Cultivated Landscapes of Native Amazonia and the Andes

William M. Denevan

OXFORD
UNIVERSITY PRESS

*This book has been printed digitally and produced in a standard specification
in order to ensure its continuing availability*

OXFORD
UNIVERSITY PRESS

Great Clarendon Street, Oxford OX2 6DP

Oxford University Press is a department of the University of Oxford.
It furthers the University's objective of excellence in research, scholarship,
and education by publishing worldwide in

Oxford New York

Auckland   Cape Town Dar es Salaam   Hong Kong   Karachi
Kuala Lumpur Madrid Melbourne Mexico City   Nairobi
New Delhi Shanghai Taipei Toronto
With offices in
Argentina Austria Brazil Chile Czech Republic France Greece
Guatemala Hungary Italy Japan South Korea Poland Portugal
Singapore Switzerland Thailand Turkey Ukraine Vietnam

Oxford is a registered trade mark of Oxford University Press
in the UK and in certain other countries

Published in the United States
by Inc., New York

ISBN 0-19-925769-8

Antony Rowe Ltd., Eastbourne

OXFORD GEOGRAPHICAL AND
ENVIRONMENTAL STUDIES

Editors: Gordon Clark, Andrew Goudie, and Ceri Peach

CULTIVATED LANDSCAPES OF NATIVE
AMAZONIA AND THE ANDES

306057182W

ALSO PUBLISHED BY
OXFORD UNIVERSITY PRESS
IN THE OXFORD GEOGRAPHICAL AND
ENVIRONMENTAL STUDIES SERIES

The New Middle Class and the Remaking of the Central City
David Ley

Culture and the City in East Asia
Won Bae Kim, Mike Douglass, Sang-Chuel Choe,
and Kong Chong Ho (eds.)

Energy Structures and Environmental Futures in Europe
Torleif Haugland, Helge Ole Bergesen, and Kjell Roland

Homelessness, Aids, and Stigmatization
The NIMBY Syndrome in the United States at
the End of the Twentieth Century
Lois Takahashi

Dynamics of Regional Growth in Europe
Social and Political Factors
Andrés Rodríguez-Pose

Island Epidemics
Andrew Cliff, Peter Haggett, and Matthew Smallman-Raynor

Pension Fund Capitalism
Gordon L. Clark

Class, Ethnicity, and Community in Southern Mexico
Oaxaca's Peasantries
Colin Clarke

Indigenous Land Management in West Africa
An Environmental Balancing Act
Kathleen Baker

Globalization and Integrated Area Development in European Cities
Frank Moulaert

Cultivated Landscapes of Native North America
William E. Doolittle

*Dedicated to John M. Treacy, 1948–89: student, colleague,
and companion in the high Andes and the far Amazon;
and to James J. Parsons, 1915–97: mentor's mentor, true friend,
and firm supporter to the end*

EDITORS' PREFACE

GEOGRAPHY and environmental studies are two closely related and burgeoning fields of academic inquiry. Both have grown rapidly over the past two decades. At once catholic in its approach and yet strongly committed to a comprehensive understanding of the world, geography has focused upon the interaction between global and local phenomena. Environmental studies, on the other hand, has shared with the discipline of geography an engagement with different disciplines addressing wideranging environmental issues in the scientific community and the policy community of great significance. Ranging from the analysis of climate change and physical processes to the cultural dislocations of postmodernism these two fields of inquiry have been in the forefront of attempts to comprehend transformations taking place in the world, manifesting themselves at a variety of separate but interrelated spatial scales.

The 'Oxford Geographical and Environmental Studies' series aims to reflect this diversity and engagement. Our aim is to publish the best original research in the two related fields and in doing so, to demonstrate the significance of geographical and environmental perspectives for understanding the contemporary world. As a consequence, its scope is international and ranges widely in terms of its topics, approaches, and methodologies. Its authors will be welcomed from all corners of the globe. We hope the series will assist in redefining the frontiers of knowledge and build bridges within the fields of geography and environmental studies. We hope also that it will cement links with topics and approaches that have originated outside the strict confines of these disciplines. Resulting studies contribute to frontiers of research and knowledge as well as representing individually the fruits of particular and diverse specialist expertise in the traditions of scholarly publication.

Gordon Clark
Andrew Goudie
Ceri Peach

The songs they recited in praise of the Sun and their kings were based on the meaning of the word *hailli*, which means *triumph over the soil*, which they ploughed and disembowelled so that it should give fruit.

(Garcilaso de la Vega 1609; 1966:1: 244)

FOREWORD

Soon after Columbus made landfall in the Bahamas in 1492 his hopes of reaching Asia by sailing west were disappointed. Instead he encountered a 'New World'. Although it was new to the inhabitants of Europe, who in the sixteenth century reacted with amazement and curiosity to its strange people, plants, and animals, we now know that this 'new' world had been occupied by people for at least 15,000 years. By 10,000 years ago some of them had begun to cultivate useful plants, and by AD 1500 many different and complex forms of agriculture were being practised in large parts of South, Middle, and North America. It is the evidence of this indigenous agricultural endeavour that concerns the authors of this trilogy of books: *Cultivated Landscapes of Native Amazonia and the Andes* by William M. Denevan, *Cultivated Landscapes of Native North America* by William E. Doolittle, and *Cultivated Landscapes of Middle America on the Eve of Conquest* by Thomas M. Whitmore and B. L. Turner II.

Of what does this evidence consist? Before answering that question, the authors' approach to the subject needs to be emphasized. As they correctly point out, research on indigenous American agriculture has been concerned more with the origins, domestication, and dispersal of crops than with the nature of agricultural landscapes and field systems. The strength and novelty of this trio of books is that they all focus on the spatial characteristics of agriculture—fields and landscapes of cultivation. It is no accident that all four authors are geographers and that they are all experienced first-hand observers of indigenous agricultural systems 'in the field'. And although they emphasize agricultural landscapes, they do not disregard the processes of agricultural production: what crops were grown, how they were planted, tended and harvested, what tools and techniques were used, as well as more general questions such as the demographic capacity and long-term sustainability of different systems of production.

To attempt a survey of indigenous agriculture throughout the Americas is a herculean task, even with four authors shouldering and sharing the burden. The range of potentially relevant evidence is prodigious. It includes descriptions by early explorers and later travellers, conquistadores, administrators, missionaries, and settlers; histories and other narratives; archaeological and biological evidence; drawings and photographs; and of course published and unpublished accounts of field projects, including the four authors' own first-hand field observations. But, despite its quantity and variety, much of this evidence frustratingly lacks detailed spatial information about the forms and functions of fields and the processes of production. It is also necessary to approach the historical sources, especially early European accounts, sceptically, prone as many of them are to exaggeration and sensationalism. These limitations have added greatly to the difficulties faced by the authors in attempting to review and synthesize knowledge of the indigenous agricultural landscapes of the Americas. That they have succeeded remarkably well will be clear to readers of the three volumes, who will also discover that the authors not only share a common aim and outlook but that their approaches differ interestingly in scope and emphasis.

The intellectual bond that links the authors in pursuit of their common goal derives from an academic genealogy that can be traced back to the influence of a group of scholars at the University of California, Berkeley, in the 1950s, particularly the geographer Carl O. Sauer. His early methodological insistence that the morphology of landscape was central to geographical enquiry, and his later explorations of New World culture history, including agricultural origins and dispersals, are seeds from which the three books can be said to have germinated. The senior author, Bill Denevan, was a graduate student in the Berkeley Geography Department in the 1950s, where he came under the influence not only of Carl Sauer but also of other scholars of Latin American geography (James Parsons), anthropology (John Rowe) and history (Woodrow Borah). I, too, was a Berkeley grad student at that time, and I remember Bill departing in 1956 to travel in South America, only to return many months later committed to what became a lifetime of research on the cultural ecology and history of Amazonia and the Andes, much of which he has distilled for us in his contribution to this trilogy. The genealogy then moves to the next generation, with Denevan supervising Bill Turner's doctoral research at the University of Wisconsin-Madison, and in due time to the next, with Turner supervising Bill Doolittle's PhD while at the University of Oklahoma and Tom Whitmore's later at Clark University, Massachusetts—a line of descent that is one of the sturdiest branches of the 'Sauer tree' (1999).[1]

It might be supposed that such direct academic descent would produce a stultifying intellectual uniformity in the four authors' approach to their joint task, but not so. Beyond their common focus on agricultural landscapes and field systems, individual differences of emphasis are allowed free rein. This adds greatly to the insights afforded by the three books, the format of which is sufficiently similar to allow easy comparison, but also flexible enough to bring out inherent contrasts in the nature and evidence of indigenous agriculture in North, Middle and South America. Thus, all the authors distinguish conceptually between rainfed, wetland, and dryland systems of cultivation, and describe the evidence of extensive pre-European landscape modelling by means of terracing, ditching, mounding, and canalization. They also give welcome attention to both horticulture and arboriculture. Those of us who have worked extensively in the tropics, whether in the New World or the Old World, know that house gardens are a widespread and important element in the domestic economy of rural comminities, but because they are relatively inconspicuous and less visually impressive than extensive field systems, they tend to be overlooked and their contribution to subsistence underrated. Likewise, forest cultivation, whether by swiddening or the more subtle exploitation of wild and semi-wild forest products, has attracted less attention than have terrace, raised-field, and canal irrigation systems. The emphasis on horticulture and arboriculture in the three volumes varies, with Whitmore and Turner devoting a chapter to them, Doolittle one to horticulture alone, and Denevan focusing on their role in Amazonian cultivation, but all the authors acknowledge the importance of these modes of production in indigenous American agriculture.

The unifying theme of this remarkable trilogy is its focus on the agricultural landscapes of the Americas 'on the eve of conquest', and the scale and comprehensiveness of the undertaking is enormously impressive. Although AD 1492 is taken as the temporal fulcrum, the authors do not allow this to inhibit them from delving deeper into the past as well as addressing the relevance of their theme to the present. Pre-European, historical, and contemporary Native American cultivation are all within their remit, although they differ in their emphasis on these three aspects of indigenous agriculture and in the extent to which they make use of archaeological in addition to historical and ethnographic evidence. They are not explicitly concerned with the origins of agriculture in the Americas, and so archaeobotanical evidence is not central to their theme, but the story they tell has fascinating implications for students of agricultural history and prehistory, and should stimulate new agro-archaeological research in the Old as well as the New World (such as the investigation of relict ditch-and-mound landscapes in lowland New Guinea and other parts of the Asian tropics). There are important implications too for agricultural development, especially among traditional communities still closely linked to the land, because many of the indigenous systems described were (and in some cases still are) both highly productive and sustainable over the long term.

Denevan disarmingly describes his contribution to the trilogy as 'to a large extent a reference work'. This is doubtless true, of his and the other two volumes (which will long be quarried by students as a guide to the vast literature), but collectively the trilogy amounts to much more than that. The authors' labours in the fields and the libraries over many years, and their determination to persevere with what must often have seemed a daunting and endless task, have resulted in a magnificent summary and synthesis of what is known, at the end of the twentieth century, about the agricultural landscapes that the Europeans encountered when they entered the 'New World' five centuries ago. It is also a testament to how far understanding of the scale and complexity of indigenous American agriculture has been transformed since the 1960s when Parsons's and Denevan's pioneering papers on ridged and drained fields in South America were published.

In 1964 Golomb and Eder (quoted by Denevan) could justly assert that 'among the surfaces deliberately fashioned by man, agricultural landforms are the most venerable, the most extensive and in many ways the least known and most ignored'.[2] Venerable and extensive they may be, but now, thanks largely to the three Bills and Tom, the cultivated landscapes of Native North, Middle, and South America can no longer be said to be either ignored or least known. With the publication of these three meticulously researched, engagingly written, and profusely illustrated books the study of indigenous American agriculture has come of age.

David R. Harris

University College London
April 2000

References

[1] Brown, Scott S. and Kent Mathewson. 1999. 'Sauer's Descent? Or Berkeley Roots Forever?' *Yearbook of the Association of Pacific Coast Geographers* 61: 137–57.
[2] Golomb, Berl and Herbert M. Eder. 1964. 'Landforms Made by Man', *Landscape* 14/1: 1–7.

PREFACE

In the Andes, *hailli* (*hayliy*) is a 'shout of triumph and happiness', a
ritual performance to celebrate the planting of maize, acknowledging 'the
victories of generations of farmers' who transformed an unfit region into
a fertile, productive landscape.

(John Treacy 1989*a*: 264)

This study is concerned with native cultivated landscapes and associated
techniques in Amazonia and the Andes. The crop field is the focus, rather than
crops themselves, which have received considerable attention elsewhere, as has
the topic of agricultural origins. I examine pre-European, historical, and contemporary Native American cultivation, based on my research over a period of
forty years, as well as the work of others.

It is often assumed that traditional agricultural technology, as distinct from
modern technology, has its roots in prehistory. However, while subsistence
knowledge has been more conservative than other aspects of culture, agricultural technology is nevertheless constantly changing in response to changing
physical environment, food production pressures, innovation, and other
factors independent of the obvious introductions of European crops, animals,
tools, and market systems. Subsequent to European contact, elements of
Indian agriculture were lost, and on the other hand, entirely new systems
emerged. Indigenous agriculture in one form or another has, however, continued with Indian people. It has generally, but not always, been ecologically
sound. Elements have been transferred to other traditional farmers, as well as
to modern farmers.

Examination of present indigenous agriculture is instructive in its own right
as well as providing insights into what the past may have been like. Likewise,
past indigenous agriculture presents options sometimes unknown to the present, with potential for the future, both indigenous and otherwise. Past agriculture can be reconstructed in part from field remnants, archaeological
excavation and analysis, crop evidence, and also from historical descriptions.
So we move backward and forward temporally from a benchmark of 1492 to
examine the agricultural technology of native South American people, without
particular concern for origins or continuity. We give attention to form and
function, distributions, change over time, and to relationships with environment, demography, and culture.

The foundation for this study was laid at Berkeley in the 1950s. Carl Sauer was
still giving seminars, and I took stimulating courses from Erhard Rostland and
James Parsons in geography, John Rowe in anthropology, and Herbert Baker
in botany, and I read the work of Sherburne Cook and Woodrow Borah on

historical demography. Senior graduate students back from tropical Latin America, still brimming with the excitement of field research, included Burton Gordon, Homer Aschmann, and Carl Johannessen. The focus of our intellectual concern as geographers was the interaction between traditional societies and their environment—the use of resources, landscape modifications, food production, plant domestication, and Indian demography. I became committed to the tropics, especially Amazonia, following travel as a journalist in Peru, Bolivia, and Brazil in 1956 and then Master's thesis research in Nicaragua in 1957. Herbert Wilhelmy, a visiting professor from Tübingen in 1959–60, convinced me that the Llanos de Mojos in Bolivian Amazonia was virtually unknown and thus a prime location for a doctoral dissertation in cultural-historical geography.

I went to Mojos in 1961 to examine how people adapted to a savanna habitat that was flooded for half the year and a virtual desert the remainder of the year. Finding remarkable linear patterns on the air photos, previously unreported, I sought them out first by bush plane and then on the ground and realized that they must be raised agricultural fields and probably prehistoric. The most exciting adaptations to the Mojos environment were not modern but pre-European. This led to a lifetime interest in indigenous agricultural fields, past and present.

Raised fields were found soon after in northern Colombia by my dissertation advisor James Parsons. Following our report in *Scientific American* (1967) and mine in *Science* (1970*a*), similar fields began turning up all over Latin America, a previously little recognized, long-abandoned form of intensive agriculture in wetlands. Later I undertook, supervised, or participated in research on raised and drained fields at Lake Titicaca, the Guayas Basin of Ecuador, highland Ecuador, and the Orinoco Llanos; on Mayan terracing in Yucatán; on riverine farming, Campa shifting cultivation, and Bora swidden-fallow management in Amazonian Peru; and on terrace abandonment in the Colca Valley of Andean Peru.

Here, I attempt to bring together this field study and associated library research, plus additional work by students, colleagues, and a host of others who have observed, thought about, and commented on Indian agriculture, with a progression of new discoveries, new ideas, and an appreciation for the capability of Indian farmers over possibly 10,000 years.

My focus is on the Andes and Amazonia, especially the Peruvian Andes and upper Amazonia, the regions I know best. As used here, the Andes includes the fringing Pacific Coast, and Amazonia is Greater Amazonia or the entire interior tropical lowlands of South America north of the Tropic of Capricorn. There was little indigenous agriculture in southern South America (see Fig. 9.1).

Of the fifteen chapters in this volume, several are based on previously published material, revised, expanded, and updated. The other chapters either represent new research or synthesize available literature. First, I present a classification of prehistoric field types and associated features in the Americas, followed by a description of crops, tools, and soft (ephemeral) technology. Three sections of chapters provide both surveys and case studies of Amazon agriculture, Andean terracing and irrigation, and raised and drained fields.

This is to a large extent a reference work, with considerable description, supported by extensive References. However, most of the chapters, including the final chapter, do examine theoretical and conceptual issues related to Indian food productivity, demography, and settlement, along with relevance for modern times. I attempt to demonstrate that South American Indian farmers were and continue to be technologically sophisticated, diverse in field systems, ecologically knowledgeable, and substantially productive, although also capable of major environmental change. Before 1492 these farmers were the equal of counterparts in Europe, Asia, the Pacific, and Africa. Today, we can profitably learn from the agricultural achievements of the pre-European Native Americans and their descendants.

This study was conceived over a decade ago as part of a larger project, which I initially directed, that includes volumes on North America (Doolittle 2000), Middle America (Whitmore and Turner 2001), and this volume. After various delays, detours, and changes in organization and publication plans, all three volumes have finally come to fruition, thanks to Oxford University Press. Following our individual inclinations, rather than a formula, we have not attempted to adhere to a common content other than a common interest in Native American cultivation practices.

William M. Denevan

Sea Ranch, California
March, 2000

ACKNOWLEDGEMENTS

Fieldwork was carried out in South America between 1961 and 1993 with funding provided by the Foreign Field Research Program, the Ford Foundation, the National Science Foundation (Grant Numbers GS-30118 and BNS-8406957), Man and the Biosphere Program, and at the University of Wisconsin by the Wisconsin Alumni Research Foundation, the Ibero American Studies Program, the Nave Fund, and the Geography Whitbeck Fund. Writing was supported particularly by grants from the Guggenheim Foundation, the National Endowment for the Humanities, and the Wisconsin Alumni Research Foundation.

Numerous graduate students in geography and anthropology at the University of Wisconsin shared in my research and/or made significant contributions on their own to Amazonian and Andean cultural ecology: Roland Bergman, Paul Blank, Sarah Brooks, Steve Brush, Marshall Chrostowski, Oliver Coomes, Hildegardo Córdova, Dan Gade, Lourdes Giordani, Greg Knapp, Robert Langstroth, Kent Mathewson, Joe McCann, Toby McGrath, Dan Shea, David Stemper, Pete Stouse, John Treacy, Jon Unruh, Robbie Webber, Stu White, Richard Whitten, and Antoinette WinklerPrins. Others worked on indigenous subsistence in Middle and North America: Bill Gartner, Laurie Greenberg, Fred Lange, Barney Nietschmann, Alf Siemens, Bill Turner, and Barbara Williams. We learned a lot together in field and seminar, and much of them is in this book. Thanks friends.

I must also acknowledge a debt to many other colleagues who have contributed, as participants on my field projects, through personal interaction, and through their writings: Janis Alcorn, Pedro Armillas, Homer Aschmann, William Balée, Stephen Beckerman, Maria Benavides, Woodrow Borah, Karl Butzer, Robert Carneiro, Robin Donkin, William Doolittle, Robert Eidt, Clark Erickson, Burton Gordon, Susanna Hecht, Mario Hiraoka, Carl Johannessen, Alan Kolata, Donald Lathrap, George Lovell, Michael Malpass, Betty Meggers, Emilio Moran, Máximo Neira, Christine Padoch, James Parsons, George Plafker, Darrell Posey, Dennis Puleston, Anna Roosevelt, Jonathan Sandor, Carl Sauer, Karl Schwerin, Clifford Smith, Pablo de la Vera Cruz, William Vickers, Robert West, Thomas Whitmore, Gene Wilken, Karl Zimmerer, Alberta Zucchi, and various others less directly. They too are part of this book.

Special thanks to graduate assistants Sarah Brooks (editing, Index), Fernando González, Laurie Greenberg, Greg Knapp, Joe McCann (Appendices), Jon Unruh, and Emily Young; to typists Sally Monogue, Nasrin Qadar, Antoinette WinklerPrins, and especially Sharon Ruch for the final manuscript; to the University of Wisconsin Cartographic Laboratory supervised by Onno Brouwer; and to Dominic Byatt and Dorothy McLean of Oxford University Press for their guidance and support.

And to Susie, who provided good meals, and much more.

W. M. D

The author and the publisher thank the following who have given permission for the use of copyright and personal illustrations.

Fig. 1.1: from John Hyslop (1990), *Inka Settlement Planning* (Austin). By permission of the University of Texas Press.

Figs 3.1 and 7.1: from Robert L. Carneiro (1974), 'On the Use of the Stone Axe by the Amahuaca Indians of Eastern Peru', *Ethnologische Zeitschrift Zürich*, 1: 107–22. By permission of Peter Lang AG, European Academic Publishers.

Figs 3.2 and 8.12: from Felipe Guamán Poma de Ayala (1980), *El Primer Nueva Corónica y Buen Gobierno* (México, D.F.). By permission of Siglo Veintiuno Editores.

Fig. 3.5: from David R. Harris (1971), 'The Ecology of Swidden Cultivation in the Upper Orinoco Rain Forest, Venezuela', *The Geographical Review*, 61: 475–95. By permission of the American Geographical Society; further reproduction is prohibited without written permission from the AGS.

Figs 4.2 and 5.8: from William M. Denevan (1996), 'A Bluff Model of Riverine Settlement in Prehistoric Amazonia', *Annals of the Association of American Geographers*, 86: 654–81. By permission of the Association of American Geographers.

Figs 5.2 and 5.3: from William M. Denevan (1971), 'Campa Subsistence in the Gran Pajonal, Eastern Peru', *The Geographical Review*, 61: 496–518. By permission of the American Geographical Society; further reproduction is prohibited without written permission from the AGS.

Figs 5.4, 5.5, 5.6, and 5.7: from William M. Denevan and John M. Treacy, 'Young Managed Fallows at Brillo Nuevo'. Reprinted with permission from *Swidden-Fallow Agroforestry in the Peruvian Amazon*, Advances in Economic Botany, Vol. 5, copyright 1988, The New York Botanical Garden.

Figs 5.8, 5.9, and 5.10: from Roland W. Bergman (1980), *Amazon Economics: The Simplicity of Shipibo Indian Wealth*, Dellplain Latin American Studies, No. 6. By permission of the Department of Geography, Syracuse University.

Fig. 5.11: by permission of Roland W. Bergman.

Figs 5.12, 5.13, 14.2, and 14.3: from William M. Denevan and Karl H. Schwerin (1978), 'Adaptive Strategies in Karinya Subsistence, Venezuelan Llanos', *Antropológica*, 50: 3–91. By permission of the Fundación La Salle de Ciencias Naturales.

Fig. 8.4: from Patricia J. Netherly (1984), 'The Management of Late Andean Irrigation Systems on the North Coast of Peru'. Reproduced by permission of the Society for American Archaeology from *American Antiquity*, 49:227–54.

Figs 8.5, 8.8, and 8.11: By permission of James S. Kus.

Fig. 8.6: from Chris C. Park (1983), 'Water Resources and Irrigation Agriculture in Pre-Hispanic Peru', *The Geographical Journal*, 149: 153–66. By permission of the Royal Geographical Society.

Fig 8.7: from Charles R. Ortloff, Robert A. Feldman, and Michael E. Moseley (1985), 'Hydraulic Engineering and Historical Aspects of the Pre-Columbian Intravalley Canal Systems of the Moche Valley, Peru', *Journal of Field Archaeology*, 12: 77–98. Reprinted with the permission of the Trustees of Boston University.

Figs 8.9, 9.4, and 10.3: photos by Robert Shippee and George Johnson (1931), Neg./Trans. numbers 334717, 334768, and 334672. Courtesy Department of Library Services, American Museum of Natural History.

Fig. 8.10: from James S. Kus (1984), 'The Chicama-Moche Canal: Failure or Success? An Alternative Explanation for an Incomplete Canal'. Reproduced by permission of the Society for American Archaeology from *American Antiquity*, 49: 408–15.

Fig. 8.13: from Gregory Knapp (1982), 'Prehistoric Flood Management on the Peruvian Coast: Reinterpreting the "Sunken Fields" of Chilca'. Reproduced by permission of the Society for American Archaeology from *American Antiquity*, 47: 144–54.

Fig. 8.14: from Karen E. Stothert (1995), 'Las albarradas tradicionales y el manejo de aguas en La Península de Santa Elena', *Miscelánea Antropológica Ecuatoriana*, 8: 113–60. By permission of Karen E. Stothert.

Figs 9.2, 10.1, and 10.6: from Sarah O. Brooks (1998), *Prehistoric Agricultural Terraces in the Río Japo Basin, Colca Valley, Peru*. By permission of Sarah O. Brooks; drawings by Douglas Tallman.

Fig. 9.5: from John M. Treacy (1994), 'Teaching Water: Hydraulic Management and Terracing in Coporaque, the Colca Valley, Peru'. Reproduced by permission of the American Anthropological Association from *Irrigation at High Altitudes: The Social Organization of Water Control Systems in the Andes*; not for further reproduction.

Fig. 9.6: reprinted from *Agricultural Terracing in the Aboriginal New World*, by R. A. Donkin, Viking Fund Publications in Anthropology No. 56, University of Arizona Press, 1979, by permission of the Wenner-Gren Foundation for Anthropological Research, Inc., New York.

Figs 11.2 and 11.3: reproduced with permission from William M. Denevan, 'Aboriginal Drained-Field Cultivation in the Americas', *Science*, 168: 647–54. Copyright 1970 American Association for the Advancement of Science.

Fig. 11.4: by permission of William A. Bowen.

Figs 11.5 and 14.4: from Charles S. Spencer, Elsa M. Redmond, and Milagro

Rinaldi (1994), 'Drained Fields at La Tigra, Venezuelan Llanos: A Regional Perspective'. Reproduced by permission of the Society for American Archaeology from *Latin American Antiquity*, 5: 119–43.

Fig. 11.9: from Stéphen Rostain (1991), *Les Champs Surélevés Amérindiens de la Guyane* (Paris). By permission of the Centre ORSTOM de Cayenne.

Figs 12.2, 12.4, and 13.7: by permission of Clark L. Erickson and Daniel A. Brinkmeier.

Figs 12.1 and 12.3: from William M. Denevan (1966), *The Aboriginal Cultural Geography of the Llanos de Mojos of Bolivia* (Berkeley). By permission of the University of California Press.

Figs 13.2, 13.4, and 13.5: from Clifford T. Smith, William M. Denevan, and Patrick Hamilton (1968), 'Ancient Ridged Fields in the Region of Lake Titicaca', *The Geographical Journal*, 134: 353–67. By permission of the Royal Geographical Society.

Fig. 13.6: reprinted with the permission of Simon and Schuster from *Time Detectives* by Brian Fagan. Copyright 1995 by Lindbriar Corp.

Fig. 14.5: from Ian S. Farrington (1983), 'Prehistoric Intensive Agriculture: Preliminary Notes on River Canalization in the Sacred Valley of the Incas', *British Archaeological Reports, International Series*, 189: 221–35. By permission of Ian S. Farrington.

For permission to reprint excerpts from previously published text we would like to thank the New York Botanical Garden (Chapter 4), the University Press of Florida (Chapter 6), Culture and Agriculture (Chapter 7), the Steward Anthropological Society (Chapter 7), the American Geographical Society (Chapter 5), the Association of American Geographers (Chapter 6), the Fundación la Salle de Ciencias Naturales (Chapters 5 and 14), the University of California Press (Chapters 5, 12, and 14), and the Royal Geographical Society (Chapter 13).

Every effort has been made to contact copyright holders; however this has not always been possible.

CONTENTS

LIST OF TABLES

LIST OF FIGURES

Part I

Introduction:
Fields and Associated
Features

The Early Spaniards were impressed by native horticulture.

(Robin Donkin 1970: 507)

1

Research on Native Cultivated Landscapes in the Americas

> Among the surfaces deliberately fashioned by man, agricultural landforms are the most venerable, the most extensive and in many ways the least known and most ignored.
>
> (Berl Golomb and Herbert Eder 1964: 5)

Until recently, there has been little scholarly interest in the techniques and ecology of native[1] cultivation[2] systems in the New World. More attention had been directed towards agricultural[3] origins and the history and nature of specific crops. It was assumed that few ancient fields still existed and that contemporary Indian cultivation was largely deteriorated and simplified and/or highly modified by European influences. Air photography, however, has revealed the existence of hundreds of thousands of early fields and associated remnant features which can be measured, excavated, and analyzed and their ecology and productivity reconstructed. Many indigenous cultivation techniques have continued relatively intact throughout the hemisphere to the present, and even where modified it may be possible to determine pre-European characteristics. Furthermore, there is much more agricultural information available in the accounts of early contacts than might be expected, as exemplified in such studies as Donkin (1979: 20–2) on New World terraces and Palerm (1973) and Rojas (1983; 1987; 1988) on Mexican fields. In addition, previously unknown systems of Indian cultivation have recently come to light, such as raised fields, sunken fields, and swidden-fallow management and other forms of agroforestry. In my time, there has been a shift from perceiving Indian cultivation as mostly technologically backward to recognizing a '*ciencia agrícola*' with pre-European roots (Antúnez de Mayolo 1986*a*).

Colonial Observations

It is remarkable how little information about aboriginal agriculture there is in the early accounts by chroniclers, travelers, and even scholars whose attention was consistently focused elsewhere, but more so for Amazonia compared to the Andes. The reasons for this neglect are not clear. Certainly most of the

Europeans who first came to the Americas were not farmers, and they were not involved in Indian cultivation practices. As European crops were introduced and commercial production increased under European management, Indian elements were modified or replaced. In any event, Indian techniques were for the most part either ignored or taken for granted. They were mentioned in passing, in some cases with awe for their size and extent, but with little description or measurement or explanation. The situation does not improve much with the hundreds of travelers' accounts published into the twentieth century, a sentence here, a paragraph there, occasionally a sketch or photo.

Several of the sixteenth-century chroniclers of Peru, such as Cieza de León, Bernabé Cobo, and Garcilaso de la Vega, did describe terraces and canals in some detail (Donkin 1979: 20–1). One of the first references to terraces was in 1535 by Pedro Sancho de la Hoz, secretary to Francisco Pizarro, who noted that 'between Tumbes and Cuzco all the mountain fields are made in the guise of stairways of stone' (Sancho de la Hoz 1917: 149; trans. by Donkin 1979: 20).

Other cultivation features received less attention. Raised fields in the tropical lowlands and around Lake Titicaca are almost unmentioned, possibly because most had been abandoned by the time of the Spanish conquest. Sunken fields on the Peruvian coast were reported by Cieza de León (1959: 337), but the artificial depressions (*cochas*) north of Lake Titicaca, still in use, were not described until recently (Flores Ochoa 1987). Reports of irrigation canals in the sixteenth century are numerous, but less detail is given than for terraces. There are even mentions of Inca and pre-Inca abandoned canals on the Peruvian coast (Garcilaso de la Vega 1966: 297). Generally, abandoned cultivation features are found in depopulated areas, and are highly eroded or buried, so it is not surprising that they did not draw more attention from early Spanish observers and subsequent travelers and scholars.

There are also some graphic portrayals of Indian agriculture from the early colonial period, including the detailed pictures by Guamán Poma de Ayala (1980) of Inca cultivation; of corn fields in Virginia by Jacques Le Moyne in 1564 and in Florida by John White in 1585 (Lorant 1946: 77, 191, 264); a garden bed (raised field) in the Caribbean by a member of Sir Francis Drake's crew in the 1590s (Pierpont Morgan Library 1996: Folio 127); an Aztec *chinampa* map (M. Coe 1964); and Aztec field implements in the *Codex Florentino* (Donkin 1979: 7–9). There are some maps carved in rock of terrace and canal irrigation systems, such as the *maquetas* of the Colca Valley and elsewhere in the Peruvian Andes (Fig. 1.1) (Brooks 1998: 287–93), and a glyph map in Sonora (Doolittle 1984: 253–4).

Pre-1961 Studies

Indian cultivation in the Americas only began to receive serious attention from scholars in the early 1960s. If a benchmark can be identified, it is the publica-

Fig. 1.1. The Sayhuite carved stone (*maqueta*) west of Cuzco, Peru, probably Inca. Canals and terraces are shown (lower left center), along with paths, buildings, humans, and animals. From Hyslop (1990: 116).

tion of 'Land Use in Pre-Columbian America' in 1961 by the anthropologist Pedro Armillas, the first major synthesis of the subject. Armillas provided a region-by-region discussion of agricultural techniques in the context of environment, culture history, and crops. The same year saw the publication of a collection of symposium papers edited by Johannes Wilbert (1961), plus articles on Mayan agriculture by Dumond (1961) and Cowgill (1961). Also in 1961 the first discoveries of prehistoric raised fields were made in the Bolivian Amazon (Plafker 1963; Denevan 1962; 1963a; 1963b). Prior to 1961, it is difficult to identify many articles, much less books, that focused on Indian agricultural techniques—prehistoric, historic, or present day.

While there were some studies before 1961 of Indian cultivation in Mexico and the Southwest United States, there were very few in South America. The *Handbook of South American Indians* (Steward 1946–59) in its 'tribe'-by-'tribe' and region-by-region treatment of archaeology, ethnohistory, and ethnography gives remarkably little information on cultivation techniques, although a chapter by Carl Sauer (1950) is devoted to crops. For example, the classic chapter on 'Inca Culture at the Time of the Spanish Conquest' by John Rowe (1946) has ten pages on subsistence, but of this only about three pages are on terracing and irrigation, tools, and the agricultural cycle. The remainder is on crops, hunting, gathering, fishing, domesticated animals, and food preparation. Other treatments of subsistence in the seven-volume study have even less on cultivation techniques.[4] The *Handbook of Middle American Indians* (Wauchope 1964–76), and the *Handbook of North American Indians* (Sturtevant 1978–), in contrast, give more attention to agricultural technology, including full chapters on subsistence, economy, or ecology.

The first detailed examination of the ecology of an Amazonian society, with data on production, labor, and diet, is Carneiro's 1957 dissertation on the Kuikuru done at the University of Michigan, supervised by Leslie White. Unfortunately, the full study has never been published.

O. F. Cook's (1916; 1920) articles on terracing and the footplow are well known, but most other early studies are in obscure South American journals. These include Latcham (1936) on pre-Colombian agriculture in Chile, Soria Lens (1954) on Aymara agriculture in Bolivia, Valcárcel (1942) on ancient Peruvian agriculture, Regal (1945) on Inca irrigation, Maldonado and Gamarra Dulanto (1978) in 1945 on abandoned terraces of the Rimac Valley in Peru, and Ardissone (1944; 1945) on terraces in Northwest Argentina.[5] Probably the earliest survey of Andean terracing is that by Swanson (1955). The German archaeologist Horkheimer (1960) described pre-Hispanic Andean food and food production. Other early studies, as well as colonial period descriptions, are cited by Donkin (1979).

To be sure, cultivation features such as terraces and canals were briefly described in archaeological reports, were mentioned by the early Europeans, and were routine but sketchy components of standard ethnographic accounts. Cultivation, however, was seldom studied for its own sake in terms of how it functioned or how it related to food production, demography, and to cultural evolution and change. The recent explosion of interest in Indian agriculture reflects developments in the 1950s and 1960s: (1) the emergence of cultural ecology in anthropology, archaeology, and geography (Steward 1955; Lathrap 1962; Denevan 1966*b*), (2) the debate over environmental limitations in the Amazon, Andes, and Yucatán (Meggers 1954; W. Coe 1957; Ferdon 1959; Carneiro 1961), (3) evidence that New World Indian populations were much larger than previously thought, requiring at times intensive, highly productive forms of agriculture (Borah 1964; Dobyns 1966; Denevan 1970*b*), (4) new discoveries of vast numbers of prehistoric field remnants throughout the hemisphere (Denevan 1970*a*), and (5) the increasing realization that traditional[6] agricultural techniques are still widely used by Indians and peasants, are productive[7] and sustainable, and are instructive for agricultural development.

Post-1961 Research in South America

The focus of recent research on Indian cultivation in South America has, first of all, been interdisciplinary, involving geographers, historians, anthropologists, archaeologists, and biologists or agroecologists. Initial efforts were primarily individual and emphasized discovery and description. Subsequent research has included major field projects, involving specialists, detailed and systematic data gathering, and theory generation. Methodology has often been *ad hoc*, developed in the field, but has also paralleled growth in the disciplines involved during the second half of the twentieth century. To a degree there was influence and stimulation from the prehistorians of Europe who have long had a sophisticated interest in relic fields there, based on air photography, excavation, and measurement (e.g., D. R. Wilson 1982: 118–27). This is part of what is now called 'landscape archaeology' (Mathewson 1986). Most of the New World research, however, has been independent of European antecedents.

Additional impetus for research on indigenous agriculture came from: (1) the development of agroecology (Altieri 1995), the study of the ecology of crops in a man-made environment (the cultivated field), (2) the recognition of the values of traditional systems in terms of ecological stability, sustainability, productivity, and cultural integrity (Klee 1980), and (3) an increased awareness of the close relationship between agricultural intensity and productivity, demographic growth, and cultural change (Boserup 1965; Cohen 1977). These currents demanded not only descriptions of agricultural variations but ecological understanding and an historical perspective.

In South America, a pioneering study of irrigation in relation to prehistoric states is the now classic *Life, Land and Water in Ancient Peru* by historian Paul Kosok (1965). This large format volume uses air photos as a primary research tool, as well as for illustration. Kosok demonstrated the utility of aerial photography for discovering and studying field remnants from the past, and he also stimulated archaeological research on irrigation. Kosok, whose studies of irrigation spanned twenty years, died in 1959, and associates assembled his book from notes and manuscripts. A more systematic monograph was intended but never completed. A variation of his aerial perspective is the use of light planes to locate old fields which are not readily apparent on the ground. The initial discoveries of raised fields in eastern Bolivia were made this way by Plafker (1963) and Denevan (1963*a*; 1963*b*).

The first serious treatment of terraces in the Andes was in an unpublished doctoral dissertation by Chris Field (1966). An excellent, comprehensive survey of terracing in the hemisphere is that by the Cambridge geographer Robin Donkin (1979). It is based on a reconnaissance by Donkin plus exhaustive search of the literature. It was the direct antecedent of a series of terrace research projects in the 1980s. Archaeologist John Rowe brought to attention sunken fields on the Peruvian coast in a 1969 article which inspired a series of other, somewhat controversial studies.

In the tropical lowlands the transition from very generalized treatments of shifting cultivation to systematic description and measurement of inputs and outputs comes with Robert Carneiro's studies of the Kuikuro (1957; 1961) in

Brazil and the Amahuaca (1964) in Peru. There is limited archaeological and ethnohistorical evidence, however, on shifting cultivation.

These publications in the 1960s led to dozens of other individual and team studies of prehistoric cultivation features by geographers and archaeologists mainly from the United States, England, and Latin America. Even more research has been carried out on present day Indian agroecology in the Andes and Amazon regions. In the 1970s the focus of this latter research shifted from description and mapping to theoretical interests in agricultural intensification in relation to demography, carrying capacity, relation of production to cultural evolution, dietary influences, and environmental limitations.

By the mid 1970s research by the individual scholar began to give way to interdisciplinary, long-term, regional team projects in which agriculture was the prime focus or a major component. In South America these have included the Moche Valley ancient irrigation project on the north coast of Peru (Moseley and Day 1982), the Peñón del Río raised field project on the coast of Ecuador (Marcos 1987), the Colca Valley abandoned terrace project in southern Peru (Denevan 1987a), the Titicaca raised-field projects (Erickson 1996; 1999; Kolata 1986; 1993; 1996b), the Bora agroforestry project in the Peruvian Amazon (Denevan and Padoch 1988), and the Kayapó ethnobiology project (Posey 1985a).

Books on the prehistory and ethnography of New World Indians still do not give particular attention to agriculture. The most comprehensive treatment of the archaeology of the Americas consists of the two volumes by Gordon Willey (1966; 1971). Although there is not a chapter on agriculture or subsistence, evidence is presented for prehistoric agricultural techniques in the regional treatments, especially for South America.

The rise in scholarly interest in ancient aboriginal fields and cultivation techniques is best reflected in several syntheses and collections. For the hemisphere as a whole there are the survey articles by Denevan (1980b; 1980c), Fedick (1995), Matheny and Gurr (1983), and R. Smith (1987). For South America, the best survey is in the textbook by Caviedes and Knapp (1995: 169–83). Topical surveys include B. Price (1971) on irrigation, Donkin (1979) on terraces, Denevan (1970a; 1982) on raised fields, and Soldi (1982) on sunken fields. Collections include Denevan (1980a) and Torre and Burga (1986). Kelly (1965) on Inca land-use regions is not very useful and at times is in error. Three valuable collections of conference papers have been published in British Archaeological Reports (Darch 1983; Farrington 1985; Denevan, Mathewson, and Knapp 1987).

Research on contempory Amazon Indian agriculture has exploded since the provocative article by Gross in 1975 on protein capture and cultural development, as exemplified by two collections by Hames and Vickers (1983a) and Posey and Balée (1989) and by the survey by Beckerman (1987). Monographs include Ruddle (1974), Smole (1976), Bergman (1980), Descola (1994), and

Balée (1994). For the Andes, published regional monographs on traditional agriculture include Basile (1974) and Knapp (1991) on Ecuador, and Hatch (1976), Brush (1977), Gade (1975), Earls (1989), Treacy (1994*b*), and Zimmerer (1996*a*) on Peru. For an overview of current traditional agriculture in South America, there is nothing comparable to *Good Farmers*, the excellent survey of Mexico and Central America by Wilken (1987).

As a result of research during the past forty years on relic field remains, on ethnohistorical evidence, and on the ecology of surviving Indian practices, it is now possible to present a fairly comprehensive survey of indigenous agricultural techniques in the Andes and Amazon. For details, however, specific studies should be consulted. What we know is very uneven, culturally, regionally, and topically. The opportunities for further research are unlimited, in terms of discovery, description, analysis, and interpretation. But there is urgency. Relic fields are rapidly being destroyed by agricultural development, urban expansion, and other activity. Also, surviving aboriginal techniques are being lost as native cultures become extinct and as modern agriculture replaces traditional methods.

Diversity

Physically, South America is one of the most diverse regions on earth (Caviedes and Knapp 1995: 21–96). This is masked by generalized maps of landforms and biomes (major bio regions) (Fig. 1.2). Two-thirds of the continent lies within the tropics, but the western fringe is in high mountains and desert. Amazonia and adjacent lowlands contain large floodplains and vast forests and savannas, each seemingly relatively uniform; however, there is considerable regional and microdiversity within each.

The Andes are obviously highly diverse biologically, both vertically as climates range from tropical to frigid as elevations change, and horizontally as precipitation and landforms vary over short distances. Thus the Andes is a patchwork of local habitats and altitudinal temperature zones, with broadly tolerant crops overlapping (Zimmerer 1999). Indigenous cultivators have adapted to this natural variability by utilizing a multiplicity of field types, field locations, and crops. A single household might have twenty or thirty different small fields located in different temperature, moisture, and soil zones, thus taking advantage of the environmental variability, reducing environmental risk, and enhancing dietary options. However, the patchy character, versatility, origin, and decline of field systems and features and crops in the Andes, as well as in Amazonia, are related not just to environmental spatial patterns but also to periodic environmental disturbance, distance factors, locational relationships, development and diffusion of cultivation technology, and various social, economic, and dietary factors, all in historical context.

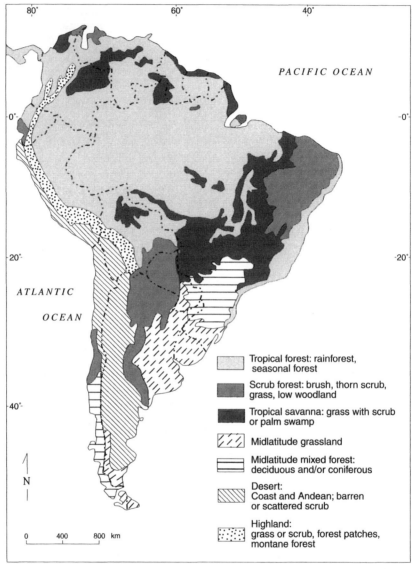

Fig. 1.2. Map of principle biomes (major habitat types) of South America. Adapted from Denevan (1980*b*: 219)

As a Geographer and Cultural Ecologist

This study falls within the conceptual discourse of cultural ecology, which is concerned with how people relate to, or interact with, their environment, especially non-modern societies. In the past, cultural-ecological studies in geography have focused on the processes involved in people-environment interactions, whereas anthropology has been more concerned with how these interactions affect aspects of culture (Denevan 1971*b*). This distinction has broken down, however. For recent treatments of the different approaches to cultural ecology (geographical ecology, political ecology, ecological anthropology, historical ecology), see Turner (1997) and Zimmerer (1996*b*) in geography and Milton (1997) and Balée (1998) in anthropology.

I have come to this study as a geographer, rather than as an anthropologist, archaeologist, historian, or agronomist, although I draw on all of these disciplines. I believe that the study of '[l]and and life is what geography has always been about' (Stoddart 1987: 334) or 'how people live in their land' (C. Sauer 1987: 156), ecology not morphology, although there are other distinctive approaches to geography. How do human beings interact with the physical environment they find themselves in by utilizing what is useful to them (resources) and in the process change that environment in ways that are positive or negative for them? Geographers examine and attempt to explain this interaction in particular places or regions and then seek to reach broader generalizations. To do this adequately, one must have competence in both the physical and social sciences, as geographers often do.

I believe that geographers must ask and pursue the 'big' questions[8] about 'human potential', about people and land in relation to the rise and fall of societies, the dynamics of food production, the growth and decline of population, and the causes and characteristics of environmental change.[9] We have failed to do so as aggressively as other disciplines, to our detriment. In this study of indigenous agricultural fields in the Andes and Amazonia, I have focused on the nature of the fields themselves; however I also raise some of these 'big questions'—questions with relevance beyond South America and its indigenous people.

Notes

[1] By 'native' (or 'indigenous') I refer to New World Indians, past and present, with recognizable cultural identities and historical roots.

[2] 'Cultivation' refers to land preparation activities for either domesticated or wild plants (sowing, planting, weeding, tillage, watering, manuring, etc.

[3] Some anthropologists use the term 'horticulture' for small fields and gardens cultivated by hand, often polycultures (multiple crops), and 'agriculture' for large fields which may involve animal traction or machinery, often monocultures (single crop). I use 'agriculture' for any type of field with cultivated crops.

[4] *Native Peoples of South America*, by Steward and Faron (1959), mainly a summary of the *Handbook of South American Indians*, has several short sections on subsistence and cultural-ecological adaptations.

[5] These are useful studies, but unfortunately information in the texts is not always tied to sources in the bibliographies.

[6] By 'traditional agriculture' I mean non-modern, that is not dependent on fossil fuels, mechanization, or chemical inputs. Today, most indigenous agricultural technology contains both modern and traditional components; however, we can still speak of traditional agriculture in a relative sense overall and we can refer to specific traditional components.

[7] Traditional agriculture is not synonymous with subsistence agriculture whereby farmers only feed their households (Angel Palerm in T. Rojas 1987: 201). Most traditional farmers, past and present, produce a surplus for storage, exchange, or market. The purely subsistence farmer is a rarity.

[8] 'We need to claim the high ground back: to tackle the real problems: to take the broader view' (Stoddart 1987: 334). See also Turner (1989).

[9] 'Land transformation' (Goudie 1993) and 'land-use and land-cover change' (Meyer and Turner 1994).

2

Classification of Field Types

Agricultural technology and the human management of the natural environment for the cultivation of domesticated plants, involves crops, tools, and the processes and forms of environmental modification. Agricultural fields can be divided into those that are artificial in terms of significant degree of surface modification and those that are not. The latter are usually rainfed or located where water is concentrated; they often have good soils; and they are mostly on gentle slopes. However, for past times our best evidence is the forms of man-made agricultural features (agricultural landforms)[1] which have survived, and these are given major attention in this study. They consist of actual fields and/or field features such as canals. In addition, there are aspects of technology that do not survive visually, but may be detected through stratigraphy, chemical analysis of soil, palynology, and other archaeological means, as well as from written accounts and observation of current practices. These include soil management, vegetation control, use of fire, etc. They are discussed in the next chapter, under the rubric of 'soft technology' (i.e., ephemeral).

Here, a classification and terminology are proposed consisting of thirty categories of agricultural fields and associated landscape features in South America (Table 2.1[2]). The primary basis of the classification is form (morphology). Form is usually related to function. Function in turn is related to environment, crops, production pressures, and tool technology. Intended function can only be inferred from the form of a relic feature, but analogy with comparable contemporary features facilitates interpretation.

Ordinary Fields

I use the term 'ordinary fields' to refer to fields in which there is no obvious human modification of the natural landscape by excavation or by earth/stone works. These fields thus tend to be invisible when not in crops, although they might be detected by soil characteristics, associated vegetation patterns, or field boundaries. Two main categories can be arbitrarily differentiated in terms of moisture sources. The first includes rainfed fields in which precipitation is sufficient for crop growth (*temporales*). These fields are mainly found in the

Based in large part on Denevan (1980*c*), which classifies Indian fields throughout the Americas.

Table 2.1. Fields and field features

Ordinary fields

Rainfed agriculture
    1. Annual cropping
    2. Dry farming
    3. Shifting cultivation
    4. Agroforestry

Water concentration
    5. Floodplain fields
    6. Seepage fields
    7. Water-table fields

Agricultural landforms: slope modification
    8. Cross-channel terraces and check dams
    9. Sloping-field terraces
  10. Bench terraces
  11. Valley-floor terraces
  12. Segmented terraces
  13. *Gradas*, or Coca terraces

Agricultural landforms: water-deficit management

Floodwater and runoff control
  14. Linear diversion embankments
  15. Semi-enclosing and enclosing field embankments
  16. Pond fields

Groundwater utilization
  17. Sunken fields

Moisture conservation
  18. Bordered gardens and lithic mulches

Canal irrigation
  19. Canals
  20. Underground canals
  21. Field ditches (furrows)
  22. Reservoirs and dams

Manual irrigation
  23. Wells

Agricultural landforms: drainage
  24. Ridges and platforms (large raised fields)
  25. Ditched fields
  26. Drainage canals
  27. Dikes and embankments

Micro management of soil, water, climate, and slope (small raised fields)
  28. Planting beds
  29. Crop mounds (maize hills, *montones*)

Field demarcation
  30. Boundary walls, fences, and markers

---

wet, lowland and montane tropics, and in the humid midlatitudes. The second category consists of fields in drylands on sites where surface or ground water has become concentrated.

Rainfed Agriculture

1. Annual Cropping Permanent cropping without major resource management techniques requires good rainfall and good soil. G. Reichel-Dolmatoff (1961: 97) quotes several sixteenth- and seventeenth-century descriptions of two to three crops per year in both highland and lowland Colombia. However, Indian cultivation was generally eliminated early from such sites and replaced with European crops and techniques. Consequently, we know very little about Indian practices on the best lands, where there are seldom any surviving field features or early descriptions.

2. Dry Farming generally refers to non-irrigated agriculture in regions where rainfall is marginal. In particular, the term is used for semi-arid farming where most years there is only enough precipitation for specific low moisture-demanding crops, or where limited moisture may have to accumulate in the soil for several years before a crop is possible. True dry farming, then, is fully dependent on rainfall (Glassow 1980: 45), but is not annual.

3. Shifting Cultivation or swidden agriculture[3] is the rotation of a few years of cropping with usually a moderate to long period of forest or bush fallow. It is the dominant form of Indian agriculture today in the tropical lowlands, but it is also important elsewhere. It is not directly identifiable archaeologically, and there are few descriptions of it in the early colonial literature.

Where soils are good or maintained by fertilization, fallowing may be for only a few years after several years of cropping (*short-fallow cultivation*). The cropping may be longer than the fallow period. With *sectorial fallowing* in the Andes, fields may be fallowed from one to thirty years after three to five years of cropping (Godoy 1988). Fallowing may be more for coping with weeds, insects, disease, or moisture deficiency than for declining fertility.

A variation is *burn-plot agriculture* or *in-field burning*. Where soil is marginal and forest fallowing is not practiced, fallowed plots covered with scrub, grass, weeds, or crop residue may be burned off in order to utilize the fertilizing effect of the ash and to clean the surface for planting. This practice is widespread.

Shifting cultivation generally involves the burning off of cleared forest. A variation, especially in areas where there is little or no dry period for burning, is *slash/mulch cultivation* (*tapado*) (Thurston 1997). The slash is not burned but rather is allowed to decompose, a slow process with less space available for crops, but a larger portion of organic material is allowed to return to the soil.

4. Agroforestry, usually rainfed, refers to the combination of annual crops with perennial tree crops and/or natural vegetation. There are infinite variations, but only recently has there been an interest in indigenous systems in the Americas. Most attention has been given to *house gardens, anthropogenic forests* (cultural forests, artificial rainforests), and *swidden-fallow management* (Denevan and Padoch 1988; Alcorn 1981; Balée 1994; Gordon 1982; Covich and Nickerson 1966; see Chapter 5). The evidence is contemporary, but many practices are very old.

Water Concentration

5. Floodplain Fields Here, we refer to plots in dry land floodplains that are naturally irrigated ('self-watered') by floodwaters when rivers briefly overflow during high water (Chapter 8). Cropping is then possible on the saturated soil. In most arid floodplains such farming preceded more reliable, managed, irrigation agriculture.

A variation is the *recessional field*, mainly found in humid tropical river valleys where high water periods lasting months alternate with low water periods lasting months. As water levels recede, crops are progressively planted on the exposed, moist sides of natural levees and emerging *playas* (beaches). This type of agriculture is risky where hydrologic regimes are uncertain, but enormous areas are planted in Amazonia and elsewhere.

6. Seepage Fields Here, the source of water is a seep or spring directly up slope from a field. In the Colca Valley in Peru, where springs are an important source of canal irrigation water, it is likely that spring-fed sloping fields preceded the development of canal and terrace systems (Chapter 10).

7. Water-table Fields The principal source of moisture for these fields is groundwater which is very close (*c.* 1 m or less) to the surface and crop roots, usually near a lake, river, or swamp. However, if the water table is too close to the surface drainage ditches may have to be dug (Chapter 14). Also, field surfaces can be dug down to water-table levels (Chapter 8).

Agricultural Landforms: Slope Modification

Agricultural landforms are features of earth and stone or excavations associated with cultivation. Many of these features require considerable labor to construct and to maintain, and they generally are or were associated with intensive (permanent or near permanent) cultivation, with means of fertility maintenance, and with at least locally dense populations.

The agricultural features under consideration are those which significantly rearrange the natural landscape, and which can be mapped, measured, excavated, analyzed, and related to other phenomena. They occur in a wide variety of habitats and in great diversity throughout the Western Hemisphere. They serve to make marginal land cultivable which is otherwise too steep, too dry, or too wet, by greatly varied methods of slope modification and water management. Millions (probably) of agricultural landform fields of pre-Columbian origin still survive on the landscape.

Agricultural landforms are grouped here according to primary function: slope modification, water-deficit management, and drainage, plus micromanagement features with variable functions. Some forms could be listed under more than one category because of more than one major function or because the major function varies under different conditions. Secondary functions may be numerous (e.g., Denevan and Turner 1974) and under certain circumstances may become primary. Field demarcations are included as a separate category; they are one of the few means of delimiting the area of many former fields.

Each agricultural landform is described and its functions are briefly discussed. The decision to combine or separate similar forms has necessarily been somewhat arbitrary. General distributions are indicated. Detail is avoided here, and only significant field studies and general surveys are cited. The typology presented is an attempt to provide a uniform terminology for the various types of fields and associated features to be discussed in this study. The classification does not parallel the organization of this study, which is based on field systems rather than field features.

Slope modification involves agricultural terraces or cropping surfaces which have been artificially leveled or reduced in slope angle, with an associated retainer wall usually of stone, but which may involve earth, volcanic ash, *tepetate* (indurated subsoil), or vegetation. Some terrace walls are of earth only, but long-term survival is then poor. There may be rock rubble below, within, or behind the wall to facilitate drainage. The major results of terracing, intended or not, are reduction of erosion, soil accumulation, and the slowing, holding (greater infiltration), and spreading of rainwater, runoff, and irrigation water. The importance of each varies regionally and with the type of terrace. Some 85 per cent of the land area with terracing has a dry season over five months long and a total annual precipitation of under 900 mm, so water conservation/management is clearly a basic function, although actual canal irrigation may not be required (Donkin 1979: 22).

Terraces are found in the New World from southern Colorado to central Chile, and are widespread in the central Andes. Most are of pre-Columbian origin. Probably at least 50 per cent of the Andean terraces are now abandoned. A comprehensive regional survey of New World terraces has been published by Donkin (1979). The main types of South American terraces (Field Types 8–11) are described in more detail in Chapter 9.

8. *Cross-channel Terraces and Check Dams* These are stone walls built across narrow valleys in dry uplands to trap soil (cross-channel terraces) and/or to

control water flow (check dams) or both. Other terms used for these features include weir terraces, silt-trap terraces, channel-bottom terraces, *trincheras*, *muros*, *presas*, and *bordos*. They are found throughout the range of terracing in the New World. However, they are much more numerous in northwestern Mexico and the southwestern United States than in the Andes.

9. Sloping-field Terraces J. Spencer and Hale (1961: 8–9) in their classification of terraces refer to 'linear, sloping, dry field terraces' where the steepness of slope has been reduced by a stone, earthen, or vegetation wall, without flattening (benching) the cultivation surface. Most are rainfed rather than irrigated. Runoff tends to be concentrated near the upslope side of the walls.

10. Bench Terraces These are the well-known Andean staircase terraces, having vertical stone retaining walls and flat planting surfaces, contoured, whose primary function is the control of irrigation water.

11. Broad-field Terraces These are a form of irrigated bench terrace with very low back walls and broad, flat cropping surfaces, often located on gentle slopes along rivers.

12. Segmented Terraces These are isolated, scattered, or discontinuous retainer walls and cropping surfaces, in contrast to parallel or serried rows across a slope (J. Spencer and Hale 1961: 9). Most are sloping-field terraces. Possibly they are incipient forms of more complex terrace systems. They are distinct from terrace walls which have been broken up by erosion. They occur in the Colca Valley, around Lake Titicaca, and elsewhere in the central Andes. Segmented terraces are also sometimes called semi-terraces.

13. Gradas, or Coca Terraces These are narrow, closely spaced, stone or earth retaining walls, about 1 m high, on steep slopes. They serve to reduce erosion and to increase settling of rainwater. The coca plants are grown in trenches between the *gradas*. Donkin (1979: 122–5) provides a brief description (with photos) of them, but otherwise they have been pretty much ignored. They are most common in the Bolivian *yungas* (eastern valleys), but have also been reported in the province of Sandia in southern Peru. None have been confirmed as pre-Columbian, but possibly some of those with stone walls are ancient.

Agricultural Landforms: Water-deficit Management

In the broadest sense, irrigation may be defined as making water available to crops by artificial means. More commonly, however, it refers to the transfer of water from source to field via man-made canals, ditches, and diversion walls. Here, canal irrigation, water diversion, and the control and conservation of flood water, slope runoff, and rainfall will all be considered as types of water management. Associated land forms include a wide variety of canals, embankments, and also excavated depressions.

An overview of prehispanic irrigation agriculture in the Americas was undertaken by B. Price (1971), but much of her emphasis is on social aspects. Regional surveys include Kosok (1965) and Park (1983) on coastal Peru. Almost all the water management literature has been on canal irrigation until recently.

The distribution of aboriginal irrigation is very similar to that of terracing. Irrigation is an arid and semi-arid land phenomenon, but it is sometimes found in wetter areas which have a long dry season, where irrigation provides greater security or makes possible a dry season crop.

Floodwater and Runoff Control

There is no clear distinction between floodwater and runoff farming. In both cases the water source is intermittent and unpredictable; temporary natural water flow is controlled, concentrated, and spread. Properly, floodwater is the flood of an intermittent stream (or flash flood) or the flood peak of a river overflowing its banks, whereas runoff is from a slope without significant drainageways. In the literature, 'floodwater' is often used to refer to both. Floodwater generally is not diverted by canals (see Number 5), but where this is done, the best characterization would be 'floodwater farming with canals' (Kirkby 1973: 38) or 'hybrid floodwater-canal irrigation' (Zimmerer 1995: 481). Here we are concerned with means of controlling floodwater and runoff.

14. Linear Diversion Embankments There is a large variety of relic embankments associated with past irrigation agriculture, especially in arid lands, and their functions are not always clear. They could have served to spread floods (spreaders), slow floodwater, stabilize fields, prevent sheetwash, prevent gullying, and to direct sheet runoff from slopes to adjacent fields (water harvesting, runoff fields, floodwater farming) or into a canal or conduit. They are not built across stream channels and hence their distinction from check dams (Number 8), in addition to usually being longer. They may occur as individual walls or in parallel groupings. When long, linear embankments direct sheet runoff to adjacent fields, they are called wings or diversion dams. Spreaders and wings may consist of stone piles, or brush, as well as earthen walls. Diversion features and canals may be associated with spillways and temporary holding basins or reservoirs. Examples of linear embankments occur in highland basins in Peru and Ecuador (Chapter 8).

15. Semi-enclosing and Enclosing Field Embankments The former are usually rectangular embankments of earth, sometimes stone, with small openings which could be readily closed. They could have served to either admit some water and then prevent further flooding, or to trap water in a man-made catchment basin or reservoir. Examples are found in Peru in the Virú Valley (M. West 1977), and in the Chilca Valley (Knapp 1982). A variation is the *albarrada*, found on the south coast of Ecuador and north coast of Peru. These are crescent-shaped ridges which trap rainfall and runoff water which is used for crops, pasture, and domestic use (Chapter 8). Enclosed embankments are completely surrounded fields. Functions are similar to those of semi-enclosed embankments. Both may include or be associated with floodwater control devices such as drops, gates, artificial depressions, canals, and raised fields. Today, fields are often enclosed by stone or adobe walls to prevent crop damage by livestock, as in the Colca Valley (F. González 1995). Discussed elsewhere are embankments enclosing raised fields (Number 27), terraces (Chapter 9), and reservoirs (Number 22).

There are also combinations of linear, irregular, perpendicular, and semi-enclosing embankments. The best example is at Chilca, where a variety of floodwater control embankments are integrated with sunken fields (Knapp 1982).

16. Pond Fields These are a form of sunken field, or artificial depression, which fill with runoff water and/or rainwater (Flores Ochoa 1987). As the water level drops the shallow margins are progressively planted in crops, similar to natural recessional fields (Number 5). They occur on the northern Lake Titicaca plain in Peru, where they are known as *qochas* (Chapter 8). They measure from about 50 to 200 m in diameter and up to 4 m in depth. This is a currently used indigenous system, but is likely a very old and continuous practice.

Groundwater Utilization

On the desert coast of Peru and northern Chile groundwater was used for agriculture by digging down to or near the water table and planting in the resulting depressions. Most of these sunken fields are located in lower river valleys where the water table is near the surface. It is not necessary to dig to the water table, which might result in water logging of crop roots, but rather only to the zone of high moisture content which lies above the water table as a result of capillary action.

17. Sunken Fields These are large depressions surrounded by earth embankments up to several meters high (Chapter 8). Individual gardens range up to between 100 and 200 m². In some instances, the surrounding embankments seem to be modified dune ridges, while the field surfaces are only slightly lowered. Floodwater management was probably involved as well as water-table farming. Relic sunken fields (*mahamaes, pukios, hoyas*) have been studied in Peru at Chilca (Parsons and Psuty 1975; Knapp 1979; 1982) and in the Virú Valley (M. West 1979). There are a variety of small depressions and trenches currently dug to tap groundwater in Peru and Chile, but antiquity is uncertain for most (*huecos*, *canchones*).

Moisture Conservation

18. Bordered Gardens and Lithic Mulches In arid lands there were several techniques of field surface modification that are apparently intended to conserve soil moisture available from direct rainfall or runoff, from fog condensation, or in some cases from canals. These features include bordered gardens edged by small stones, and lithic mulches of stones, sand, gravel, or ash. They occur in Arizona and New Mexico but are rare to the south (Lightfoot 1994). In South America, bordered gardens have been reported in northwestern Argentina and in coastal Peru; these may have served for moisture retention (see Chapter 8).

Canal Irrigation

Canal irrigation involves not only surface canals but a number of other landscape features which have survived from the past. These include field ditches, drops and headgates, aqueducts, dams, reservoirs, underground canals, and covered canals.

19. Canals Irrigation canals are highly diverse in terms of size, length, and construction. They are lined or unlined, artificial (or at least improved) channels of rock, earth, or stone, which tap water at higher points and direct it to fields at lower points. They may be fully excavated or partly excavated with canal banks raised above ground level by the excavated material. The water sources may be rivers, lakes, floodwater, ice melt, bogs, groundwater, springs, or reservoirs. Canals can be subdivided into primary canals, carrying water cross country from the source; secondary or distributary canals, providing subregional distribution; and tertiary or feeder canals, leading water into field furrows or flooding a field. Associated with canals are drops (spillways) usually with terracing, and gates (headgates, outlets) through which water is released into secondary canals or into fields. Drops and gates may survive and be identifiable, especially if made of stone.

Some pre-Columbian canals are still in use, but many are damaged and abandoned. Many have been completely filled in or buried by sand or sediment. The lines of former canals can often be identified, especially from the air, by soil, moisture, and vegetation patterns. Otherwise, trenching can be undertaken where canals are suspected. The best-studied canals are on the Peruvian coast (Chapter 8).

Aqueducts are artificially elevated canals to facilitate a steady gradient over uneven terrain. Most in the New World were built by the Spaniards, but a few are pre-Columbian. In coastal Peru there is a 3 km long aqueduct in the Pampa de Zana (Kosok 1965: 137). In the Andes, there is a small aqueduct in the Colca Valley (Denevan 1987*a*: 27). Field (1966: 460) mentions small aqueducts in northern Chile.

20. Underground Canals These are very common in the drylands of the Old World, where they are usually called *qanats*. They occur in northern Chile and southern Peru (Troll 1963; Barnes and Fleming 1991). Apparently most are of Spanish origin, but some may be indigenous (Chapter 8).

21. Field Ditches (furrows) Distinct from actual canals are the irrigation furrows within fields that are fed water from the canals. These ordinarily do not survive from the past; however, on the desert coast of Peru a variety of pre-Columbian furrow patterns can still be observed.

Furrows may be straight, with crossing ridges to pool water, E-shaped to pool water, S-shaped or serpentine where there is more of a slope gradient, or a combination of E-shaped and serpentine. Also, there may be stone piles, of unknown purpose, with or without the furrows.

22. Reservoirs and Dams Reservoirs (or tanks), usually earthen, are for water storage for canal irrigation (as well as for domestic use). They may be depressed (excavated), embanked, or both depressed and embanked. They were rare in pre-Columbian times in South America, but there were some in the Andes and Peruvian coast (Chapter 8).

True dams are earthen or rock constructions to block water flow in order to form reservoirs behind them, in contrast to check dams (Number 8) and water diversion embankments (Number 14). Andean reservoirs are mostly filled via canals, rather than by blocking channels. In contrast, there are large dams of pre-Columbian date in Mexico, such as the Purron dam in the Tehuacán Valley, which is 18 m high and 300 m wide (Woodbury and Neely 1972: 82–99).

Manual Irrigation

Manual irrigation refers to the transfer of water from source to field by hand methods, either by splashing (splash irrigation), whereby water is splashed directly from a stream or canal by hand or scoop, or by a container (gourd, bowl) which is filled and then emptied on to a field. Container or pot irrigation, in particular, is more water-efficient but less labor efficient than either basin flooding or furrow irrigation. Both methods were probably widespread prehistorically in conjunction with various forms of canal irrigation and raised-field agriculture. Terrace irrigation by means of pots of water (*cántaros*) has been reported for Peru in the sixteenth century (Antúnez de Mayolo 1980: 40). Container irrigation of plants in house gardens is common today.

23. Wells Vertical shafts dug to the water table are not common in pre-Columbian South America, probably because of a lack of technology for digging deep wells and also because most old wells which have been filled in are not readily identifiable. Walk-in wells have been described at Chan Chan in Peru (Day 1974). It is believed that their main purpose was for residential use, but they may also have been used via containers for house gardens.

Agricultural Landforms: Drainage

In contrast to terracing and irrigation, pre-Columbian drained-field agriculture received little attention until recently. Since 1961, great numbers of raised fields in poorly drained tropical savannas and in highland basins have been discovered, primarily with the aid of aerial photography. Less common are ditched fields, drainage canals, dikes, and embanked fields. In Latin America there was more poorly drained land in cultivation in pre-Columbian times than at present.

24. Ridges and Platforms (Large Raised Fields) Raised agricultural fields include any prepared land involving the transfer and elevation of soil above the natural surface of the earth in order to improve cultivating conditions on that surface (Denevan and Turner 1974: 24). They have a variety of functions, but the primary one for large raised fields in South America is drainage. (These are considered here, and small raised fields are described under Numbers 28 and 29.)

There is a great diversity of relic raised, drained fields in terms of size, shape, and patterns. One major distinction can be made between platforms, which are low and very wide, and ridges, which tend to be higher and narrower and usually occur where flooding is deep. Raised fields range up to about 2 m in height, 25 m in width, and 500 m in length. Most are separated by ditches, which served several potential functions besides the source of earth for the fields: subsoil drainage, water removal, irrigation, canoe travel, fish culture, and a source of soil nutrients.

As with other agricultural landforms, many raised fields have been destroyed by erosion, or buried by sediment accumulation, or are invisible under water, tall grass, or a forest cover. They can nevertheless be detected by remote sensing techniques, soil and vegetation marks, vegetation removal, and, where buried, by exposure in river cuts or by trenching.

Large raised fields have been found throughout South America (Chapter 11). Almost all are on seasonally inundated sites where wet-season cultivation would not be possible without some form of drainage. All the raised fields are prehistoric in origin and are now abandoned, with the exception of a few which have been restored or built anew.

25. Ditched Fields Drainage by ditching only, without an accompanying build-up of raised fields, is rare; the earth from ditches has to go somewhere.

Possibly this reflects the fact that ditches fill rapidly with sediment and are difficult to identify. One good example occurs in the Llanos de Mojos of Bolivia (Chapter 12). The ditches are about 1 m wide and 20–35 cm deep and 2–10 m apart. Present day aboriginal ditching is found in the Andes and with the Karinya Indians in the Orinoco Llanos (Chapter 14). It was probably an ancient practice in these areas as well as elsewhere.

26. Drainage Canals Whereas ditching with or without raised fields serves for subsoil drainage or localized water removal, further removal of excess water from a field area requires larger canals. Such canals have been identified in association with fields in the Llanos de Mojos (Denevan 1966a: pl. 18) and at Lake Titicaca (Kolata 1993: 224–8). Some of these canals, however, may have served other functions such as for canoe travel and transport and fish ponds. Drained fields are also irrigated fields in that a portion or even most moisture is obtained through subirrigation or capillary action as water moves from ditches through the subsoil to crop root levels.

27. Dikes and Embankments Fields were also protected against poor drainage by the construction of dikes (stone or earth walls), or by semi-enclosing or fully enclosing embankments. Single walls occur at Lake Titicaca, often paralleled by drainage ditches. Full and semi-enclosing embankments around raised fields also occur at Lake Titicaca (Chapter 13). These are irregular in shape, even circular. They seem to occur in particularly wet areas. They probably represent individual efforts to control flooding, in contrast to a more regional effort to remove water by canals or to protect fields by long continuous dikes. In the Llanos de Mojos, Plafker (1963: fig. 4) shows full and semi-enclosing embankments around clusters of raised fields, apparently to keep excessive flood waters out of the field areas.

Micro Management of Soil, Water, Climate, and Slope

There are a variety of forms of small raised fields, past and present, developed to modify soils physically, to conserve moisture, to improve drainage, to change slopes, or to modify microclimate (temperature, wind). From region to region, form may be similar but function different. The two basic forms are planting beds and mounds. Given their small size and construction of earth, survival is poor but there are relic examples of each.

28. Planting Beds In South America, narrow (1–1.5 m) planting beds are often called *camellones*, although the same word is also used for larger raised fields. Cobo (1956: 2: 253) used this Spanish term in 1653. Beds are very common in the Andes where they are called *eras* in Colombia and *huachos* in Peru. They are still constructed with the footplow in Peru, mainly for potatoes, and they are very similar to the 'lazy beds' (narrow crop ridges) of Ireland and northwestern Europe. They loosen the grass sod, control weeds, improve aeration, drain soil subject to waterlogging, reduce evaporation, and increase ground temperatures. They occur both on level surfaces and on slopes. Types of Peruvian *huachos* are described by Sánchez Farfan (1983: 167–9). Abandoned *eras* in Colombia may be pre-Columbian according to R. West (1959).

29. Crop mounds (maize hills, montones) Mounding, especially for maize and for manioc, is and was a widespread practice throughout the Americas (Gallagher 1992). Most of this has involved only a slight raising of earth for one or two plants, and survival is brief. However, somewhat larger mounds have persisted for a long time in a few places in South America, including the Colombian Llanos and Salta in Argentina (Denevan 1980*c*: 643). The Spaniards reported hundreds of thousands of large manioc and sweet potato mounds (*montones*) on Hispaniola in the early sixteenth century (large enough for six to ten manioc cuttings), but few remains have been found (Sturtevant 1961). Mound functions are similar to those of planting beds, and crop mounds around individual maize plants also protect from wind throw.

Field Demarcation

Many of the features described here are associated with agriculture but are not cultivated fields, with the major exceptions of terrace floors and raised fields. Thus it may be very difficult to identify actual former fields. In some instances, however, the fields may be outlined by fringing canals, dikes, or drainage ditches. There are, in addition, several types of protective field enclosures which also were or are intended as field boundaries.

30. Boundary Walls, Fences, and Markers Prehistoric stone walls primarily serving as field boundaries occur, but most of the walled fields in Latin America are post-Columbian. One reason, in contrast to the Old World where field walls serve to keep domesticated animals in or out of fields, is that there were no significant domesticated animals in the New World outside the Andes.

There are prehistoric rock-walled fields on high slopes in the Colca Valley of Peru and elsewhere in the Andes, often associated with terraces. On the Pacific Coast of Peru, prehistoric field fences were constructed of spiny branches, such as *algarrobo*, and of *adobe* (Latcham 1936: 294). Netherly (1988: 271, 274) for the Río Moche basin describes three Inca coca fields walled with *tapia* (mud) to keep out wild animals or possibly to display the Inca presence in a hostile area. Non-permanent fences constructed from a variety of materials are common around Indian fields today, but may not have been so in the past. Living fences, containing trees and shrubs, are important for several economic and ecological functions in addition to field boundaries (Gade 1975).

Stone boundary markers are common in the Andes but are seldom reported. The importance of field boundary markers for the Incas was indicated by chronicler Bernabé Cobo (1979: 211): 'The boundaries of the lands and fields . . . were kept so exact, and the care and protection of these markers of the fields . . . were so impressed upon the Indians that it was one of the most important religious duties that they had.'

Field fences, walls, and boundary markers, even living fences, may persist long after field abandonment and thus delimit those fields. For a general discussion of the archaeology of field boundarie see Gleason 1994: 2–11.

Thus, thirty categories of Indian agricultural fields and field features have been differentiated in South America, some minor or unique to be sure, and some have not clearly been identified as pre-European. These categories can be further subdivided into distinct variations, and additional forms will surely be described. For relic fields, functions need to be clarified, dating techniques better developed, and methods for locating and describing need to be refined. For practices that persist, systematic investigation is essential before they are engulfed by modern agricultural technology.

Notes

[1] Terms used for agricultural landforms include 'agro-engineering works' and 'geointensive agriculture' (Atran 1993: 636, 695).

[2] Some additional forms have been described in Mesoamerica and North America (Denevan 1980c; Wilken 1987; Glassow 1980).

[3] 'Swidden', meaning burned field in Old English, has become a generic term for all forms of shifting cultivation.

3

Crops, Tools, and Soft Technology

The focus of this study is on the visible landscape features created by or for cultivation. As background, however, I first provide a brief listing and description of the associated aspects of indigenous agricultural technology in South America—the crops grown, the tools used, and soft technology or non-landscaping techniques of cultivation, such as fertilization and crop rotation. These aspects will be elaborated on in subsequent chapters where appropriate.

Crops

New World crops (cultivated plants) have received much more attention than cultivation techniques, and information about crops—origins, history, characteristics, distribution—will not be covered here.[1] C. Sauer (1950) provides a good but dated treatment of the native cultivated plants of South America. Useful regional studies are given at the end of Appendix 1A. Entire books are devoted to individual species, such as Salaman (1949) and Ochoa (1990) on the potato, Yen (1974) on the sweet potato, Heiser (1979) on the gourd, and Andrews (1984) on chili peppers. Good general inventories of crops are by Schery (1972) and by J. Sauer (1993). A large portion of the articles in the journal *Economic Botany* are concerned with New World crops.

Crops are, of course, the reason for agricultural fields and associated cultivation techniques which are designed to enhance ecological conditions for crop growth. Technology is often crop or crop-group specific; hence crops will frequently be mentioned in the chapters that follow. For that reason, an inventory of South American crops (over 200 species), with identifications, is provided here (Appendix 1A, 1B). The listing of major and secondary crops should be fairly complete. A large number of minor crops are also listed.

Three types of 'crops' can be differentiated. The first are cultigens or true domesticates, which have undergone morphological or physiological change to the extent of dependency on people. The second group might best be termed 'semi-domesticates', in which there has been a degree of modification as a result of human selection and/or management, intentional or unintentional. A third group would be wild plants which are at times planted or protected;[2] several useful tropical palms fall into this category. Native people, however, may make no distinction between these categories (Descola 1994: 167).

In Appendix 1A, the first common name listed is the term used in the text.

Field Tools

A history of pre-Columbian field implements in the highlands of Middle and South America is provided by Donkin (1970; also 1979: 3–16). Other good sources are Latcham (1936: 304–30), Rivero (1983), Gade and Rios (1972), F. Kramer (1966), Bullock (1958) on Chile, and Rojas (1984) on Mesoamerica. Some information on tropical lowland tools is also included in F. Kramer. Knowledge of pre-European cultivation tools is still inadequate, however (Rojas 1984: 175). A summary is presented here of South American agricultural tools. These tools were used to clear vegetation, for weeding, for excavating and raising earth, and for planting. Most were made of wood or stone and occasionally copper or bronze.

Stone Axes A primary means of removing trees for cultivation in tropical forests (and elsewhere) was the stone axe. This tool is clearly quite old and quite widespread, and occurs in archaeological sites. Early European observers, however, provided little information on form, making, or use. Possibly the earliest description of stone axes is by Pedro Vaz de Caminha, who was with Cabral on the Brazilian coast in 1500: 'And they cut their wood and boards with stones shaped like wedges put into a piece of wood, very well tied between two sticks' (Caminha 1938: 26).

Hans Staden (1928: 136–7) in 1557 mentioned the stone axes of the Tupinamba of the Brazilian coast: 'Before the ships began to arrive the savages had (and they have even now in many places where the ships do not come) a certain bluish-black stone shaped like a wedge, which they sharpen at its broadest end. These stones are about a span long, two fingers thick, and as broad as a hand, but some are larger and some smaller. They then take a thin reed and bend it round the stone, binding it with bast.' Padre Cristóbal de Acuña (1942: 54) in 1639 observed 'axes of stone' which could cut down trees of any size, used by Amazonian Indians.

Stone axes consisted of a shaped, sharpened stone hafted to a wooden handle. The type of stone was determined by what was durable and was available locally or could be imported, usually stream cobbles already water smoothed. Stone is rare in much of Amazonia, and long-distance trading for stone axe heads was important. Metal (bronze and copper) axes were used in the Andes, but generally not as agricultural tools.

Amazonian axes are described by Carneiro (1974). They consist of three types: (1) the celt, which has a smooth butt, is the most common, (2) the T-shaped axe from western Amazonia, which has a notched butt for lashing, and (3) the bent-over or split-branch axe of central Brazil (Fig. 3.1). All these axes

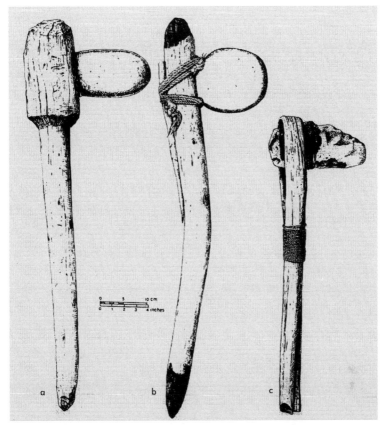

Fig. 3.1. Types of stone axes in Amazonia: (a) imbedded celt (Héta), (b) T-shaped axe (Amarakaeri), (c) axe with bent-over handle (Nambicuara). From Carneiro (1974: 120)

are subject to fracturing and dulling and are by no means an easy means of felling large trees. They might be expected to smash rather than cut fibers, but a pure 'cut' is possible (Carneiro 1979*a*: 40). Their edges are blunt, thus reducing their bite compared to metal axes. Their use was supplemented by girdling, firing, charring the trunk base, and driving tree falls; often the largest trees were left standing (Carneiro 1974: 114–15).

Carneiro (1974) reported on four Amazonian societies still using stone axes in the twentieth century: the Héta in Paraná, Brazil; the Akuriyo in southeastern Suriname, the Kreen-a-kore in northern Mato Grosso, and the Amarakaeri in southeastern Peru. His description of Amahuaca axes (southeastern Peru) is based on stories of its use and surviving specimens. The one detailed description of stone axe making is by Vladimír Kozák (1972) for the Héta.

Stone axes are, of course, much less efficient in felling trees than are iron and especially steel axes. This has important implications for the feasibility of different forms of forest site cultivation (see Chapter 7).

Macana The term *macana* (Quechua) refers to a wooden club. It is used here for a tropical forest wooden tool for slashing forest undergrowth and for weeding. They have been replaced by the steel machete almost everywhere. One version is that of the Shipibo on the Río Ucayali in eastern Peru which is made of hard assai palm wood, with a fiber handle and a notched end, at times decorated with geometric designs (Steward and Métraux 1948: fig. 78e). Shipibo told me that the tool was also used as a war club. They are primarily made for the tourist trade today. The Amahuaca have a similar 'sword club' called a *windó*, used to fell bamboo and other small vegetation (Carneiro 1974: 116). Gumilla (1963: 429) described clearing of brush with a *macana* or 'hardwood sword' in the Orinoco Llanos in 1745. Descola (1994: 185) maintains that the wooden broadsword of the Achuar is almost as effective for weeding as the metal machete. The Kuikuru prior to 1900 used piranha mandibles to clear undergrowth (Carneiro 1961: 47, no source). This is the only mention I know of for such a tool.

The term *macana* is also used for a paddle-shaped spade (see below), and some forms may have served both clearing and digging functions. In the Andes and in Panama a *macana* was a wooden, sword-shaped, double-edged war club (Rowe 1946: 276; Oviedo 1959: 26).

Garabato Another simple weeding tool used today by the Karinya in the Orinoco Llanos is the *garabato*. This is the fork of a branch with one arm cut short and the other left long as a handle. It is used as a hook to pull vegetation toward a worker who then cuts it with a *macana* or machete (Denevan and Schwerin 1978: 17–18).

Digging Stick (Planting Stick, Dibble, Palo) This is the basic agricultural tool found all over the New World. Most are branchless wooden poles, sharpened and fire-hardened at one end. Some may be tipped with stone or metal, especially in the Andes (Donkin 1979: 5–6). They range up to about 2 m in height, with considerable thickness (see Fig. 5.3). They are used primarily for planting seed. The stick is plunged into the soil to make a hole several cm to 0.3 m or so deep, into which three to ten seeds are inserted, and then the hole is filled with dirt. Oviedo (1959: 14) in 1526 described the use of the digging stick on Hispaniola: 'An Indian takes in his hand a stick as tall as he is, and plunges the point into the earth, then he pulls it out, and in the hole he has made he places with his other hand about seven or eight grains of corn.'

Lampa The *lampa* in the Andes (*coa* in Mexico) is a short, metal-bladed digging tool, triangular with a pointed end, used for digging, hilling, planting, weeding, breaking up clods, and directing irrigation water. It lacks a foot rest and so is not properly a spade or shovel but rather intermediate between a planting stick and a shovel. Cobo (1990: 214) described copper-bladed *lampas* in Peru *c*. 1650. At the time of the conquest they seem to have been mainly found in northern Peru and Ecuador, as well as Mexico (Donkin 1979: 9). In Chile, they were constructed of algarrobo wood (Latcham 1936: 310). A *lampa*-like triangular bladed tool (*chuzo*) is used in highland Colombia to dig *eras* or lazy beds (R. West 1959: 279).

Another, longer variation of the *lampa*, or planting stick/shovel, is the *barreta* in the southern Andes (Latcham 1936: 307–9). This is a sturdy stick with a chisel-shaped end, used to break and turn the soil. On Chiloe Island the *barreta* was made with two sticks (*luma*) worked by two Indians together (Tschudi 1852: 10; Donkin 1979: 13). The Araucanians of central Chile had a *barreta* in the form of a trident or pitchfork. Three pronged *barretas* are found in the Colca Valley (S. Brooks, pers. comm.). All of these may be post-Columbian.

Spade or Shovel These are digging tools, usually wooden, but in parts of the Andes they have metal blades lashed on. Nordenskiöld (1919: 29) shows a distribution map for western South America. There are several forms. One is the true spade, for which there is a foot rest or shoulder on one or both sides. The other form lacks the shoulder but could still serve as a shovel for scooping up earth. These are paddle-like tools, usually from a single piece of wood. They are widespread in South America, including the Andes, central Chile, the Chaco, and Amazonia (F. Kramer 1966: 59–64). Padre Gumilla (1963: 429) described Indians in the Orinoco Llanos in the eighteenth century making raised fields with *palas de macana*, which seemed to be wide-ended digging sticks or spades. The Yanomami of Venezuela used a palmwood shovel (*fimo*) until recently (Smole 1976: 127). The Shipibo were using a manatee-shoulder blade hoe for weeding in the nineteenth and early twentieth century (Myers 1990: 36).

Footplow (Chaquitaclla, Taclla) The Andean footplow has been called the most advanced agricultural tool of native America (Gade and Rios 1972: 3; also see Donkin 1970: 515–19; 1979: 10–14; O. F. Cook 1920; and Rengifo Vásquez 1987). It dates to at least AD 1300 and extended from Ecuador to Bolivia. The *taclla* consists of a stave with a flat end or a metal share to which a wooden foot rest is lashed. Higher up is a handle or grip. Several excellent sketches from the early seventeenth century appear in Guamán Poma de Ayala (1980: 3: 1044, 1050, 1062) (Fig. 3.2).

The *taclla* is especially used for breaking compacted earth for the making of low, narrow potato ridges (*huachos*), at times by teams of men in tandem, with women following to turn the sod (*champas*). It is also used with other crops and for other purposes such as planting, harvesting, cutting sod for fences, and digging irrigation and drainage ditches. Farmers in Puno at Lake Titicaca who are reconstructing raised fields today use the *taclla* and probably did so in the past (Erickson 1988*b*: 13; Garaycochea 1987: 390). There are a wide variety of

Fig. 3.2. Harvesting potatoes using the *taclla* footplow and the *liukana* hoe, *c.* 1615. From Guamán Poma de Ayala (1980: 3: 1044)

footplow forms. One is the *huiri*, which is short and has a curved handle. It is preferred on relatively flat terrain (Gade and Rios 1972: 9).

Bernabé Cobo (1990: 213) described the Inca use of the *taclla* as follows: 'The manner in which they broke and plowed the land with this *taclla* was

Fig. 3.3. Simple footplow used in the 1960s by the Campa in the Gran Pajonal, Peruvian Amazonia. Photographer unknown

by raising the top of it up to the right shoulder and the point about two or three spans from the ground, and they would drive it down with all the force of their arms and their left foot which bore down on the above-mentioned rest; they did this so that they could make the plow strike with force and penetrate deep.'

The *taclla* has been replaced in many areas by the ox-driven plow, but it is still used on high, steep slopes or where oxen are expensive or where potato ridges are desired.

Amazonian Footplow There is a curious version of the footplow in Peruvian Amazonia where some of the Campa Indians still use it (Fig. 3.3). It is a very simple pointed or flat ended pole, with a flat rock lashed on for a foot rest. It is the only advanced agricultural tool reported in Amazonia, and was probably derived from nearby highland Indians. It was observed in the 1920s by Tessman (1930: 90), but is likely much older. The use of such footplows and also crude hoes for weeding may reflect a more intensive cultivation in Amazonia in the past.

Mattock *(Liukana, Raucana, Azadón)* This is a short-handled hoe still widely used in the central Andes (Guamán Poma de Ayala 1980: 3: 1029; Cobo 1956: 2: 252). Copper and bronze blades have been found in archaeological sites from Ecuador to northeastern Argentina. Blades of wood, stone, and bone were probably also used (Donkin 1979: 15). The blades are tied to a hooked stick

(Fig. 3.2). These tools are used for working the soil, weeding, and harvesting. Long-poled true hoes were absent in pre-Columbian South America or rare at best (Donkin 1979: 14).

Clod Crusher (Clod Breaker) This was and is used in the Andes to break up clods of earth, usually after plowing with the *chaquitaclla* (Guamán Poma de Ayala 1980: 3: 1062). The crusher is of hard wood (*huyni*) or a perforated, rounded stone tied to a shaft (*huypu*); they are called *mazas* in Chile (Latcham 1936: 323). Stone crushers are common in archaeological sites, although some may have been used as weapons or for other purposes. One was recently found within a raised field in the Casma Valley of coastal Peru (J. Moore 1988: 273). Broken doughnut-shaped stones are common in prehistoric fields in the Jequetepeque Valley (Eling 1987*a*: 168–71).

These, then, are the basic, traditional cultivation tools, with numerous variations of each. The distribution of some of these tools (*coa, taclla, liukana*) is mapped by Donkin (1979: 4).

Soft Technology

'Soft technology' is a term used by Hecht and Posey (1989: 177) to refer to relatively invisible but essential field technologies which have little or no long-lasting impact on the landscape, in contrast to 'hard technologies' consisting of permanent field features, our primary concern in this study. It must be emphasized that fields can not function well and for long (sustainable) without these soft technologies for manipulating the field environment, based on a deep knowledge of nature, crop ecology, and local conditions. Soft technology includes soil management, spatial and temporal crop patterns, fallowing, use of fire, and pest control. These techniques are difficult to detect archaeologically, and there is little mention of them in the early literature. We know of them mainly from ethnographic accounts, for which there may be uncertainty as to indigenous origins.[3] Here, I will briefly outline some of these techniques and provide some historical and contemporary examples. There is further discussion in later chapters where appropriate. Systematic treatment of most methods mentioned is provided by Altieri (1995).

Organic Fertilizers

Intensive forms of agriculture were widespread in South America and were often on marginal soils. Fertility was maintained by the application of organic material, fallowing, crop rotation, and mixed cropping. The most systematic application of organic material that we know of was in the central Andes.[4]

Guano Bird *guano*, rich in nitrogen and phosphorus, from the offshore islands of Peru, was the most important source of fertilizer for the coastal Incas according to several sixteenth-century chroniclers. Apparently the Incas recognized bounded claims or rights to *guano* on the islands (Julien 1985: 221). Garcilaso de la Vega (1966: 1: 246), not always a reliable source, reported that there were sanctions against killing the birds, going to the islands during nesting, and mining *guano* from the claims of others. There was even a god of *guano*, Huamancántac (Arriaga 1968: 52).

Cieza de León reported *guano* use as maize fertilizer on the south coast of Peru in 1548–9: 'Its use is advantageous to them because the earth where they plant is made rich and fertile instead of sterile, because if they had not used this dung, they would harvest very little maize . . . Because of the value of this dung, the Indians trade it among themselves like a precious substance' (Julien 1985: 189). *Guano* was incorporated into the soil a handful at a time with seeds and/or later after plants had sprouted (Julien 1985: 190). The *guano* was obtained in enormous quantity from the offshore islands and moved by boats (of sea lion skins?) to the coast (Cobo 1956: 1: 85).

Diez de San Miguel (1964: 245) in 1567 reported that the Indians of Moquegua on the south central coast of Peru brought *guano* 20 leagues (*c.* 100 km) from the coast. Vázquez de Espinosa (1942: 519) in 1628 said that llama trains carried *guano* along the coast. A. Wright (1962: 99–100) mentions the use of llama-transported *guano* on upland terraces in Arica, Chile. Julien (1985: 192) suggests that a farmer would require more than 15.5 kg of *guano* for his farm, or about one-third to one-half of a llama load. Thus, enormous caravans of llamas carrying *guano* would be necessary to have an impact on altiplano farming where there were hundreds of thousands of family farms. Consequently, *guano* use far from the coast was probably limited and localized. However, western mountain towns in Arequipa had rights to coastal *guano* in the eighteenth century and probably in the sixteenth century. Production in these towns, in part based on *guano* use, helped support large Indian populations.

Fish Several early writers mention the use of fish or fish heads (anchovies) as maize fertilizer in the coastal valleys of Peru (Rowe 1969: 321; Donkin 1979: 1). According to Cieza de León (1959: 337): 'But the seed of the corn in no wise sprout nor yield if they did not put with each a head or two of the sardines they take in their nets.' Sardines were burned and placed with maize seeds in sunken fields on the coast of Peru (Garcilaso de la Vega 1966: 1: 247). Some fish fertilizer may have been transported to the highlands. A mural at Pachacámac shows a maize plant growing out of a fish (Fig. 3.4). Kolata (1993: 187) indicates the probable importance of fish fertilizer for the Titicaca raised fields.

Dung The use of llama dung for potato crops in the Andes was reported by Garcilaso de la Vega (1966: 1: 246) in the sixteenth century and was probably of

Fig. 3.4. Inca mural showing maize growing out of a fish, suggesting use of fish as fertilizer, Pachacámac, coastal Peru. From Bonavia (1985: 177)

major importance. Dung is the main source of organic fertilizer today in the Andes (Winterhalder *et al.* 1974). Different dungs were analyzed by Winterhalder. Generally, sheep and cattle manure have a somewhat higher nutrient content than llama dung, and llama dung production per animal is less; however, llama dung is clearly productive and is still widely used. Guinea pig droppings, mixed with floor refuse, are also used today. Different dungs and forms of dung are used for different crops (Sánchez Farfan 1983: 169–71). The only early mention of the use of human waste as fertilizer was on maize fields at Cuzco by Garcilaso de la Vega (1966: 1: 246).

Latcham (1936: 300, no source) said that for northern Chile and Peru maize and potato planting would be preceded by digging trenches which were filled with dung and then covered with a soil layer. For other crops manure was equally distributed on the surface. If in limited supply, then only individual plants were fertilized.

Folding This is a European term, referring to the construction of temporary corrals of branches and vines or rope, which are moved often over fields in order to fertilize them with manure. They have been reported at Acora on the Peninsula of Chuquito on Lake Titicaca (Tschopik 1946: 517), and at Pampallaqta and Espinar in Cuzco (Sánchez Farfan 1983: 171; Orlove 1977: 95). I have observed them in the Vilcabamba region north of Cuzco. In highland Ecuador such moveable pens are called *talenqueras*. To fertilize just one hectare, as many as 500 sheep may be bedded for one hundred nights (Basile 1974: 108, 113). Folding may or may not be indigenous in the Andes.

Anthropogenic Top Soil Settlement sites concentrate organic material resulting from human activity—manure, ash, garbage, bones, etc.—and soils around old sites are enriched accordingly. Such abandoned sites may be intentionally sought out for agricultural fields. One example is the so-called black soil or *terra preta do índio* of Amazonia, scattered patches of rich soil of pre-European cultural origin (see Chapter 6).

Other potential sources of organic fertilizer include ash, garbage, weeds, leaves, and field residues. However, there are few early ethnohistorical reports of their use specifically as fertilizer. Oviedo (1959: 14) in his *Historia Natural de las Indias* in 1526 mentioned the use of ash as a field dressing on Hispaniola. The Aymara in highland Bolivia applied ash to soil, as well as *salitre* or nitrate according to Soria Lens (1954: 92), who also said that 'manured or organic soils' were transferred to sites of poor soil.

Mucking This is the addition of rich organic, aquatic deposits to field surfaces,[5] and is mainly associated with raised fields, both in the tropical lowlands and in Andean basins. It is a current practice with Mexican *chinampas* (Wilken 1987: 82–8) and New Guinea raised fields (Steensberg 1980: 85–7), and field excavations at Lake Titicaca confirm mucking of prehistoric raised fields (Kolata and Ortloff 1989: 242–6). Restored raised fields in Puno at Lake Titicaca currently use muck, with excellent results (Erickson 1988b; Garaycochea 1987). In wetlands with low soil fertility, mucking was probably essential to make raised fields productive on a relatively continuous basis.

Composting Compost is a mixture of partially decomposed, piled organic material, including dung, ash, garbage, leaves, etc., which is applied to fields as fertilizer, either as a surface dressing (or mulch) or mixed with soil.

The Kayapó of the Brazilian Amazon prepare compost heaps of mixed organic material which is mixed with soil from termite and ant nests (with living termites and ants included) and carried into savannas and made into small mounds. These sites over time may become larger forest islands (*apêtê*) dominated by useful wild and cultivated plants (Posey 1985a: 142). The sunken fields of coastal Peru were fertilized with amendments of decomposing vegetation according to Latcham (1936: 286, no source).

Nitrogen Fixation Green manure refers to a nitrogen-fixing legume crop which is rotated with primary crops and plowed in to help restore fertility. Indigenous examples in South America are uncommon. Beans and peanuts are important legume crops, but they are intercropped rather than rotated in the tropics. The

European broadbean (*Vicia faba*) is important today in rotations in the Andes, and New World beans likely served that role in earlier times. Intercropping maize with beans was widespread aboriginally throughout the Americas for nutritional (complete amino acids) as well as for fertility enhancement purposes. In maize/bean intercropping, Gliessman (1982) found that maize yields were increased up to 80 per cent more than with monoculture maize.

Leguminous trees may be left or planted as individuals, in clusters, or intercropped for nitrogen fixation. This is reported in Mexico (Wilken 1987: 66–8), and I have seen it in both the Andes and Amazonia.

Donkin (1979: 2) mentions chemical analysis of soil from abandoned terraces at Choisica (near Lima) indicating that beans were planted on higher terraces so that soluble nitrates would flow down slope to other fields.

Llakoshka This is an old, but continuing fertilizer practice in the province of Anta (Cuzco), which increases yields by 20 per cent or more (Anonymous 1985, no source). Seeds are dipped in a putrefying and fermenting mixture of dried llama dung, salt, and *chicha* (maize beer), and sometimes juice from the fruit of the molle tree. Resulting biochemical processes make inorganic elements in the soil more easily assimilated, parasites and aerobic organisms are destroyed, an anaerobic bloom is created, the dung provides nutrients for the seedlings and root system, yeast from the *chicha* turns seed starch to sugar which is advantageous to root development, plus there are other positive effects. At Chilca on the coast a similar process used putrefied anchovies instead of dung.

Inorganic Soil Additives

There are several sixteenth-century reports of top soil for terraces being brought from a distance (Matienzo 1910: 12; Sarmiento de Gamboa 1907: 98). Cieza de León (1959: 193) reported that the soil of the Valley of Cuzco was so unproductive that thousands of loads of dirt were brought from the Andes and spread out to make the valley fertile 'if this is true'. Garcilaso de la Vega (1966: 242) also mentioned soil transfer to terraces in rocky places. The soil scientist S. W. Buol (J. Treacy, pers. comm.) on a visit to Machu Picchu examined terrace top soil whose source he believed to be the banks of the Río Vilcanota below. Terraces above the Cusichaca Valley in Peru (Fig. 14.5) probably contain soil brought up from the valley below (Keeley 1985: 563). The Kayapó in Amazonia today carry rich top soil to rocky sites and place it in cracks which are then planted (Posey 1985*a*: 152). It is unlikely, however, that long distance soil transfer was a common practice anywhere, given the weight involved and the consequent high labor inputs required.

Wilken (1987: 70–5) discusses the management of silts, mucks, and sands in field applications in Mesoamerica by contemporary traditional farmers. For South America, such practices undoubtedly are and were common.

Silting or silt trapping involves the slowing down of silt-laden water so that the silt is deposited in cropping areas. The main mechanism is the silt-trap terrace or check dam built across slope or in a channel. (These may also serve to reduce erosion and conserve water.) They are common in northwestern Mexico and the southwestern United States, but not in South America. Donkin (1979: 92, 108, 126, 130) describes old silt-trap or cross-channel terraces in western Ecuador, near Sicuani in southern Peru, at Socaire in northern Chile, and at Ancasti in northwestern Argentina. Brooks (1998: 130–1) describes them in the Colca Valley of Peru. The highland Indians were well aware of the value of silt deposited by rivers or carried in irrigation canals (Cobo 1956: 1: 85; Soria Lens 1954: 92). In Peru such silt was referred to as *ihuanco* or *colmataje*, meaning silt fertilizer (Antúnez de Mayolo 1980: 40). A *cacique* (chief) in 1555 reported that the Inca Cusi Guallpa (Huascar) diverted a river in order to deposit silt for a new maize field (Rostworowski 1962: 135, 142). The Aymara in Bolivia also used silt traps (Soria Lens 1954: 89).

Fire

The use of fire as a tool in preparing land for cultivation is worldwide and well known and well studied. Land cleared of forest is invariably burned before planting, not only for shifting cultivation but for more permanent forms of agriculture. Cleared debris is thus removed, microflora and microfauna are reduced, and a nutrient-rich (carbonates, phosphates, and silicates) layer of ash is left behind (Nye and Greenland 1960: 66–73). At San Carlos in the Venezuelan Amazon, soil charcoal samples dating from *c.* 4000 BC to AD 1750 are believed to represent both agricultural burns and wild fires (Clark and Uhl 1987: 3–4). Fire also warms the soil, beyond the burn itself, by removing litter which blocks solar radiation and by blackening the surface of fields and thereby increasing absorption of solar radiation. Surface temperatures of burned sites may be several degrees higher than otherwise, and this can be important for crops in cold lands (Gallagher 1989: 580).

The use of fire, however, can be much more sophisticated than the simple burning of cleared vegetation. The Kayapó use small, localized fires (in-field burning) of weeds, crop residue, and forest litter for managing soil within their swidden plots throughout the cultivation cycle: 'The control of the volume of biomass, seasonal timing, diurnal timing, and the temperature of the burn' (Hecht and Posey 1989: 180). The Kayapó have many descriptive terms for types of ash, as well as songs and rituals, and shamans specialize in burning techniques. In addition, burning can help control disease pathogens.

Kayapó burning may take place after some crops have been planted as a form of phyto-sanitation which is crop and pest specific; for example, for sweet potato virus disease control (Kerr and Posey 1984: 393). Controlled burning is also important for the concentric structure of Kayapó swiddens (see below), and probably helps explain plot architecture of other Indian groups. Such complex use of fire and ash seems beyond the knowledge of most non-Indian settlers in Amazonia.

Mulching

Mulching involves the surface spread of straw, ash, garbage, leaf litter, manure, aquatic plants from ditches, cut weeds, and crop residue. Functions include protection of new crops from direct sun and rain, conservation of moisture, reduction of soil temperature, erosion, and runoff, a slow-release nutrient source, and improvement of soil structure. Mulch material may be brought in from adjacent forest or developed in place as with crop and weed residues. Specific crops may be planted. Material may originally be a surface dressing but later buried or mixed with soil.

Wilken (1987: 62–6) describes mulching techniques by Indians in Mexico and Guatemala, and the practice undoubtedly was and remains widespread in South America, but is seldom described. One example is the Kayapó who use palm leaves, banana leaves, crop residue, and cut weeds (Hecht and Posey 1989: 180–1); also see J. Kramer (1977) on the Urarina in Peru.

Cropping Patterns

Indigenous planting patterns are highly varied in terms of crop combinations, zonation (space management),[6] location, and sequence (scheduling). The functions of these patterns relate to soil fertility, pests and weeds, shade control, crop preference, and field mobility.

Mixed Cropping (Polyculture) The practice of planting different crops together has been widespread in both the highlands and the tropical lowlands. The advantages are multiple: reduction of diseases, insects, and animal pests; maximum soil protection from sun and rain; maximum utilization of soil nutrients by roots reaching different depths and by different plants taking up different quantities of specific nutrients; complementarity through vertical zonation in terms of sunlight and shade requirements; efficient use of solar energy; the combination of nitrogen-restoring legumes (beans, peanuts) with nitrogen-demanding seed crops (quinoa, maize), dietary diversity, and reduction of environmental risk (Cox and Atkins 1979: 670–2). Even without these advantages, overall productivity tends to be higher for several crops than for a monoculture because of a more efficient use of space (Altieri 1995: 206–8).

Variations of mixed cropping include: interplanting, 'the practice of planting two or more species with different patterns of seasonal activity, so that while one species is reaching maturity a second is undergoing early growth to mature at a later time' (Cox and Atkins 1979: 670); intercropping, 'the cultivation of two or more species whose growth and maturation tend to be synchronous' (Cox and Atkins 1979: 670; also see Innis 1997); agroforestry, the combination of annual or perennial crops with forest plants on the same land; and house gardens (kitchen gardens, dooryard gardens), which are highly diversified plantings in the house vicinity. An Achuar garden may have 100 different species, of which 62 cultigens appear in most gardens (Descola 1994: 160). There are numerous recent accounts of mixed cropping by Indians in South America, but few early reports of specific combinations other than the ubiquitous maize, beans, and squash.

Another variation of mixed cropping is companion cropping, or an association of crops in which one crop provides an ecological benefit to another. The combination of leguminous beans with maize to improve nitrogen availability for maize is an Indian practice throughout the hemisphere. The maize stalks in return provided support for the bean vines. Usually a third crop, squash, is part of the mixture, providing a protective ground cover. Crops may be mixed and spaced so as to provide appropriate shade or sunlight conditions. The Kayapó know of some two dozen tubers plus medicinals that grow well in the shade of bananas (Posey 1985a: 150–1). These plants are known to the Kayapó as 'companions of bananas'. Also see allelopathy below under Pest Control.

There is a general impression in the literature that mixed cropping dominates native swidden agriculture in lowland South America—the Geertzian model (Geertz 1963) of swiddens that mimic the tropical forest. This may have been more true in the past. While polycultural swiddens still exist [e.g. the Waika (Yanomami) of the upper Orinoco (Fig. 3.5) described by D. Harris 1971], more common are monocultural fields dominated by a single crop with a few individuals of other crops (see Beckerman 1983a and associated articles in *Human Ecology*, 2/1 1983). Thus a field may have many different species, but 80 per cent or more of the total crop may be manioc, as with the Jivaroan groups (Boster 1983: 50). Or crops may be zoned, as described below. Some of the functions of mixed cropping, such as pest protection, may be accomplished by polyvarietal interplanting (many varieties of the same crop), or by other characteristics of crop patterning. In the Peruvian Andes individual farmers may cultivate a dozen or more varieties of potatoes in a single field, with over one hundred varieties known to a single valley (Brush 1980a: 40; 1992: 163). The main advantage of monocropping would seem to be that labor time is reduced (Stocks 1983a: 81; Beckerman 1983a: 4–6), as well as a dominating dietary preference. One problem with layered polyculture is that it tends to shade out sun-loving annuals (the main staples) and may increase pests and disease due to more moist soil conditions.

Patch Farming Traditional agriculture usually involves family fields and house gardens which are located near settlements. Crops, however, may be planted individually or in clusters in many other places. The Kayapó provide one of the best studied examples. These people make plantings along trails, campsites,

ARROWROOT		TOBACCO		COTTON	MAPUEY	MANIOC		SUGARCANE
Maranta arundinacea	LECHOSA OR PAPAYA *Carica papaya*	*Nicotiana tabacum*	OCUMO *Xanthosoma sagittifolium*	*Gossypium barbadense* var. *brasiliense*	*Dioscorea trifida*	*Manihot esculenta*	BANANA *Musa paradisiaca*	*Saccharum officinarum*

Fig. 3.5. Waika (Yanomami) polycultural swidden in southern Venezuela illustrating crop diversity and layering. From D. Harris (1971: 480)

forest gaps, and graves. Since they trek enormous distances over a network of thousands of kilometers of trails in an area the size of Western Europe, the Kayapó are dependent on this 'nomadic agriculture' (Posey 1985*b*: 165–9). Mekranotí-Kayapó Indians exploit forest gardens far from the nearest village (Werner 1983: 232). Fruit trees either in groves or as individuals may be planted or protected. Planting along trails and camp sites may be either intentional or accidental via seeds spat out and from defecation. Also, 'While squatting to defecate, the Kayapó often plant tubers, seeds, or nuts they have collected during the day and stored in a fiber pouch or bag' (Posey 1985*a*: 149–50). In one 3 km section of trail, admittedly near a major village in this case, Posey (1985*a*: 149) reported 185 planted trees of fifteen different species, *c.* 1,500 medicinal plants from an unknown number of species, and 5,500 food-producing plants (domesticates, semi-domesticates, wild) from an undetermined number of species.

Forest openings created by tree falls or by trees felled by honey collectors, allowing for sunlight penetration, provide for another type of small isolated forest field for the Kayapó (Posey 1985*a*: 148). These various types of forest fields do not get the cultivator's attention a swidden field receives, but are visited frequently enough for some weeding and other minor management; losses to animal and bird pests are tolerated. The Kayapó farming techniques probably have considerable antiquity and were likely present among many other groups in Amazonia.

Concentric Rings The Kayapó (Hecht and Posey 1989: 182, 185), Bari of northern Colombia (Beckerman 1983*b*), and the Candoshi in northern Peru (Stocks 1983*a*: 77–8) arrange plantings in concentric rings within a swidden. The Kayapó fields have three rings in which different crops dominate and which are managed differently. Trees are felled outward from the center of the swidden. The central plot is fully burned and kept clean and is intensively planted in sweet potatoes. The middle ring contains the large tree boles, which when burned leave deep, scattered ash layers. There is a wide variety of crops, with specific crops being located in relation to variation in post-burn fertility. Nutrient demanding maize, beans, tobacco, cotton, and yams are planted in ash concentrations and by logs and slash where there is gradual nutrient release. Manioc is planted in the least favored sites. The outer ring contains less biomass (from the canopies of the felled trees) and hence less ash. Low nutrient-demanding plantains and bananas are the main crop and persist for twelve years or so (while the other rings are in fallow). A major source of nutrients is decomposing debris from the adjacent forest.

Stocks believes that the advantages of concentric rings are reduction of overshading, reduction of moisture conditions conducive to diseases, and maximally dispersing plants of one species along the ring thus inhibiting disease and pests. Hecht and Posey emphasize differences in soil fertility between the rings.

Other Intra-field Zonation Patch farming within shifting cultivation fields is widespread in Amazonia. The Yanomami in Venezuela have sectors of plantains/bananas/ocumo (cocoyam), manioc, plus a waste or non-cropped zone (Smole 1976: 130–1). The sectors are based on soil and moisture differences on slopes. The Yekuana, also in Venezuela, plant large monozones in manioc, with patches of other crops on the outer margins as monocrops or mixed crops, based on environmental variation (Frechione 1982: 66–7). Block and angular zonation have been described for the Yekuana (Frechione 1982) and the Ka'apor (Balée and Gély 1989: 137, 140–1). Crops are thus matched to microenvironments and angles provide longer ecotones. The Kayapó may plant specific crops in 'nutrient hot spots' enriched by ash or mulch or compost, where nutrient concentrations are favorable to those crops (Hecht and Posey 1989: 182).

Stocks (1983*a*) for the Candoshi of Peru and Beckerman (1983*b*) for the Barí of Colombia suggest that there are various reasons for systematic spatial patterning of crop zones, including shading characteristics and reducing the spread of disease and insects. In any event, numerous investigations indicate that planting crops in zones and in specific microsites is not random but highly deliberate based on sophisticated site, crop, and pest knowledge. This indigenous practice was probably widespread in the past and likely still persists more than is recognized.

Crop-growing units within fields are also common in the Andes (see Gade 1975: 92–3, for examples). Secondary crops (root crops, quinoa, vegetables) may either be dispersed within the staple crop or located in separate patches within or at the margins of the staple.

Environmental Zonation (life zones, verticality, agroclimate belts, production zones, ecological complementarity).[7] Fields are commonly dispersed across a mosaic of microhabitats in terms of specific crops, crop varieties, and scheduling. This may be inefficient spatially, with greater travel time, but environmental risk is minimized (Soria Lens 1954: 86–7). Individual farmers in the Andes have been known to have from twenty to as many as ninety separate plots (Godoy 1988: 16). In her study of field scattering in the District of Cuyo Cuyo in the southern Andes of Peru, Goland (1993: 317) concludes that: 'Pooling harvests of dispersed fields buffers households from production shortfalls in an environment characterized by temporally and spatially unpredictable microclimatic and agroecological factors.' Seldom will there be crop failure from both climate or pests all at the same time.

Field dispersal may be in a systematic pattern based on changing elevation (vertical zonation). Where slopes are fairly steep (steep environmental gradient) and temperatures and precipitation change rapidly, a farmer may find ecological zones close enough together so that he can farm different crops in several of them. A cold zone for potatoes, a temperate zone for grains, and a warm zone for subtropical crops may be within a day's walking distance and can be exploited accordingly. Where distances between zones are substantial, multiple zone exploitation may be on a marketing or colonization basis (Brush 1976; 1977: 10–16).

In some regions, a relatively flat or horizontal zonation of farm fields may occur, where the environment changes rapidly and systematically. One example is the Amazon floodplain where there are complex microhabitats based on microrelief with differential flooding and soil conditions. Site use by crop and season vary accordingly (see Chapter 4; Bergman 1980).

Scheduling This involves the time a crop is in a field and the sequence of crops in a field. Sequential planting or relay cropping is the planting of crops in sequence in order to utilize as much of the growing season as possible. Double cropping refers to two crops on the same site in sequence over a year. Multiple cropping refers to three or more crops on the same site in sequence over a year. Cropping periods can be accelerated by the use of short growing period crops, or by transplanting seedlings from seed beds. The former is common, as in floodplains during low water, but there are few indigenous examples of the latter in South America.[8] Latcham (1936: 300, no source) said that in Peru tree crops were transplanted from seed beds (*mallqui*), as well as several annuals, including cotton, chili pepper, and tomato.

Phased Planting 'Crops with different growth and maturation periods are planted together or at staggered planting dates so that crops in the same field are at different stages' (Wilken 1987: 254). This maximizes the use of space and time and spreads out harvesting.

Crop Rotation This refers to different crops being grown in a regular sequence on the same land over several years. Crop rotation can have positive effects on soil fertility, nutrient use, plant pathogens, physical properties of soils, erosion, microbiology, insects and other pests, weeds, earthworms, and other aspects of cropping (Altieri 1995: 233). Rotation may incorporate fallowing and/or pasture phases. Rotation may also reflect changing soil, microclimate, and other environmental conditions following forest clearing for swidden agriculture. The Amuesha, for example, in the upper Amazon of Peru, have a distinct succession. On low-lying plots this is from maize to manioc to plantains to other fruit trees, to utilized forest fallow (Salick 1989: 194–7).

Fallowing

Fallowing has multiple functions. Alternating cropping with periods of fallow is 'the most elementary method of restoring soil fertility' (Donkin 1979: 2). Depleted soil nutrient levels recover fully or partly under a regrowth vegetation of grass, bush, or forest. Lengths of cropping versus fallowing vary considerably. Tropical shifting cultivation systems usually have fallows of over ten years. However, even fields that are considered permanent and intensive, such as terraces, usually have periodic short periods of fallow.

Pest and weed control may be equally or more important as a reason for fallowing than fertility decline. Generally, animal pests, disease, and weed build-ups increase with the age of a field and the dominance of a few crop species. Thus even on good soils, a field may be fallowed in forest, bush, or grass when labor inputs for weeding become excessive or when crop losses to pests become excessive.

In the central Andes,[9] the sectorial fallowing (*laymi*, common field) system is one form of fallowing (Orlove and Godoy 1986; Godoy 1991). Here, sectors of fields under communal control are planted in the same crop and fallowed as units, with a cropping period of a few years being followed by a longer period of fallow which often consists of pasture. Fertility recovers, pests are reduced, pasture is provided, manuring results from livestock, and a diversity of crops is produced. Since households have plots in different sectors, environmental risk is reduced. Timing is staggered so that different sectors start the sequence in different years. The family plots in each sector pass through the same sequence. A typical sequence consists of a potato crop the first year, other Andean tubers the second year (oca, ulluco, añu), and then eight years in communal pasture (Orlove and Godoy 1986: 170–1). The system was introduced from Spain; however, some elements were clearly prehistoric in the Andes (Godoy 1991: 397–401).

Garcilasco de la Vega (1966: 1: 242) mentioned fallowing of poor land after a year or two of cropping in the Andes. The Araucanians of central Chile fallowed two or three years or more after one season of cropping (Latcham 1936: 292–3). A *cacique* of Chuquito, Lake Titicaca, reported in 1567 that people in that province rested their lands for 'four years as ordered' (Diez de San Miguel 1964: 36). In prehistoric Amazonia, shifting cultivation may or may not have been the dominant technique of native cultivation (see Chapter 6). There are few early descriptions of fallowing.

From what we are beginning to learn from several contemporary Amazon societies, such as the Bora (Denevan and Padoch 1988) and the Kayapó (Posey 1985*a*), Indian fallowing was not a period of plot abandonment. Instead, fallows were managed to produce an anthropogenic forest in which there was a higher than natural proportion of useful species, including fruits, tubers, fibers, medicinals, and construction materials. This management includes planting domesticates and semi-domesticates and protecting useful wild species (see Chapter 5).

Afforestation

Latcham (1936: 301) reported that the advanced Indian cultures in Chile planted trees on slopes in order to prevent soils from drying out, and presumably for erosion control. No source is given, so this example of planned conservation cannot be confirmed. In the pre-Hispanic Andes people planted trees such as molle, quishuar, and aliso (*Alnus* spp.) for fuel, construction, and implements (Gade 1999: 60–3).

Pest Control

The problem of crop pests—weeds, insects, viruses, animals, birds, and even humans (crop theft)—is the subject of billions of dollars of research and control mechanisms in modern agriculture. It is often assumed that traditional farmers have had few and only poorly effective means of reducing crop losses to pests. This is not so for many indigenous farmers; however documentation of specific and intentional practices has been scant.

As mentioned above, one of the main functions of swidden burning is to reduce insect pests, and the Kayapó example of using spot fires to control sweet potato virus was given. Mulching, crop rotation, temporary flooding, interplanting, barrier crops, time of planting, organic additives, cover crops, and green manure all help control plant pathogens (Altieri 1995: 309–19).

Crop species and varietal complexity reduce vulnerability to insects and diseases. Intercropping reduces the susceptibility of a particular crop to insects by lowering that crop's density. Certain crops repel certain insects chemically. Some crops or associated weeds provide a habitat for insect predators. Fields that progress through successional stages create changing habitats and thus reduce pest and disease build-ups.

The Bora soak peanuts in a solution of crushed basil leaves before planting to prevent ant predation (Denevan *et al.* 1984: 348). The Kayapó are aware that the *Azteca* sp. ant, apparently via a pungent smell, repels the leaf cutter ant which is destructive to many crops. They transfer pieces of *Azteca* nests around the forest and near fields in order to discourage the leaf cutters (Posey 1985*a*: 143–4). The Kayapó also bathe maize with a paste made from the bulb of *Costos warmingi* to prevent pest attack (Kerr and Posey 1984: 395).

Sectorial fallowing in the Andes has, as one of its primary functions, the control of nematodes which attack potato roots. Fallowing or alternative crops for several years reduces the nematode populations (Orlove and Godoy 1986: 179–80; Brush 1980*b*: 163).

It is difficult to demonstrate examples of pest control in pre-European times. However, on the north coast of Peru an insect called mariquita may have been managed to control a leaf worm that attacks cotton, as it is used today. The mariquita appears on a tapestry from the Jequetepeque Valley, *c*. AD 1250, suggesting that it had some perceived value (Vreeland 1986: 307). Today in the Piura Valley, people plant the shrub *Lippia* sp., the smoke from which, when ignited, repels harmful insects; and in Chimú field ditches the plant *Dysdercus peruvianus* apparently served as an insect repellent (Vreeland 1986: 289–90). One of the functions of the prehistoric raised fields, intentional or not, was to reduce root rot caused by *Phytophthora* spp. and other pathogens (H. D. Thurston, pers. comm. 1988).

Mature crops are best protected from animals, birds, and robbers by dogs or human presence. Guamán Poma de Ayala (1980: 3: 1032, 1056), in sixteenth-century Peru, shows the use of a drum to scare away foxes in a corn field, and in another sketch a farmer with a rattle, tassels, and a sling is scaring away birds and a skunk. Other than such methods, the use of scare devices, such as scarecrows within fields, seems to be largely European.

Crop theft did occur in earlier times, but may not have been common. Guamán Poma de Ayala (1980: 3: 1038) shows a man stealing maize ears, hidden in a field from the eyes of a watchman (*arariwa*) guarding the crop. Gade (1970: 9–10) shows huts used today by crop guards in the Cuzco region. In another study he reports a Catholic religious rite of excommunicating crop pests (Gade 1975: 58). This raises the whole realm of ritual to ward off pests, as well as natural disasters and to assure good yields; however, ritual is not examined here.

Weed Control

Weeds were controlled by burning, by mulching, by shading via multiple canopies, and by companion cropping of allelopathic species. The degree of weeding with cutting tools, such as the mattock and *macana*, in pre-Columbian times is uncertain. Undoubtedly, much weeding was done by hand and with hand tools, particularly in the tropical lowlands. This has the advantage of being able to select individual plants for destruction if undesirable or for survival if useful. After Europeans arrived, the machete rapidly became the almost universal weeding device. A major function of shifting cultivation is the control of pests and weeds, which tend to become progressively more serious as a field ages.

Allelopathy refers to the suppression of plant growth (and insects) by toxic chemical compounds in other plants through root exudation or from decaying plant material (Altieri 1995: 291–2). Certain cultivated crops, such as tobacco, are thus able to reduce competitive weeds.

Other Procedures

Ground preparation was highly varied and is most obvious with landform modification, as will be discussed in the chapters on irrigation, terracing, and raised fields. Actual plowing was mainly confined to the Andes and involved the use of the footplow and spades or digging sticks and clod crushers. Elsewhere, the ground was broken for individual plants. Potatoes were usually banked or planted in narrow beds (*eras, huachos*). In the Andes and on the coast, the soil around individual maize and other plants was and is banked up (*aporque*) to prevent stalk collapse and wind throw, and to reduce drying and frost (Gade 1975: 41).

Planting was by individual plant, or a few seeds would be dropped in a shallow hole, or a few cuttings of tubers would be stuck into loosened earth or in a small, low mound. There is no clear record of broadcast seeding until European influence and the introduction of the animal-drawn plow. Latcham (1936: 300) said that quinoa and a few other crops were broadcast, but he gives no sources; there are Indian names, however, for this type of sowing.

Harvesting procedures, transport, storage, and preliminary crop processing varied with crop, tools, and region and will not be treated here.

An important conclusion regarding the efficacy of 'soft' agricultural technologies is that reached for the Kayapó by Hecht and Posey (1989: 186): 'The idea that inherent chemical features of soils determine human population densities [thus] becomes meaningless.'

Notes

[1] For excellent, recent treatments of crop domestication in South America, see Pearsall (1992) and Piperno and Pearsall (1998).

[2] This categorization corresponds roughly to Rindos' (1984) agricultural continuum of 'agricultural domestication', 'specialized domestication', and 'incidental domestication'.

[3] Wilken (1987) provides considerable detail on soft technologies of contemporary Indians in Mexico and Central America; however there is no comparable study for South America.

[4] Descriptions of the pre-European use of organic fertilizers are provided by Latcham (1936: 294–301) and Rostworowski (1987). An article by Antúnez de Mayolo (1980) on agricultural fertilizers in ancient Peru has disappointingly little on actual fertilizer use. A second part of the article was never published. See Miller and Gleason (1994*b*) for a general survey of prehistoric fertilizers.

[5] Muck is defined by Jacks *et al.* (1960: 102) as 'partially decomposed organic matter, plant remains not discernable, accumulated in a wet place and mixed with some mineral matter'.

[6] For a good description of traditional crop space management in Mexico and Central America see Wilken (1987: 240–61).

[7] For various approaches to this much discussed concept in the Andes, see Troll (1958; 1968); Murra (1972); Brush (1976; 1977); Gade (1975: 95–107); Masuda, Shimada, and Morris (1985); Tosi (1960); Mayer (1979); and Zimmerer (1999).

[8] The use of seed beds (*chapines*) in the *chinampas* of the Basin of Mexico is well documented (Wilken 1987: 258–61).

[9] Central and southern Peru and western Bolivia; see map in Orlove and Godoy (1986: 187). Most of these fields are neither terraced nor irrigated.

Part II

Amazonian Cultivation

There is a methodological problem with interpreting prehistoric Amazonian lifeways mainly through ethnographic projection ... present-day Indians' resource management modes may not be representative of prehistoric ones.

(Anna Roosevelt 1989*b*: 31)

.

4

A Diversity of Habitats and Field Systems

Greater Amazonia[1] (Fig. 4.1) is enormous, and rather than being a uniform region of tropical rainforest and impoverished soil, as often thought, there is a considerable diversity of environmental conditions at both local and regional scales. Likewise, there has been a diversity of indigenous subsistence systems ranging from non-agricultural hunting and gathering to intensive swamp reclamation, with population densities ranging from under $0.1/km^2$ to several hundred/km^2, and with settlements ranging from small, dispersed, and unstable to relatively large and permanent.[2] Long-fallow shifting cultivation is the prevailing form of Indian cultivation today; however, there are many variations. In prehistory, more permanent systems of cultivation were probably common (Chapter 7).

When I studied Campa agriculture in the mid-1960s, few specific accounts of indigenous cultivation were available for tropical South America. Most notable was the work on the Kuikuru and Amahuaca by Robert Carneiro (1957; 1961; 1964). Several theoretical statements, however, already had been put forth on the relationship of subsistence patterns to population density, settlement stability, food productivity, and cultural evolution. To test these theories and to make comparisons possible, ecological fieldwork on varied societies needed to be carried out in conjunction with ethnographic and demographic surveys. Detailed measurement was needed of labor inputs (both time and energy), production, location factors (distance and site), carrying capacity, and dietary intakes and nutrition, preferably over a full year food-obtaining cycle. The need was urgent in view of the rapid cultural changes being experienced by tropical peoples and the population decline and actual extinction of many societies. Salvage cultural ecology was needed not just for understanding Indian groups but to provide insights into how Amazonian settlers could come to terms with the tropical environment in productive and non-destructive ways. Indeed, numerous studies of Indian cultivation in Amazonia were made in the 1970s and 1980s, and considerable quantitative and qualitative data is now available for groups throughout Amazonia (see, e.g., the collections edited by Hames and Vickers 1983a, Posey and Balée 1989, Beckerman 1983a, and the books on the Shipibo by Bergman 1980, on the Achuar by Descola 1994, and on the Ka'apor by Balée 1994). What has emerged is not just more information

54   *Amazonian Cultivation*

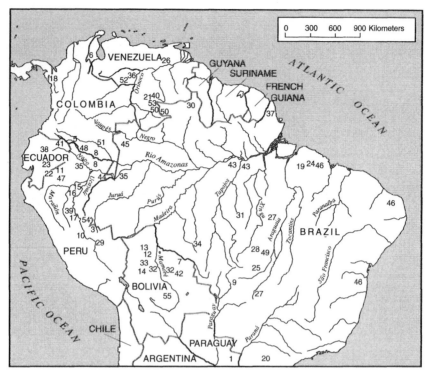

Fig. 4.1. Map showing locations of Indian cultures in Greater Amazonia. By W. M. Denevan. *Key*:

1. Ache	— Guayakí (Ache)	38. Runa
2. Aguaruna	20. Héta	39. Shipibo
3. Amahuaca	21. Hoti	40. Sanema-Yanoama
4. Amuesha	22. Huambisa	— Shuar (Jívaro)
— Asháninca (Campa)	23. Jívaro	41. Siona-Secoya
5. Andoke	24. Ka'apor	42. Sirionó
6. Barí	25. Kalapalo	43. Tapajó
7. Baure	26. Karinya	44. Ticuna
8. Bora	27. Kayapó	45. Tukano
9. Bororo	28. Kuikuru	46. Tupinambá
10. Campa	29. Machiguenga	47. Urarina
11. Candoshi	30. Macusi	— Waika (Yanomami)
12. Cayuvava	— Makiritare (Yekuana)	48. Witoto
13. Chácabo	31. Mekranotí (Kayapó)	49. Xavante
14. Chimane	32. Mojo	50. Yanomami
15. Cocama	33. Movima	51. Yapú
16. Cocamilla	34. Nambicuara	52. Yaruro
17. Conibo	35. Omagua	53. Yekuana
18. Emberá	36. Otomac	54. Yora
19. Guajá	37. Palikur	55. Yuquí

and variation on a normative pattern of traditional indigenous shifting cultivation, but an awareness of more complex, ecologically sustainable, and more
productive agroecosystems than had been assumed previously.

The issues include environmental limitations on culture (Meggers 1957),
the origins of advanced cultures (Lathrap 1970; Roosevelt 1980; 1987;
Carneiro 1961), population size and density (Denevan 1976; 1992a; 1999; Meggers 1992a), the *várzea/terra firme* (floodplain/upland) settlement dichotomy
(Gross 1975), agricultural origins (Lathrap 1977), settlement patterns (location, size, duration) (Myers 1990; Meggers *et al.* 1988), optimal foraging
(Beckerman 1983c), and the processes of change and stability (adaptation,
population pressure, migration, carrying capacity, risk avoidance) (various
Amazonianists).

For a discussion of the conceptual thinking of Steward, Meggers, Lathrap,
Carneiro, Gross, and others, see especially Roosevelt (1980). Also on theory
and concepts see Sponsel (1986), Lathrap (1970), Myers (1990; 1992a), Hames
and Vickers (1983b), Meggers and Evans (1983), Balée (1989), A. Johnson
(1982), Gibbons (1990), Roe (1994), and Viveiros de Castro (1996). My emphasis here is on fields and field technology; however, I will use the perspective
gained to consider several of the above themes in this chapter and in Chapters
5, 6, and 7.

Pre-European Conditions

What was Amazonia like at the time of initial European contact in the early sixteenth century? Unfortunately, eye-witness descriptions from that time are few
and lack detail. Instead, reconstruction can be attempted using: (1) archaeo-
logic information, which is especially fragmentary for Amazonia, (2) ethnohistoric information for the time of early contact, but mostly later, and (3) recent
ethnographic information. Prehistoric patterns, however, were not necessarily
still in place at the time of initial description. For ethnohistoric and ethnographic information it is necessary to filter out post-contact influences, which
is difficult (see Chapters 6 and 7).

The Natural versus the Anthropogenic Environment

The forests and grasslands of Amazonia were long considered to be among the
last pristine environments in the world until recent development activity and
associated deforestation. Indigenous people seemingly had had little impact.
Scattered shifting cultivation plots supposedly quickly returned to original forest, and large sectors were believed to have been inhabited by 'food gathering
peoples with no more influence on the vegetation than any of the other animal
inhabitants'.[3] We now know otherwise (Balée 1987; Denevan 1992b; Raffles
1999; Roosevelt 1999; 2000).

First of all, it may take cleared tropical forest hundreds of years to return to pre-disturbance biomass and diversity conditions; Riswan and Kartawinita (1988) calculate 150 to 500 years for East Kalimantan, Indonesia; Beckerman (1987: 72) gives 500 years for Amazonia. Second, much forest never recovers but is replaced by savanna which is maintained by frequent burning. It is impossible to say how much of the roughly one-third of Greater Amazonia that is savanna is natural and how much is not, but it is certain that natural savannas have been expanded by human disturbance. Finally, a large portion of what appears to be undisturbed tropical forest is not and, as result of human activity, may contain a larger number of useful species than is natural. Swidden fallows are managed; useful plants are spread around trails, campsites, villages, and other places; the foraging for wild plants in the forest modifies the distribution of those plants; and useful wild plants are protected that might otherwise vanish in the process of forest succession. Also, swidden fires escaping into forest and game exploitation affect vegetation. The result is the creation of anthropogenic forests which may appear to be natural. Balée (1989: 15) estimates that at least 11.8 per cent of the Brazilian Amazonian *terra firme* forest today is anthropogenic to some extent as result of peasant/Indian manipulation. The portion was undoubtedly much greater in the past when Indian numbers were greater and more dispersed than are populations today.[4] Vegetation modification, of course, also results in changes in soils, wildlife, microclimate, and hydrology, especially where forest is converted to scrub and grassland.

The important point here is that many habitats in which Indian people obtain their subsistence are not natural but rather have been slightly to considerably modified, both intentionally and unintentionally, in ways which can either enhance or depress subsistence productivity.

Subsistence

Amazon societies experienced deculturation not only from the moment of first contact but indirectly even before direct contact. Major changes occurred in belief systems, language, art, clothing, and tools to the point of detribalization. Even very isolated groups have experienced this to some extent. Subsistence knowledge, however, seems to have been more resistant to change despite the introduction of new crops (plantains, taro, sugar cane, citrus, rice), animals (pigs, chickens, cattle), and metal tools (machetes, axes, guns[5]).

Basic techniques of shifting cultivation, forest management, hunting, gathering, and fishing remained essentially indigenous, although not necessarily unchanging. The most significant impact by a European introduction was that of the metal axe, which rapidly replaced the much less efficient stone axe (see Chapter 7). This may mean that prehistoric shifting cultivation in the *terra firme* forests was rare or confined to easily cleared vegetation, with a resulting greater emphasis on hunting and gathering, with consequently lower population densities. Another possibility is that fields were cropped for longer periods of time, using soil maintenance techniques, polycropping to reduce pests, plus a high reliance on products from anthropogenic forests, thus reducing the need for frequent forest clearing.

Anna Roosevelt (1989*b*: 31) emphasizes the unreliability of 'ethnographic projection' of resource management patterns to the past and the resulting disagreement between scholars on many aspects of subsistence. 'Where Indians roam "virgin" forests today, prehistoric people farmed and built mounds' (Roosevelt 1999: 385). Betty Meggers (1992*a*: 199), in contrast, believes that the present 'way of life' has existed for at least 2,000 years. Demographic decline alone would have brought about significant changes in subsistence *strategies*, although not necessarily in specific resource management *techniques*. 'The densely-settled prehistoric groups would be expected to have used the land much more intensively than do the sparse Indian populations of today' (Roosevelt 1989*b*: 32). Also, the relative importance of different crops may have been different in the past.[6] In her article on resource management before the conquest Roosevelt (1989*b*) indicates that by 3000 BC early horticultural villages were root crop-based, whereas by AD 1000 agricultural chiefdoms were maize-based. Most Indians today have a manioc staple; for some, however, the staple is maize, plantains, or rice.

Certainly in prehistoric times riverine (floodplain and bluff) habitats were more attractive than was the interior *terra firme*. However, few of the riverine groups have survived, so we know little about their subsistence, whereas we have considerable information on *terra firme* people.

Population

I have previously estimated aboriginal (1492) population densities for Greater Amazonia, deriving total populations of 5.75 million, 5.1 million, and 5.7 million and overall densities of 0.6–0.7/km^2 (Denevan 1970*b*; 1976; 1992*a*). The methodology used, given very little data for contact times, was to estimate reasonable or potential densities for thirteen habitat types and then multiply those densities times the total area of each habitat type. The determination of densities was based on archaeological, ethnohistoric, and ethnographic evidence of different dates and kinds, so there was no consistency or control for aboriginal versus modified conditions.

This habitat-density method, however, assumes a relatively uniform distribution of people within each habitat. I am now convinced that populations were clustered, thus making this method problematic (Denevan 1996; 1999). We do not know how many villages there were or how large they were, so overall estimates are very difficult to derive. Five to six million can remain a working total and is probably conservative. Meggers' (1992*a*) estimate of a density of only 0.3/km^2 for both *terra firme* and *várzea* is too low for the *várzea* in my opinion, as is the total of 2,931,000 derived for Greater Amazonia using her density.

Riverine villages contained up to several thousand people along bluff tops (see Chapter 6). There were also concentrated populations in parts of the interior *terra firme* (see Chapter 7). Reports of the Jivaro uprising in 1599 mention mobilization of over 20,000 warriors (Harner 1972: 21). Ethnohistoric accounts indicate Kayapó settlements periodically numbering over one thousand (Posey 1987: 139, 147). The subsistence base for such numbers is not clear. *Terra preta* (black, anthropogenic soils) habitation sites as large as 50–200 ha or more have been observed, but these may not have been single villages and probably included agricultural sectors.

The Indian population was rapidly reduced after European contact, mainly from introduced diseases. In the 1970s, the total was about 500,000 (Denevan 1976: 232), and is probably similar today; while some groups have declined in numbers, others have grown in size.

We can proceed then, with qualifications: (1) the resource base has been modified by Indian activity, (2) subsistence has been changed nearly everywhere by European contact and other factors, in terms of both crops and technology, and (3) present populations are greatly reduced from those of prehistoric times.

Riverine Habitats (*Várzea*)

Floodplain Heterogeneity

> One does not farm the Amazon Plain, but [rather] hundreds of thousands of particular tracts of ground, each endowed with . . . a specific local climate and many other equally localized and specific environmental components.
>
> (Hilgard O'Reilly Sternberg 1964: 323)

The concept of ecological zonation as a guiding principle of traditional land-use management in Latin America has been primarily considered in terms of altitudinal variation (verticality, complementarity) in the central Andes (Murra 1972; Masuda *et al.* 1985). Different altitudinal zones are utilized in order to spread risk, to diversify crops, to spread demands on labor, and to utilize land that is close at hand. A related concept, environmental gradient, refers to how close biotopes (microenvironments) are to one another (Porter 1965). It is with such units of nature that individual humans interact, not with a polymorphous 'savanna' or 'tropical forest'. On a mountain slope, farmers have easy access to a variety of biotopes. We say that the environmental gradient of a mountain slope is steep because of its spatially rapid changes in temperature and in associated moisture, vegetation, and soils.

The concept of zonation, however, with a steep environmental gradient and the exploitation of multiple biotopes, need not be limited to mountain slopes.

Zonation may likewise occur in aquatic situations and elsewhere. A good example is provided by Nietschmann (1973: 98) for the Miskito Indian habitat on the Caribbean coast of Nicaragua. A wide variety of biotopes exist along a sea-to-shore-to-inland transect and are recognized and utilized. The Karinya in the Venezuelan Llanos exploit a horizontal sequence of biotopes with specific types of agriculture for each (Chapter 5). It seems appropriate, then, to speak of horizontal zonation where there is a marked environmental gradient but little or no change in elevation except at a micro scale.

Whereas the Amazon Basin in the past has been thought to be ecologically uniform, with recognition mainly of forest, savanna, and floodplain, Amazonian scholars are increasingly aware of considerable heterogeneity. Frequently, this heterogeneity exhibits regularity or zonation, which is perceived and utilized by local people in their gathering, hunting, fishing, and agricultural activities (Denevan 1984a). The floodplain habitat probably has the greatest heterogeneity.

Floodplains occupy only a small portion of the Amazon Basin, an estimated 154,400 km^2 (Sippel *et al.* 1992) or 300,000 km^2 (Ohly and Junk 1999: 284) for the main river and major tributaries in Brazil, or about 5–10 per cent of the region. The proportion of floodplain to *terra firme* apparently is greater (*c.* 12 per cent) in Peruvian Amazonia (Salo *et al.* 1986). It is in the floodplains, of course, that most of the best Amazon soils occur.

A transect across the floodplain of the Amazon River or one of its tributaries will intersect various biotopes in a regular and somewhat predictable sequence. Some portions of the sequence may be repeated several or even many times along the transect. A complex mosaic thus exists within a floodplain, which may be enormous in width. The floodplain biotopes are created by or influenced by a number of micro- and macrofactors, including geology, geomorphology, hydrochemistry, climate, soils, and biogeography. Figure 4.2 shows a simplified floodplain transect.

The Amazon floodplains in Brazil are often differentiated between whitewater high-sediment-load rivers mostly originating in the Andes, and blackwater low-sediment-load rivers originating in the Brazilian and Guiana Highlands. The whitewater floodplains generally contain good quality alluvial soils, and rich aquatic life, in contrast to the low fertility, dispersed game, and sparse human populations of interfluve forests.[7] The whitewater floodplain soils do tend to decrease in fertility from the upper to the lower Amazon (Zarin 1999: 316).

Várzea is used here to refer to any floodplain in Amazonia, a common practice in Brazil, and this is how it is used by Sternberg (1995: 115, 124), with *igapó* being a 'more or less permanently inundated' variant. However, *várzea* is used by some scholars to refer to floodplain forests that are seasonally flooded, while using *igapó* for floodplain forests that are permanently flooded (see Chernela 1989: 23). Others use *várzea* for whitewater floodplain forests, and *igapó* for blackwater floodplain forests (Prance 1979: 29; Chernela 1989: 23). Brazilians

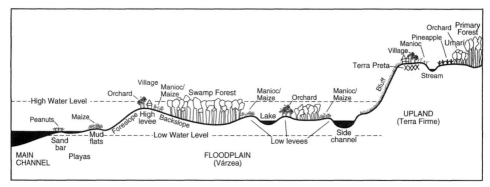

Fig. 4.2. Cross-section of the upper Amazon floodplain showing the main channel, water levels, natural levees, bluffs, and a *terra preta* site. From Denevan (1996: 657), adapted from Coomes (1992*b*: 164)

also distinguish between high *várzea* (mainly natural levees) and low *várzea* subject to lengthy and deep flooding (Parker 1981: 55, 57).

River Rise and Fall

> Each year there is a flood and all life adjusts to it. It comes not as a disaster, but as a season. Day by day the river rises a few inches, drops a little, then rises higher . . . every ten years or so the floods cover even the highest levees.
>
> (Roland Bergman 1980: 53)

Unusually high floods covering all or most of the floodplain occur every five to ten years. The duration of exposure of agricultural land at low water is of course relative to the relief of the land as well as to the rise and fall of the river. Of all the land exposed during lowest water, as little as 10 per cent or even less will be exposed during highest water. For the amounts of land above water at different times of the year at specific locations along different rivers, almost no data are available. Adaptation is further complicated by the instability of the land itself as a river swings back and forth across its floodplain, destroying land in one place, creating it in another (Sternberg 1975: 17–18). A farm or village may disappear overnight as banks cave in. A large *playa* fronting a village during low water one year may not reappear a year later.

For the farmer, annual flooding means the annual deposition of fertile sediment, which makes annual cropping possible. While floodplain soils are far superior to most *terra firme* soils, considerable variation in soil texture and fertility create options for different crops (Ohly and Junk 1999: 284–5). A second benefit of annual flooding for the lowest land is the destruction of pests and vegetation. Little or no land clearing is necessary, and crop losses to pests are minimal. The floods themselves are the main hazard. On the higher, forested ground of the levees the reverse is true, and short-fallow shifting cultivation is practiced, not so much because of deficient soil fertility as because of weed invasion and pest problems. There is little data on river-levee cultivation cycles. On the Río Alto Pachitea of eastern Peru, Campa Indian swiddens average one to three years of fallow and three to four years of cropping for manioc and up to twenty years for bananas (Allen and Holshouser 1973). However, at San Jorge near Iquitos, swiddens on high levees are cropped for two to four years followed by four to twenty years of fallow prior to reclearing (Hiraoka 1985: 16). High levee soils only infrequently receive flood silt deposits and hence fertility is less than for the low levees and *playas*, and this results in field fallowing according to Hiraoka (1985: 8, 15).

The major disadvantage of floodplains, of course, is the irregularity and variability of flooding. On the tributaries destructive floods can occur without warning, even during low water seasons. The rise and fall of the main Amazon is more regular, but periodic extreme floods occur, filling the entire floodplain, topping the natural levees, and destroying all the crops. Thus floodplains are a rich but high risk habitat.

Agricultural patterns, including labor scheduling, amount of land cropped, and type of crop, are all linked to water-level variation. Short growing season crops, such as maize, beans, peanuts, and rice, are grown on the lowest lying ground, especially on sandy soils. The fine-tuning of planting is best exemplified by rice, which is characteristically planted (or transplanted) in progressive steps as the river level drops. Manioc, bananas, and plantains dominate the higher levees, which are above flood levels most years. Permanent tree crops are located only on the highest ground where there is seldom or only brief flooding. On the central Amazon, levees have growing periods of eight months or more and *playas* four months or less (Ohly and Junk 1999: 287).

Thus, there are numerous environmental variables causing ecological heterogeneity or biotope diversity in the Amazon floodplains. These combine to create both regional variations, including significant differences along the course of the same river, and local variations, which occur in regular patterns that generally are linear and parallel to the main channel. Associated with this regularity is a horizontal zonation of agriculture, which is a major concern here. The *várzea* could also be viewed as an archipelago, consisting of isolated units with different possibilities for aquatic and terrestrial wildlife and for human utilization. Another relevant spatial concept is that of 'edges'. Settlements tend to be located on the edges of certain types of biotopes adjacent to certain other types so that both can be readily exploited.

River Terraces

Várzea agriculture, while very productive, is at risk because of periodic high floods that fill the floodplains. This is not entirely so, however. There are large river terraces of varied origin within the *várzea*—surfaces of Pleistocene and

Holocene age, which are higher than the highest levees and are seldom or never flooded. Ages range between about 5,000 and 100,000 years in upper Amazonia. Dumont *et al.* (1990: 128, 131) map terraces in the Río Ucayali floodplain with dates of 8520 BP and 32750 BP. Thus river terrace soils are relatively young and less weathered than *terra firme* soils and have a relatively high nutrient content (Räsänen *et al.* 1993: 211). Other terrace soils, however, are poorer colluvial material washed down from uplands. Floodplain terraces cover 0.7 per cent or about 4,800 km² of the Peruvian Amazon, based on measurement on Landsat photography (Räsänen *et al.* 1993: 211). In Brazilian Amazonia, Pleistocene terraces, *c.* 4 to 100 m high, are mostly reworked Tertiary and older deposits (Sombroek 1966: 29, 34–5).

There is a river terrace near the community of San José (pseudonym) on the Amazon near Iquitos (Coomes and Burt 1997: 29–35). The old Tertiary alluvium soils of the local *terra firme* are heavily leached of minerals and are acidic. In comparison, the terrace soil is siltier and less acidic, and it has a higher nutrient content. The vegetation is dominated by the ivory nut palm (yarina), which may reflect past human disturbance. The fields on the alluvial terrace are particularly favored for maize and rice because of their good fertility.

Some river terraces have a considerable potential for settlement and cultivation because they do not flood, but only if they are readily accessible. They do not seem to be utilized much today, possibly because they are distant from the main channels, not easily reached by canoe, and because they are fragmented. The same difficulties probably prevailed in prehistory. River terraces have seldom been examined by archaeologists, so their past agricultural importance remains little known. Mora *et al.* (1991: 6) map river terraces along the Río Caquetá in the Colombian Amazon and show several archaeological sites on them.

Floodplain Cultivation

Considering the general optimism about the potential of the Amazon floodplain for agriculture, past and present, it is remarkable how little is cultivated and how little attention has been given by both agronomists and scholars (Padoch *et al.* 1999). In the 1970s it was estimated that less than 1 per cent of the full Amazon *várzea* was cultivated (Petrick 1978: 33). Eden (1990: 127), in his book on land management in Amazonia, stated that: 'There is less dependence on *várzea* cultivation itself than was the case in pre-contact times.' Whether this is true has not been demonstrated, but it is quite possible. Certainly, much more research is needed on the archaeological and ethnohistorical evidence for indigenous floodplain cultivation.

The keys to indigenous adaptation were the canoe and flexible exploitation of productive micro niches, with numerous settlements resulting. A prehistoric agricultural model can be proposed based on ethnographic example. On the natural levees there was short-fallow shifting cultivation, with fallowing resulting from flooding or weed/pest invasion rather than fertility decline. As much as twenty years of continuous cropping is possible followed by just a few years of fallow (Lathrap 1968*b*: 27; Allen and Holshouser 1973). Río Pataza (Ecuador) alluvial soils are cultivated almost continuously with little decline in crop yields (Descola 1994: 176–7). As river level recedes during the low water period of the year, *playas* and islands are progressively exposed and can be planted in what might be called 'recessional' agriculture, using fast-growing crops such as maize, beans, peanuts, and melons. Crop selection varies with soil texture and length of growing season as determined by slope position and thus length of flooding. Similar conditions and opportunities are presented in those back swamps which dry out periodically. Evidence of former floodplain settlement is meager because of destruction of sites by river action. Most sites are found on the fringing bluffs, and it is there that most permanent settlements were located rather than within the floodplains.

Reports from expeditions down the Amazon in the sixteenth and seventeenth centuries described the cultivation of *playas*, natural levees, and bluffs, and also mention fruit orchards. Most of the riverine societies, however, were quickly destroyed. One of the few surviving to the present is the Shipibo, described in Chapter 5. Others are the Isconahua (Momsen 1964) and the Cocamilla (Stocks 1983*b*), all in the Peruvian Amazon.

Early Descriptions

Floodplain agriculture for the Omagua and Tapajó Indians is described by Meggers (1971; 1996), based on sixteenth- and seventeenth-century accounts. The primary crops were maize and manioc. For the Omagua, other crops were sweet potatoes, peanuts, beans, tobacco, achiote, cotton, gourds, peppers, pineapple, cacao, avocado, and various other fruits. Carvajal (1934*a*: 172–96) reported that Orellana's men in 1542 obtained very large quantities of food in some villages, allowing fifty or so Spaniards to stay for twenty days in one and thirty-five days in another, with enough food to feed 1,000 men for one year in another. However, very little information is given on the forms of agriculture practiced. It is not clear whether the primary production was occurring on *playas* and islands, on levees, or on *terra firme* bluffs, or on all, which is likely.

There are early reports of maize on islands and *playas* (see Roosevelt 1980: 153, 155). Gumilla, in 1745, described what was clearly a form of recessional cultivation by the Otomaco on the Orinoco: 'When the waters are drying up after the rains, all the Indians that live near the large lagoons sow all the bare earth from which the water has withdrawn. There they get an abundant crop because the soil is very rich . . . in the space of a year they get six crops [of two month maize] by searching out the appropriate terrain' (trans. by Roosevelt 1980: 156–7). Sugar cane, root crops, calabash, and watermelons were closely intercropped with the maize. Roosevelt (1980: 148–57) makes a good case for maize often being the floodplain staple, in contrast to manioc which requires a much longer growing season and which usually was the dominant crop on *terra firme*. The early reports of riverine manioc probably referred to bluff and high-levee cultivation. For example, the Omagua had 'great quantities of large cakes made out of cassava baked hard like bisquit' (Carvajal 1934*b*: 424–5).

Food was stored in order to feed people during the high water and flood periods. The storage of manioc in pits in the floodplain, keeping for up to two years without rotting, is mentioned by Fritz (1922: 50) and Acuña (1942: 35–6). (The Ticuna today bury manioc during high floods, Bolian 1971.) The Tapajó stored maize in baskets buried in ash for protection, according to Meggers (1971: 134, 141, no source). Maize was stored in rafters and in raised cribs, along with bitter manioc flour, mixed maize-manioc flour, and also dried game and fish (Carvajal 1934*b*: 425, 432; Oviedo 1934: 398). Manioc flour (*farinha*) today keeps indefinitely. In addition, fish were smoked and turtles were kept live in pens (Acuña 1942: 38–9). Orellana's men obtained 1,000 turtles in a single village according to Carvajal (1934*a*: 193), and men of the Ursúa-Aguirre expedition in 1560–1 reported a village with 4,000 turtles (in *corrales*) caught in the dry season for eating in the wet season (Vázquez de Espinosa 1948: 385). The capacity for food storage alleviated the problem of flooding, but may not have been sufficient to counter high floods of long duration or high floods occurring several years in sequence. The loss of crops to flooding was mentioned by Fritz (1922: 50): 'when the River is in high flood, they are left without a *chagra* [field] and not seldom without anything to live upon.' Likewise, Acuña (1942: 35) said that: 'the Indians are exposed to great loss [of crops], on account of the powerful floods.' In addition, fish and game availability are greatly reduced in the floodplains during high water.

Storage of food was mainly seasonal, and at best only provided for a year or two of non-production. More than food was involved, however. Seed and tuber cuttings must be preserved in adequate quantity for future plantings. Seed can be stored, if not eaten in emergencies, but I know of no long-term storage of manioc cuttings. If manioc plants were destroyed by flooding, new cuttings would have to have been obtained from *terra firme* fields. Thus, non-floodplain (*terra firme*) sources of food would have been essential for the support of large numbers of people. Today, of course, floodplain farmers and fishermen have access to market sources of food.

Blackwater Floodplains

The blackwater rivers of Amazonia are more transparent than whitewater rivers, carry dissolved organic compounds, have a high electrolyte content and a low pH value, are brown in color and carry a low sediment load, and have a relatively low mineral nutrient content; primary phytoplankton is low and thus aquatic biomass productivity is low (Chernela 1989: 239–40). The surrounding flooded forest is a more important source of nutrients than the rivers themselves, and thus more productive of fish. The soils of blackwater floodplains are poor for the same reasons, with limited nutrient renewal from suspended sediments. They are extremely acid with very low fertility. Blackwater rivers have a reputation of being 'rivers of hunger'; that is, they support relatively low human populations. Agriculture is concentrated in forest, with long-fallow shifting cultivation, with a minimal use of *playas*, which are much more restricted than along whitewater rivers. However, there have been no comparisons of population densities of blackwater rivers with whitewater rivers. Early accounts of Indians along the blackwater Rio Negro do not mention great numbers such as were reported along the whitewater Amazon. For discussion of the nutrient-poor characteristics of blackwater ecosystems, see Moran (1991) and Coomes (1992*a*). However, not all blackwater floodplain soils are unproductive, and they may well be more fertile than adjacent *terra firme* soils (Coomes 1998: 51).

The Tukano are an example of a blackwater society. They are located in Colombia and Brazil on the Río Vaupés, a tributary of the Rio Negro. The study of Tukano subsistence by Chernela (1989; 1994) suggests that tribal populations of blackwater rivers may not have been as sparse as assumed. Present Tukano villages range from 30 to 150 people each and are located along the river 3 to 24 km apart. Long-fallow shifting cultivation is practiced on forested levees or on upland soils of low fertility. Cropping is for two to four years, with yields declining rapidly. Fallowing is for up to twenty years, but primary forest is preferred over secondary forest for new plots because of much higher crop yields (double) in the former. However, even primary forest first-year plots have much lower productivity than plots on whitewater floodplains. Manioc is the staple (85 per cent of calories), with fish the main source of protein (Chernela 1989: 242).

Fish consumption is very high for the Tukano (68 g per day per person), and Chernela (1989: 242) believes this permits a higher population than would otherwise be possible. By far the largest portion of fish by source (about half the total) comes from the flooded forest, which is where fish congregate to feed. Recognizing the importance of the flooded forest as a fish resource, the Tukano avoid clearing this forest for agriculture, thus maintaining a riparian reserve. The Tukano example is instructive, but more data on adaptations to blackwater habitats are needed in order to better evaluate associated indigenous productivity and settlement.

The Estuarine Floodplain

The Lower Amazon floodplain and delta is characterized by daily tidal floods which exceed the seasonal variation in flood levels. This combined with poorer quality sediments (low suspension loads) than in the Middle and Upper Amazon, limits agriculture to the higher levees and river banks (Anderson 1990). *Caboclo* (non-Indian Amazonian peasants) land use today is a combination of house gardens and small swiddens on the river banks and forest product extraction and management in the floodplain forest. Given waterlogged conditions, these forests have low biological diversity; many of the trees, however, have subsistence and economic value, such as *Inga, Spondias, Mauritia*, and *Euterpe* (Anderson 1990). Populations today are relatively dense and may also have been large in prehistoric times; however, there has been little archaeology in the estuarine zone, except on Marajó Island, nor is there much historical information on Indian agriculture.

The Upland Forest Habitats (*Terra Firme*)

The *terra firme* of western Amazonia consists of primarily Pleistocene fluvial deposits, now dissected and highly weathered, with very poor soils, covered by a biologically diverse tropical forest. In eastern Amazonia these deposits are interrupted by sedimentary and metamorphic exposures of the Brazilian and Guiana Highlands, with diversified soil and vegetation that includes savanna and dry *caatinga* scrub. In western Amazonia, there are sectors of better limestone and volcanic soils in Peru and Ecuador. The *terra firme* may be 25 m or more above the *várzea* and by definition is not subject to flooding. Following are brief discussions of aspects of contemporary indigenous cultivation in the upland interfluve forests. Additional information is given in Chapters 3 and 5.

Swidden Characteristics

A survey of the general characteristics of swidden cultivation of contemporary Amazonian groups has been provided by Beckerman (1987). He draws on descriptions for thirty-four groups, most of which are on *terra firme*, with a few being riverine. Summaries of portions of his survey follow.

Crops generally appear as single species stands, with a few scattered individuals of other species (but see D. Harris 1971). Some form of crop zonation is common. The staple crops are manioc (sixteen groups), plantain (five groups), and maize (one group, the Amahuaca). Maize is the most important secondary crop. The crop food proportion of the diet ranges from 45 per cent (Siona-Secoya) to 90 per cent (Barí). Mean field size ranges from 0.14 ha (Yaruro) to 2.6 ha (Yanomami[8]), with an overall mean of about 0.4 ha.

Field architecture (layering, zonation, intercropping, spacing) varies considerably (Chapter 3). Most fields are located close to the house or settlement. For fields further removed, the maximum walking distance is about 7 km. Fields are seldom immediately adjacent to one another, the buffer zone in between facilitating fallow recovery. The mean time from initial planting to 'abandonment' (cessation of weeding) is three years, but the time may be as short as one year (Cocamilla) and as long as fifteen years (Barí). Ecological reasons for abandonment of annual crops are soil fertility decline, weeds, disease, and pest invasion. Weed invasion is considered most important by some Amazonianists, but on poor soils fertility may be most critical (Descola 1994: 185). Manioc is usually replanted once after the initial manioc crop is harvested, whereas maize is not. The length of fallowing ranges from about eight (Urarina) to seventy years (Kuikuru).

Productivity varies with environment, field stage, soil additives, and other factors. For manioc, Beckerman's nine samples range from 4,500 (Yekuana) to 24,700 kg/ha/yr. Banana yields given are 2,800 (Yanomami) and 4,800 (Aguaruna) kg/ha/yr. Maize (not in Beckerman) ranges from *c*. 1,000 to 3,000 kg/ha/yr.

Annual labor times range from 63 (Siona-Secoya) to 200 (Machiguenga) hours/ha for clearing; 166 (San Carlos) to 193 (Barí) hours/ha for planting manioc, 75 (Shipibo) to 109 (Emberá) hours/ha for planting plantains, and 40 (Machiguenga) hours/ha for planting maize; and 75 (Barí) to 600 or more (Machiguenga) hours/ha for weeding. The overall labor demands given range from 188 (Shipibo) to 2,600 (Machiguenga) hours/ha/yr—a range that seems unlikely.

Population density ranges from 0.01 (Mekranotí Kayapó) to 1.0 (Campa) per km^2 for thirteen *terra firme* societies, and is 4.0 per km^2 for a *várzea* group (Shipibo). The average for the *terra firme* groups is 0.39 per km^2. This may be too high. Meggers (1992*a*: 203) using a different sample gives a range from 0.04 to 1.0 per km^2, with a mean of 0.30. These densities are much lower than for tropical forest swidden farmers elsewhere in the world (up to forty or even more per km^2). Beckerman (1987: 87) does not believe that the Amazonian densities are related to agricultural activity.

Forager/Cultivators

There are nomadic groups, such as the Ache, Guajá, Héta, Hoti (Hödi), Makú, Nambicuara, Yuquí, and Yora (Yaminahua), which subsist largely from hunting and gathering. These people are constantly moving from camp to camp, rather than having a permanent village. They can be distinguished from trekking groups, such as the Yanomami and Kayapó, who make trips over long distances and long periods but return to their villages and fields. Most of the nomadic groups are not pure foragers, however. They may practice periodic cultivation of small plots, and/or they may have some interaction with settled farmers, including exchange and crop theft, so they do know about cultivation. Furthermore, these nomadic groups rely strongly on food products from trees dominating abandoned settlement and agricultural sites rather than purely on 'wild' resources plus limited crops (Balée 1992). Most of these groups have probably experienced 'agricultural regression' (Balée 1992: 37–41) or 'cultural devolution' (Lathrap 1968*b*: 29) since earlier times. They are agricultural refugees, probably not incipient farmers, who have fled from more powerful Indians, from Europeans, or from demographic pressure along the rivers, an argument made in 1968 by Lathrap and even earlier by Claude Lévi-Strauss (1963) (also M. Martin 1969; Henley *et al.* 1994–6; Headland and Bailey 1991). Roe (1994: 198) refers to the 'illusion of pristine survivors'.

The Sirionó of eastern Mojos (Beni) in Bolivia were the subject of a classic field study by Holmberg (1969) in 1941–2. (For a recent re-examination see Stearman 1987.) I briefly visited the group at Ibiato in 1962. Holmberg described the Sirionó as nomadic hunters, and so they are generally thought of, but they did practice some crude horticulture.[9] They utilized both savanna and forest resources, but agricultural plots were in the forest. Manioc, maize, sweet potatoes, papaya, cotton, and tobacco were planted in small natural clearings of *c.* 30 m^2, probably due to tree falls. Fields were left unattended but were returned to after a few months to harvest what had not been lost to insects, birds, rodents, and weeds. After harvest, a site was abandoned indefinitely if not permanently. The Sirionó in 1940 numbered some 2,000 Indians (now a few hundred), with a very low population density, living in small camps, with a simple material culture.

Intensive Swidden

Contemporary indigenous swidden farming is usually extensive in the sense that the cropping to fallow ratio is high, typically one to three years of cropping followed by twenty or more years of fallow, a ratio such as 1 : 20 or 3 : 70. I have suggested that such high ratios were probably not representative of prehistoric agriculture (Denevan 1992*c*; 1998; see Chapter 7). Given the inefficiency of stone axes for clearing large trees, once a clearing was available it was probably farmed nearly continuously (i.e., intensively) or at least for many years followed by a short fallow, a low ratio such as 12 : 5 or 20 : 10. The archaeological, ethnohistoric, and ethnographic records are not very helpful, however, in confirming this.

The Kayapó are one of the few well-studied examples of relatively intensive shifting cultivators in *terra firme* forest, thanks to years of research by Darrell Posey and colleagues (Posey 1984*b*; 1985*a*; Hecht and Posey 1989). Cropping for five to six years and fallows as brief as eight to eleven years reflect site crop-specific planting, soil fertility management, micromanagement via in-field burning, composting and mulching, and both polycropping and concentric ring zonation (see Chapter 3). Even with this agriculture, the Kayapó still trek for long periods for long distances, in part to obtain forest foods.

Today, some 2,500 Kayapó are scattered over two million ha in eastern Brazil (Posey 1984*b*: 113). Indications that they were once very numerous, with large villages, include a report of 5,000 in four villages in 1896 and a village of such a size that it may have contained 3,500 to 5,000 Indians in 1900 (Posey 1987: 139, 147). The village of Gorotire still has 600 people which is very large for forest Indians (Posey 1985*a*: 140). If Kayapó villages did once number in the thousands, then highly productive agriculture was necessary, and present techniques may be survivals of that.

Another example of present-day intensive swidden is that of the Waika (Yanomami) at Ocamo on the Upper Orinoco in Venezuela. These people cultivate polycultural fields for five or six years. As weeds increase, root crops give way to fruit trees and then abandonment, with some subsequent harvesting of surviving useful plants in the fallows (D. Harris 1971: 480–1) (Figure 3.5).

Slash/Mulch Systems

A variation of slash-and-burn shifting cultivation is 'slash/mulch' or *tapado*. The vegetation cut for a clearing (slash) is not burned but rather is allowed to decompose (mulch), a slow process with less space available for crops, but a larger portion of the organic material is allowed to return to the soil than with burning. This system protects the soil, reduces erosion, conserves moisture, reduces soil temperatures, and prevents weeds and some diseases; however, yields are low. The system can be sustainable with periodic addition of new mulch from weeds and brush, with only short fallows. Most examples occur in very wet regions where there is an inadequate dry period to get a good burn. Contemporary examples, both Indian and non-Indian, occur in the Chocó of Pacific Colombia, Guanare in Venezuela, coastal Ecuador, and in some parts of Amazonia, and there are also early colonial descriptions (R. West 1957: 129; Thurston 1997: 30–6).

The Achuar of the Ecuadorian Amazon at times practice slash/mulch cultivation not because of a short dry season but because for one reason or another there is insufficient time to wait until cleared vegetation is dry enough to burn. The Achuar also practice slash/mulch for small maize gardens. Only a portion of the trees are felled. The slash is left on the surface as a mulch compost and the maize seed is broadcast into it. As maize grows rapidly, there is not time for the young plants to be choked by weeds. Only a portion of the maize seeds germinate, so density is low; however, the technique is very labor-efficient (Descola 1994: 158, 180).

The Canelos Quichua in Ecuador meticulously clean all the vegetation from the surface, leaving the felled logs unburned. The cut material is piled at the edge of the clearing to dry as a compost. After one or two plantings of manioc, maize, and plantains, other crops are planted or broadcast into the new brush, which is then cut down and left in place. Then the compost mulch previously piled at the edge is spread over the field (N. Whitten 1976: 70–6).

Slash/mulch cultivation may have been more common in pre-European times when it was difficult to continuously clear new forest.

Agroforestry Systems

Agroforestry is the combination of annual crops with perennial tree crops and/or useful forest species, planted either simultaneously or sequentially. There is a tremendous diversity of agroforestry systems, ranging from house gardens which are completely managed to swidden fallows or altered primary forest in which there may only be a few managed plants present.[10] Indigenous forms in Amazonia include anthropogenic forests with a high proportion of useful plants (Balée 1989: 6–15); forest patches of planted or spontaneous cared for crops and fruit trees along trails, campsites, and tree falls, as with the 'forest fields' of the Kayapó (Posey 1985a); and swidden-fallow management in which the number of useful fallow plants is intentionally increased and managed, as with the Bora described in Chapter 5.

Only recently has it become clear that various forms of agroforestry were widespread in prehistoric Amazonia and have continued to the present with both indigenous and mestizo societies. Over large areas the composition of mature and especially secondary tropical forest reflects past and present human management.

The nomadic Nukak (Makú) in Colombian Amazonia create 'wild orchards' as the result of the concentration of seeds from consumed fruits in their camp sites. These resource patches are frequented in the Nukak's 'cycles of mobility'. Also, thinning of vegetation for trails and camp sites favors some sun-loving economic species (Politis 1996: 504–7).

One of the best descriptions of indigenous forest management practices is by Balée (1994: 116–65) for the Ka'apor in eastern Amazonia. He defines management as: 'the human manipulation of inorganic and organic components of the environment . . . [that] involves direct and indirect human interference in species' populations, distribution, and behavior . . . some species may become locally extinct [but there] may be a net increase in the ecological and biological diversity' (Balée 1994: 116). The Ka'apor swidden-fallow forests are the most managed, and sampled plots indicate nearly half of the ecologically most important species are significant food species, whereas in mature forest only 20 per cent of the main species are food species (Balée 1994: 137). Many of these useful plants are semi-domesticates, particularly fruit trees and palms (Balée 1994: 215–19). For Amazonia, of the 138 cultivated or managed 'crops' listed by Clement (1999: 192), eighty-six are incipient domesticates and semi-domesticates, and seventy-nine are trees and woody vines, mostly fruit and nut trees. These may occur in gardens and fallows as well as in disturbed forest. Thus, forest succession can be a human-managed process (Balée and Gély 1989).

House Gardens

House gardens (dooryard gardens, kitchen gardens) are common in Amazonia but are not as large or as well studied as in the Caribbean, Central America, and Mexico. These gardens are highly complex mixtures of perennials and annuals, planted and spontaneous, food crops including medicinals, ornamentals, fuel, and artisanal species. They are invariably polycultural. Some house gardens are very large and very diverse, with dozens of different plants present, providing for many household needs. Others may consist of just a few fruit trees.

Fertility is maintained by decomposing plant refuse, cooking fire ash, garbage, and human and animal waste.

In Amazonia today, most Indian house gardens are not very impressive. They are small and impermanent and do not receive much attention, the plants being almost incidental. Bora and Karinya gardens are described in Chapter 5. Other descriptions, invariably brief, exist for the Siona-Secoya in Ecuador (Vickers 1983*b*: 37–8, 41), the Ka'apor (Balée 1994: 148–54), and the Amuesha (Salick 1989: 201–5). Some of the oldest and largest house gardens in Amazonia are in Moyobamba and Rioja in the Río Mayo Valley in Peru (Works 1990). These are mestizo, but certain indigenous (Aguaruna) elements persist.

Gardens take several years to develop, and so are not important to cultivators like the Campa who not only abandon their fields every two to three years but also move their houses as frequently. The Siona-Secoya house gardens are of particular interest because of their large size (one-third to half a hectare) and hence their major economic role (Vickers 1983*b*: 37–8, 41; also 1976). Between thirty and seventy-nine species were present in each of four gardens studied. Staples (manioc, maize, plantains) dominate early stages but later decline as non-staples shade out and replace them. Other food plants continue to be important, however.

The probable reason for poorly developed Indian house gardens in Amazonia is that most Indians today move their houses frequently, so that there is no time for gardens with perennials to develop. Large, permanent house gardens are associated with stable dwellings. Houses and villages are now moved frequently given short-cycle shifting cultivation, game depletion, village fissioning, disease, and flight from other people. In prehistory, however, gardens may have been much more important.

When inefficient stone axes were used to clear forest, agriculture was probably much more permanent than at present, permitting more stable settlements with large house gardens, fruit orchards, and other forms of agroforestry, as well as small, fixed fields. The presence of *terra preta* soils on both bluffs and interfluves, the product of semi-permanent prehistoric settlement, is supporting evidence. These issues are discussed in Chapters 6 and 7.

Savanna Habitats

Lowland Savanna

Most of the lowland or 'wet' savannas are subject to seasonal flooding of several months duration from overflowing rivers or standing rainwater. The largest are the Llanos de Mojos in northeastern Bolivia, the lower Orinoco Llanos of Venezuela and Colombia, the Pantanal of western Mato Grosso in Brazil, and along the Atlantic Coast from Guyana south to Marajó Island.

Much of the soil of these savannas is alluvial but usually heavy clay which is difficult to cultivate. Agriculture is further complicated by the growing season also being the time of flooding. There are few examples of any kind of agriculture today and even fewer for Indians. However, some of the savannas were farmed in prehistoric times by means of large man-made raised fields which provided drained planting surfaces during flooding (see Chapters 11 and 12). Fertility apparently was maintained by transferring rich organic muck from the intervening ditches to the field tops.

The largest surviving clusters of these fields in Greater Amazonia are in the Llanos de Mojos, but others occur in the Orinoco Llanos and Guianas. All the raised fields are prehistoric, and none have continued in use to the present. The Karinya still drain savannas by ditching in the Orinoco Llanos (Chapter 14). The Macusi (Makushi) (Blank 1976) farm the Rio Branco savannas of northern Brazil by means of minor mounding of clusters of manioc plants.

Upland Savanna

The well-drained upland 'dry' savannas have much poorer soil and wildlife resources than do the lowland savannas and forests.[11] They include some of the oldest, most weathered soils in the world. Particularly extensive are the enormous *campos* (grassy savannas) and *cerrados* (scrub savannas) of the central plateaux of Brazil and the eastern Orinoco (Llanos Altos).

The *cerrados* are considered to have had limited Indian settlement, with agriculture restricted to the forest islands and gallery forests. Surviving groups such as the Nambicuara are primarily trekking hunters and gatherers, with only sporadic plantings of crops which are left unattended for long periods. Their population density in 1907 was only 0.2/km^2 (Steward 1949a: 659). On the other hand, some Indians, such as the Kayapó, did manage the savanna to make it more productive.

Even without savanna cultivation, savannas often were preferred settlement locations for foragers and cultivators of adjacent forest: Chácabo and Sirionó in the lowland Mojos savannas; Yaruro (Pumé) in the lowland Orinoco Llanos (Leeds 1961); Nambicuara, Xavante, Kayapó, etc., in the upland scrub savannas (Moran 1993: 123–36). Because groups located in savannas utilized both forest and savanna resources, population densities were not determined just by savanna resource management. Savanna people hunted and gathered and cultivated in forest islands and gallery forests and into the fringing forests. The savanna/forest boundaries were highly disturbed by these activities, and swiddens cut into the forest edges, combined with burning, were capable of expanding the savanna at the expense of the forest.

Apêtê

The Kayapó are generally thought of as tropical forest shifting cultivators. However, their territory also includes expanses of upland scrub savanna which they exploit, particularly by creating and managing small forest islands (up to

2 to 4 ha in size) called *apêtê* (Posey 1984*a*; 1985*a*; Anderson and Posey 1989). These islands, or many of them, apparently are initiated from compost heaps in nearby forests consisting of sticks and leaves plus soil from termite and ant nests. The rotting mulch is used to make mounds (1–2 m diameter, 0.5 m high) in the savanna which are planted with forest species and then enlarged over the years with additional plantings at the edges as well as by natural growth. Some 85 per cent of the species present are planted or plantable. About 98 per cent of the species present are useful, including domesticates (Posey 1985*a*: 141). The savanna vegetation and soil are thus modified and made more productive.[12] Posey (1984*a*: 32) believes that such savanna management was practiced by other Indians. There are thousands of forest islands in the central *campo cerrado* region of Brazil which may have originated as *apêtê*, or at least be partially anthropogenic in terms of plant composition. If *apêtê* are indeed widespread, then savanna population densities probably increased significantly over what they would have been otherwise.

Multiple Habitat Exploitation

We have examined cultivation and population in floodplains, forests, and savannas, the three principle biomes (macrohabitats) in Amazonia. Each could be subdivided into biotopes (microhabitats). An important point to make is that while some groups are confined to a single biome, other groups exploit multiple biomes and/or biotopes, with appropriate technologies for each. As a result, higher populations are potentially possible than with single biome utilization. We saw this for the Kayapó who farm both forest and savanna, including a range of microhabitats. The Shipibo of Peru, who farm *playas*, levees, and backswamps are discussed in Chapter 5. The Karinya of the Orinoco Llanos in Venezuela, who farm floodplains, levees and islands, house gardens, *terra firme* forests, and palm swamps, are described in Chapters 5 and 14. Such adaptive variability is probably typical, if not as dramatic, of many Amazonian Indian groups.

Population densities may be greater than with single agrosystems, not just because a combination of systems may be more productive, but also because subsistence risk is reduced. This kind of flexibility counters the thesis of Meggers (1992*a*) that because of unpredictable crop-destroying floods in the *várzeas* population densities would have been kept at low levels. The *várzea* farmers had crops elsewhere: unflooded river terraces, fringing *terra firme* forest (bluffs) (Chapter 6), plus storage capability and the option of periodic greater reliance on wild plants and animals.

Notes

[1] Amazon Basin, Orinoco Basin, Guiana Highlands, northern and central Brazilian Highlands, and northeastern Brazil, an area of *c*. 9.8 million km^2 (Denevan 1976: 206, 230).

[2] For a general survey and bibliography of Amazonian cultural ecology, see Sponsel (1986). For

archaeology, see Lathrap (1970), Meggers (1987), Meggers and Evans (1983), Roosevelt (1980), Raymond (1988), and Myers (1990). For general discussions of subsistence see Meggers (1971) and Eden (1990: 62–85).

³ Richards (1952: 404). Also see the discussion on this statement by Sauer and Richards (C. Sauer 1958).

⁴ For the Caribbean mainland region from Yucatán to Colombia, Gordon (1982) believes that probably almost all of the prehistoric forest was anthropogenic.

⁵ Shotguns are now widespread, but shells are expensive and many Indians with access to guns still rely more on bows and arrows, spears, and blowguns in hunting.

⁶ Roosevelt (1989*b*: 34) suggests that biological and chemical techniques in archaeology can help reconstruct prehistoric subsistence. Thus far, such techniques have been used mainly to determine crops and diet.

⁷ Chernela (1989: 246) provides Amazonian data showing first-year maize yields that are 40 per cent to nearly 500 per cent greater on alluvial soils than on upland soils, and *manioc* yields that are nearly 50 per cent higher on alluvial soils.

⁸ The Yanomami are the largest of several linguistically related groups in northern Brazil and southern Venezuela. Other terms for the Yanomami, or subgroups, in addition to Waika, include Yanomamö, Yanoama, and Yanomama.

⁹ The related Yuquí further west near the Río Chimoré practiced no cultivation when first contacted in the 1950s, but likely did so in the past given linguistic evidence, and they have recently relearned how to farm (Stearman 1989: 73).

¹⁰ Peters (2000) makes a distinction between managed forest fallows and managed forests which are mature forests that may or may not once have been swidden fallows.

¹¹ Roosevelt (1980: 184) estimates that the carrying capacity for animals on the dry savannas of Parmana on the Orinoco is no more than one-quarter that of the gallery forest.

¹² Parker (1992; 1993) has questioned whether *apêtê* are indeed artificial forest islands; see Posey's (1992; 1998: 112–14) response.

5

Fields of the Mojo, Campa, Bora, Shipibo, and Karinya

In the previous two chapters, material on indigenous cultivation in Amazonia was taken from many sources. Following are five case studies based on my own field research, plus collaboration with others, and use of historical documents. The reader is referred to our previous publications for more details.

These case studies provide examples of different historic and contemporary agricultural systems in Greater Amazonia: the colonial Mojo (relatively intensive swidden) in Bolivia, the Campa (extensive swidden) in Peru, the Bora (managed fallows) in Peru, the Shipibo (floodplain cultivation) in Peru, and the Karinya (multiple biome fields) in Venezuela. An additional group of relatively intensive swidden farmers, the Kayapó, has already been described in several sections of Chapters 3 and 4.

The Mojo: A Colonial Reconstruction

For the colonial period, seldom is there enough information available to reconstruct agriculture for a specific tribe or region. Following are some early descriptions of forest cultivation in the Jesuit Province of Mojos in northern Bolivia (Fig. 5.1) (Denevan 1966*a*), a region better known for pre-European drained, raised fields in seasonally flooded savanna (Chapter 12). Shifting cultivation was and continues to be practiced in gallery forest and in forest islands within savanna, both habitats tending to have semi-deciduous forest relatively easily cleared with stone axes. The main Indians involved were the Mojo and the (related) Baure chiefdoms.

Swidden (shifting cultivation) fields were called *chacras* (or *chácaras*) during the colonial period and are now referred to as *chacos*. There are many early references to forest cultivation but few details. Several of the descriptions of Mojos by members of the Solís Holguín expedition in 1617 mention fields: '*chácaras* in carefully cleared *montaña*'; and 'great *chácaras* of maize and other vegetables' (Lizarazu 1906: 149, 195). The Jesuit, Altamirano (1891: 30) wrote in 1710 that the Mojo first felled the trees and rooted out the useless weeds and grasses. Coronel Aymerich, Governor of Mojos from 1768 to 1772, wrote that because of flooding, crops were grown in the forests on *alturas* (high ground)

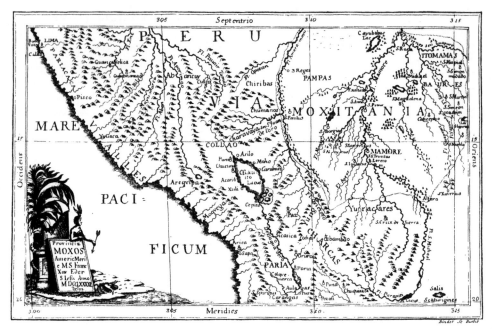

Fig. 5.1. Map of colonial Peru *c.* 1767 showing Mojos (Moxitania). From Eder (1985: 428)

next to the rivers (gallery forests on natural levees), where transport was easy to villages that were sometimes two or three days away (René-Moreno 1888: 68). Padre Eder (1985: 75), in 1791, said that cultivation of the *campos* (grasslands) was impeded by flooding, but that they were not inferior to the forests because the ashes of burnt grasses provided good fertility.

The process of clearing forests for fields in Jesuit Mojos has been described by Métraux (1942: 59, based on Eder 1985: 73–4):

The Mojo and Baure cleared fields in the forests, which were not flooded during the rainy season. At the end of August they first destroyed the underbrush, then cut the base of large trees by alternately charring and hammering the wood with stone axes. They waited until a strong wind blew down the undermined trees, or else felled selected trees, which knocked down all the others. The dry trunks were burned and their charred remains left on the field to protect young maize stalks. The Spaniards who penetrated the country with Solís Holguín were amazed at the size of the Mojo fields, which were crossed by wide roads.

The Mojo and Baure used digging sticks for planting, and presumably so did the other tribes in Mojos. Stone axes were used by the Mojo to clear forests. Alonzo Soleto Pernia, a member of the Solís Holguín expedition, said that: 'we found trees cut, as if by stone axes; they have mines where they obtain stone for axes for cutting trees, and their edges are like iron' (Lizarazu 1906: 211). Rock outcrops may not have been far from the area visited by the expedition; however, in the central savannas most stone tools were obtained in distant trade and were a valuable commodity. Possibly in much of Mojos there were not enough axes to clear sufficient forest to support large numbers of people. The Mojo had cutting and sawing tools made from bone, teeth, and chonta palm wood, but these would not have been very effective for clearing mature forest. The change to metal tools undoubtedly had revolutionary effects for shifting cultivation (Chapter 6). By 1676, and probably much earlier, Mojo Indians were traveling south to Santa Cruz to trade cotton goods for 'machetes to cut and clear their *chacras*' (Marbán 1898: 148). The Jesuits used axes and machetes as major gift items in gaining the friendship of the native people.

In the early seventeenth century the Indians of southeastern Mojos had large farms producing great quantities of food. These must have been forest fields, assuming that most raised fields in the savannas had been abandoned by that time. This is indicated by the reports of the members of the Solís Holguín expedition. The relation of Juan de Limpias (Lizarazu 1906: 170) states that a Captain Diego Hernández Vexarano saw a large number of *percheles* of maize and other legumes and told Juan and another soldier to count them. Juan de Limpias counted over 700 *percheles* in one group. There seemed to be 20 to 30 *fanegas* (30–45 bushels) of food in each *perchel*, and the men were much impressed. The other soldier counted over 400 *percheles* in a group. Captain Gregorio Jiménez said that Captain Diego Hernández Verarano reported seeing a *chácara* with over 500 *percheles* of maize, which formed one of the granaries (Lizarazu 1906: 158).

Métraux (1942: 59) translated *percheles* from these accounts as 'probably the forked sticks used to support maize'. This usage is apparently based on one meaning of *percha* as perch or pole and *perchonar* meaning to leave shoots on a vine stock. However, it would be impossible for a single 'forked stick' to support a harvest of 30 to 45 bushels of maize. A unit of land is a possibility, but most likely *perchel* refers to a maize crib on pilings for protection against flooding and animals (the term has been so used in Portugal). Such cribs were still being built by the Chácobo in Mojos when visited by Nordenskiöld (1920: 3) in the early twentieth century. Seven hundred *percheles* each holding 30 to 45 bushels of corn would total 21,000 to 31,500 bushels—a sizable amount for what was presumably one village.

Padre Castillo (1906: 309–10), in 1676, listed the following Indian crops: manioc, maize, beans, squash, sweet potato, peanuts, papaya, chili pepper, cotton, arracacha, tobacco, and plantain. Most of these were also listed in the reports of the earlier Holguín expedition. The Old World plantain was apparently pre-Jesuit, being cultivated in 1677 according to Castillo. The staple seems to have been manioc. Padre Marbán (1898: 139) wrote that: 'yuca [sweet manioc] is the common bread of the land,' and Padre Orellana (1906: 13) in 1687 said that: 'yuca is their principal food.' On the other hand, the numerous maize *percheles* suggest that maize was at least locally important. In the reports of the Solís Holguín expedition in 1617, maize is mentioned ten times and manioc only once. But Marbán (1898: 138) later said that: 'there is not much maize because these Indians do not use it for *chicha* [maize beer], except once in a while.'

The Jesuits introduced rice, and they said that rice produced as many as five crops a year and that harvesting was done in canoes (Eder 1985: 75–6). This indicates that the Jesuits were growing wet rice on the savannas; however the Indians may also have been harvesting a wild rice.

The colonial fields contained both indigenous and Jesuit techniques and crops. Raised fields in the wet savannas were largely abandoned. Land use today is quite different, with subsistence fields being smaller. Steel axes and machetes are now used for clearing and weeding. *Percheles* no longer exist. Manioc is the dominant crop, and maize and rice are rare. The savannas are used for livestock. Food is imported.

Thus, it has been possible to extract considerable information on Mojo forest agriculture from sixteenth- and seventeenth-century exploration and missionary accounts. However, our information on the Mojo is from different places and times and may pertain in part to tribes other than the Mojo. And we are told little or nothing about field types, cropping/fallowing patterns, and agroecological techniques. For most other Indian groups in Amazonia, even less early information is available, and we have to rely on twentieth-century reports long after initial contact, or on what can be learned from archaeology.

The Campa

Unstable Swiddens

> I wouldn't want to live in a world without lions. These people [Campa] are like lions.
>
> (Werner Herzog 1982)

The Campa (Asháninca, Ashéninka) live in the eastern foothills of the central Andes of Peru. They number at least 30,000 and possibly 45,000, making them the largest indigenous group today in Amazonia (Hvalkof 1989: 128).[1] In the 1960s I made a field study of Campa agriculture in the Gran Pajonal, a dissected plateau with scattered patches of savanna (Fig. 5.2) (Denevan 1971a).[2] These Campa are best described as extensive swidden farmers with a strong emphasis on hunting and minimal gathering and fishing. While males spend most of their working time hunting, most household food by weight and calories comes from cultivated plants. Manioc is the staple, but each field is initially planted in intermixed manioc and maize (Fig. 5.3).[3]

Fields are usually cropped for only two years, with a fallow of ten years or more. In a typical year, a new field is cleared and planted, the crop in the previ-

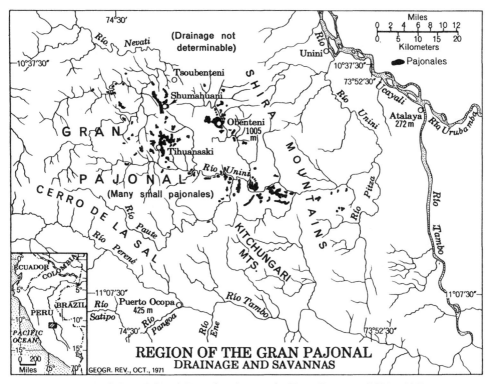

Fig. 5.2. Map of Gran Pajonal, Peruvian Amazonia. From Denevan (1971*a*: 497)

ous year's clearing is harvested, and residual crops in a two-year-old field are also harvested. There is relatively little management and use of either swidden-fallow fruit trees or house gardens, given frequent movement. This contrasts with the Río Pachitea Campa to the north, who have longer cropping and shorter fallows (Allen and Holshouser 1973: 143), and the Bora, who have managed fallows (see below).

I was impressed that many Campa are not systematic land-use planners. Clearing and burning can take place well into the wet season, with a poor burn resulting, leaving little fertilizing ash and much debris. Planting is done immediately after burning, regardless of the season, even weeks before the rainy season begins. Resulting yields, therefore, may be poor for most crops, although manioc does fairly well. The Campa are certainly aware of the ideal timing of clearing, burning, and harvesting, but other factors intervene.

A field that is still producing may be abandoned prematurely because of a house shift so that a family is located too far away to be able to utilize an old field. Relatives may move in and help to deplete a field. Because of bad weather, an extended trading or fishing trip, or sickness or an injury, a man may not clear a field some years. As a result, some families have plenty of food and their

Fig. 5.3. Campa boy planting maize with a digging stick on a recently burned swidden, Peruvian Amazonia. From Denevan (1971*a*: 506)

future needs are assured, whereas others are caught short with a field depleted before another comes into production. Then they either move in with relatives or live off the land, hunting and gathering, until the new field is producing. Such instability of production seems to be common in the Gran Pajonal and helps explain why the Campa frequently clear, burn, and plant out of season. They fully realize that they will not get a good crop, but a poor crop planted in the wrong season is better than no crop at all. Such seasonal and managerial irregularity in swidden systems may be more widespread than we realize (Brookfield 1968: 421; Salick and Lundberg 1990).

A new field may be adjacent to a current one or as much as tens of kilometers away. Given plentiful land, the reason for distant moves of houses and fields can possibly be explained (not always) by game depletion rather than by soil depletion, pest invasion, or social factors, since game is the primary source of protein. Game is rapidly hunted out or scared away around a settlement; hence the frequent moving to new locations. This so-called 'protein thesis' of Amazonian settlement is discussed below.

Semi-nomadic tribes with a strong emphasis on hunting, such as the Campa, are often thought of as being incipient or primitive agriculturalists. However, the large number of crops cultivated[4] and the many varieties of each, plus the utilization of a wide range of microecological conditions, suggest considerable agricultural sophistication, more so than that of the average non-indigenous colonist in tropical Peru. The strong emphasis on manioc cultivation is misleading. It is a labor-saving way to provide calories, while the Campas' greater physical efforts are directed toward hunting for the more basic food element, protein. Although a labor/time study was not made, agricultural productivity seems to be high in terms of time expended, whereas hunting productivity is low for time expended.

The Protein Thesis

It has been postulated that the greater population densities and more developed cultures have been located along the large Amazonian rivers, in contrast with sparsely populated interfluves, because of the availability of protein-rich aquatic resources which supplement the protein-poor root crops that dominate the diet of Amazonian people (Lathrap 1962: 547–9; 1968*b*; Carneiro 1970*a*; Denevan 1966*b*; Gross 1975). Away from the rivers, game is the main source of protein, but it does not seem to exist in quantities large enough to support large social units, nor is it permanent enough to support long-enduring settlements. Hence, the aboriginal pattern in the upland forests is usually one of small semi-nomadic social groups with a limited material culture.

One of my purposes in studying the Gran Pajonal Campa was to test the above argument with a brief dietary study (Denevan 1971*a*). These Campa clearly fit the non-riverine model. Their total protein intake is low (less than 50 g most days), and they seem to go to great efforts in hunting, with small returns, to stay above the minimum. The result is settlement instability. The Gran Pajonal Campa move about once every two years, to average distances of about 8 km.[5] This is not to say that other factors besides game depletion are not important, and a careful sociological study is needed to define these factors, their causes, and their relative significance compared to the dietary argument. In contrast with the Gran Pajonal Campa, the riverine Campa are much less nomadic.

Certainly diets can change, and a greater use of protein-rich maize and beans would reduce the need for animal protein.[6] The emphasis on protein-poor root crops in Amazonia is rational given the great productivity and ease of ground storage of root crops, so long as the population is concentrated along the large rivers where protein-rich fish and game are found in great numbers. It is interesting to note that the missionaries have been able to establish much larger and more permanent Campa villages along the lowland rivers. On the other hand, the Seventh Day Adventists prohibit meat eating, or at least restrict the varieties of game and fish that the Campa in their missions can eat. As a result, severe nutritional problems arose in the past at some missions, as on the Perené (Paz Soldán and Kuczynski-Godard 1939); elsewhere, the Campa have often ignored the restrictions. The Adventists have made major efforts to shift the Campa diet from traditional manioc toward maize, beans, peanuts, and other crops with a relatively high protein content.

Robert Carneiro (1960) and others, have shown that quite substantial settlements and population densities can be supported by indigenous shifting cultivation in Amazonia. On the other hand, if shifting cultivation does not supply sufficient protein, the availability of unevenly distributed animal protein becomes a limiting factor. Such a limitation, apparently applicable to the Campa, must be viewed as culturally determined in so far as the dietary pattern responsible for it is culturally determined.

The Campa obtain at least 90 per cent of their food, by weight eaten, from agriculture and at most 6 per cent from hunting, 3 per cent from gathering, and 1 per cent from fishing. The evidence for the Campa, though not precise, indicates that a group consistently thought of as being hunting-oriented actually may obtain no more than 5 to 10 per cent of its total food (by weight consumed) from hunting.

The argument that the availability of protein from game in Amazonia not only influenced shifting cultivation but determined low population density, small and unstable settlement, and cultural development, became a major debate in cultural anthropology in the 1970s well after my Campa research and Carneiro's 1970 paper, and it has continued (Denevan 1984a; Hames 1989). The catalyst was an essay on 'Protein Capture' in the *American Anthropologist* in 1975 by Daniel Gross. Numerous field studies followed which attempted to support or disprove the thesis, one direct result being the stimulation of research on Amazonian cultural ecology.

Following Gross, the topic was picked up by Marvin Harris (Harris and Ross 1987) and the cultural materialists as a prime example of how human behavior can be explained as invariably functional in some way in terms of survival, in contrast to the structuralists who argue that much behavior can be explained in other ways or is inexplicable. The materialists further claim that limited game availability can also explain other aspects of behavior in Amazonia, such as warfare (competition for game territory), food taboos (restrictions on large game animals), and sex (greater availability for successful hunters).

Many of the leading Amazonian scholars became involved in the debate. Opponents of the thesis include Beckerman (1979); Chagnon and Hames (1980); and Diener, Moore, and Mutaw (1980). Beckerman and Diener argue that there is little evidence of protein scarcity. Brokers include A. Johnson (1982) and Sponsel (1983). Renewed support has come from Baksh (1985), Good (1987; 1995), and Frank (1987), even though Chagnon (1992: 96) maintains that 'the protein debate has now pretty much been laid to rest'. A. Johnson emphasizes perceived versus actual protein scarcity. Good emphasizes greater labor inputs to procure adequate protein from progressively smaller game as population size grows, rather than absolute scarcity.

Thus we have a situation today in the Amazonian interfluves where Indian settlements, almost without exception, are small and moved often, with low population densities (mostly $1.0/km^2$ or less; Beckerman 1987: 86). Why? Limited game seems to be more critical than poor soils. Despite evidence that game is more plentiful than previously thought, there is some threshold where it will not be adequate, possibly at 2 persons/km^2 or even 20, but certainly before 200.

However, this is not a direct environmental limitation. It is the result of a diet dominated by manioc with most protein coming from animal sources, and this particular diet is a cultural choice given options available, pressures present, and history. A diet based on maize and legumes, as in tropical Mesoamerica, is nutritionally well balanced and not dependent on protein from fish and game. The Amazonian Indian diet works because population densities are low and because labor inputs are low. A denser population could be supported without game if the diet were changed to emphasize seed crops and legumes, but the labor inputs would be high; there are numerous examples in Asia and Africa.

In my opinion, protein scarcity is one but not the only explanation for the consistently extremely low densities of historical *terra firme* Indian populations (Denevan 1984*b*). The argument is less valid, however, for prehistoric times when shifting cultivation was limited by stone axes and cultivation and settlement were probably more permanent.

The Bora: Swidden-fallow Agroforestry

The use of swidden fallows is a widespread but little-studied practice in the tropics.[7] When we undertook the Bora Agroforestry Project in 1981–3 in the village of Brillo Nuevo (Denevan and Padoch 1988), we believed that fallow management was unique in Amazonia. 'Forest gathering from abandoned polycultural swiddens has seldom been commented on in the literature on shifting cultivation, though it increases significantly the overall productivity of the system' (D. Harris 1971: 482). We subsequently learned that fallow management was common not only to Indian societies but also to *caboclo* (peasant), *ribereño* (river settlers), and other mestizo groups, but less so to new colonists.

Brillo Nuevo is a Bora mission village located on the Río Yaguasyacu, a small branch of the Río Ampiyacu which joins the Amazon at Pebas, 120 km northeast of Iquitos in Peru (Fig. 5.4). The village fields are on unflooded *terra firme*, with humid tropical forest and low fertility soil (ultisols). The village contained forty-three families, all in the process of being assimilated into Peruvian society. Bora subsistence, however, retains many of its traditional elements, with a reliance on swidden cultivation, house gardens, fallow management, and collecting, hunting, and fishing. The young Bora fallows (three to nineteen years) were studied by John Treacy and Janis Alcorn and the older fallows by agronomists Salvador Flores Paitán and Wil de Jong.

Bora shifting cultivation can be described as a short-cropping system followed by a fallow of about twenty years. Manioc, the staple, is intermixed with a wide variety of other crops. In the initial clearing various fruit trees and other useful trees are spared, and new trees are planted which may survive into the next fallow. The swiddens include patches of peanuts, coca, and pineapple.

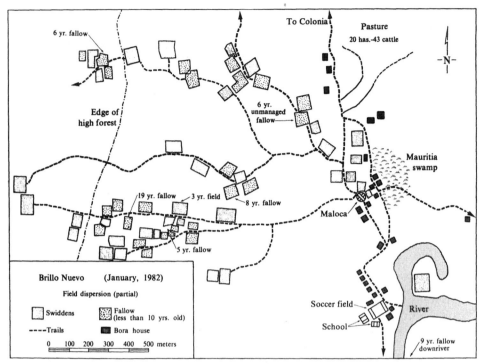

Fig. 5.4. Map of Brillo Nuevo village, swiddens, and fallows, Peruvian Amazonia. From Denevan and Treacy (1988: 9)

After two or three years, the manioc stage (*yucal*) is replaced by a brief pineapple stage (*pinal*), and other annual crops are abandoned. The early fallow stage is dominated by individual and clusters (orchards) of managed fruit trees.

Fallow management involves both purposeful and unintended human manipulation of both individual plants and groups of plants, both wild and domesticated or semi-domesticated. Practices include saving and protecting useful plants either from the original forest or invaders in swiddens or fallows, weeding to eliminate undesirable species, shade/sun light control, plantings in the original field or in a fallow, transplanting, and sometimes fertilization. Secondary effects of human activities (unintended manipulation) include seed dispersal through spitting out seeds, from defecation, from garbage, and from scattering via indiscriminate weeding; provision of sunlit openings at house clearings, trails, and campsites, which are all places where useful species are likely to be introduced either intentionally or accidentally; and the improvement of soil fertility around habitations and other sites of human activity due to accumulation of garbage, bones, weeds, human and animal waste, and ash. Such unintentional manipulation can give an advantage to useful species appearing spontaneously in the secondary vegetation, and over time such plants may become more numerous and valuable than intentionally managed plants. It is during the initial ten to twelve years of fallow that management is

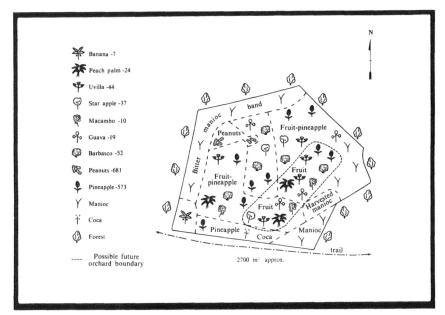

Fig. 5.5. Three-year-old transitional swidden/fallow, Brillo Nuevo, Peruvian Amazonia. From Denevan and Treacy (1988: 17)

most active. Older fallows still contain useful plants and are harvested for up to thirty-five years or more, but management is minimal.

Fallows of three, five, six, eight, and nine years, plus another of nineteen years and an unmanaged fallow of six years were examined by means of transects and sample plots to determine how the relative frequency of useful plants changed with fallow age. These are not replicated samples and thus only suggestive of successional sequence. There can be considerable difference in composition between fallows of the same age depending on their history, including the age of the original forest, proximity to forest, composition of the swidden, differences in soils, drainage, pests, etc., and especially management patterns. These factors, as well as age, determine when a fallow is most productive of useful plants as well as what those plants are. Despite this sampling problem, some important conclusions can be drawn from our data. We found that the transition from swidden to fallow, or field abandonment, does not occur at a point in time but is a gradual process over years, partly regulated by human populations and partly by ecological processes.

Each of the younger fallows is described in Denevan and Treacy (1988). A summary appears here in Table 5.1. A sketch map of the three-year-old fallow is shown as Fig. 5.5. Figure 5.6 shows a fifteen-year-old fallow, and Fig. 5.7 shows a fallow being harvested.

The Bora recognize that two ecological processes, soil nutrient depletion and secondary succession, must be confronted. They acknowledge that manioc is not sufficiently productive to merit harvesting after three or four years, mainly,

Amazonian Cultivation

Table 5.1. Succession of harvestable plants[a] in Bora fields and fallows

Stage	Planted harvestable	Spontaneous harvestable
High forest	None	Numerous high-forest construction, medicinal, utilitarian, handicraft, and food plants available
Newly planted field; 0–3 mos	All species still developing	Dry firewood from unburnt trees for hot fires
New field; 3–9 mos	Maize, rice, cowpeas (*Vigna* sp.)	Various early successional species
Mature field; 9 mos–2 yrs	Manioc, other root crops, bananas (*musa* spp.), cocona, and other quick-maturing crops	Abandoned edge zone has some useful vines, herbs
Transitional field; 1–5 yrs	Replanted manioc, pineapples, peanuts, coca, guaba, caimito, uvilla, avocado, cashew, barbasco, peppers, miscellaneous root crops; trapped game	Useful medicinals, utilitarian plants within field and on edges. Seedlings of useful trees appear. Abandoned edges yield straight, tall saplings, including *Cecropia* spp. and balsa (*Ochroma* spp.)
Transitional fruit field; 4–6 yrs	Peach palm, banana, uvilla, caimito, guaba, annatto, coca, some root crops, pineapple; hunted and trapped game	Abundant regrowth in field. Many useful soft construction woods and firewoods. Palms appear, including chambira and huicungo. Many vines; useful understory aroids
Orchard fallow; 6–12 yrs	Peach palm, some uvilla, macambo; hunted game	Useful plants as above; self-seeding *Inga* spp. Probably most productive fallow stage
Forest fallow; 12–30 yrs	Macambo, umarí, breadfruit (*Artocarpus* sp.), copal (*Dacryodes* sp.)	Self-seeding macambo and umarí. High forest species appearing. Early successional species in gaps. Some useful hardwoods becoming harvestable. Many large palms: huicungo (*Astrocaryum huicungo*), chambira, assai, ungurahui
Old fallow; high forest; over 30 yrs	Umarí, macambo	Only a few residual previously managed trees

[a] Plants not identified here are in Appendix 1A.

they say, because of soil depletion, but also because of weed invasion. Termination of manioc production occurs within the space of one to two years in fields planted almost entirely in manioc. However, if fields are planted with trees, weeds may be the major obstacle to extended field use. Management

Fig. 5.6. Peach palms in a Bora fallow aged about fifteen years, Peruvian Amazonia. From Denevan and Treacy (1988: 38)

Fig. 5.7. Bora man carrying peach palm fruits and chambira fronds harvested from a ten- to fifteen-year-old fallow, Peruvian Amazonia. Chambira fiber is used for making hammocks and other materials. From Denevan and Treacy (1988: 43)

shifts from replanting and weeding manioc to dealing with encroaching secondary vegetation threatening tree crops. With periodic weeding, planted trees can remain productive for several years before their yields decline or they succumb to the effects of shading, competition for nutrients, or other factors.

Our observations of the sampled fallows plus numerous other fallows indicate that the most productive fallow period is between about four and twelve years. Before four years, fruit trees are not yet producing or have limited production. After twelve years, management is minimal and many of the smaller useful plants are shaded out. Harvesting of some trees continues, however, for up to twenty to thirty or more years. Another important characteristic is seasonality. The various Bora tree species bear fruit sequentially allowing a spread of produce throughout the year.

A number of tree species planted in Bora fields frequently grow in secondary forests. Umarí and macambos are common cultivated trees found in old fallows, either growing alone or in clusters.[8] These survivors of swidden orchards are valued components of Bora fallows. At twenty to thirty years of age most fruit trees cannot be easily harvested; however, the Bora occasionally gather the fallen fruits. Fallen fruits also attract game animals. It is common to find an umarí fruit on the forest floor with toothmarks of a majás (a rodent, *Cuniculus paca*) or other browser. For this reason older fallows are good hunting grounds.

The process from field to fallow to forest regeneration clearly has a spatial aspect. While successionary processes are complex (Uhl *et al.* 1981), there is a pattern of centripetal forest regrowth explained largely by the history of weeding. Harvesting and weeding of manioc holds regrowth at bay. As a manioc zone is abandoned, terrain is gradually surrendered to the forest, usually from the edges inward, and the field shrinks in size.

The life of a field is one of sequential overlapping utilization rather than simply planting-harvest-abandonment-fallow. Harvesting proceeds from grain-producing annuals (rice and maize) to root crops and pineapples to planted and unplanted fruit trees and other useful trees and vines. The concept of 'sequential variation' is probably more appropriate here than 'phased abandonment'.

There are similarities between complex swidden systems and agroforestry systems (Hecht 1982). Agroforestry combines the production of trees and other crops on the same unit of land, a strategy characteristic of swidden-fallow management. Both systems rely on the succession of tree crops following the harvests of short-term cultigens.

Bora agriculture becomes an agroforestry system during the early stages of forest fallow. The enriched swidden to fallow sequence closely resembles the natural succession analog approach to tropical agroforestry outlined by Hart (1980; also Uhl 1983). Hart suggests that appropriate cultigens be placed in the niches otherwise occupied by common early successional species. These 'analog' plants should have growth structures and resource requirements similar to those of their weedy counterparts. Thus, rice or maize replaces early annual grass species, bananas replace wide-leafed *Heliconia*, and late appearing tree crops mimic early successional tree species. Whether by accident or design, the Bora seem to follow this approach. Bananas do well in low shady areas, where *Heliconia* plants are also common. The most obvious example is uvilla which matches the ubiquitous *Cecropia*. Guaba is also in the same genus (*Inga*) as its semi-domesticated analog, the shimbillo. Further research may reveal other similarities between naturally appearing species and cultigens, and whether or not productivity is greater in analog situations.

Another feature of Bora swiddens in relation to agroforestry is the use of space. Trees are clustered to take advantage of variation in slope and other conditions. More important, slowly abandoning ground to secondary forest seems to be a sound strategy for tropical farming. Agroforestry plots do not have to be entirely planted. Managed forest regrowth can provide useful products, as well as canopy cover for the soil and a source of stored nutrients for when the forest is cleared to begin the swidden and agroforestry cycle anew.

Swidden-fallow agroforestry, enriched with tree crops planted in areas of forest regrowth, approximates the 'tree garden' model of silviculture which may have been a pre-Columbian agricultural adaptation in the Caribbean lowlands (Gordon 1982). This model involves a combination of overstory fruit trees and subcanopy woody shrubs, interspersed with open areas of swiddens containing maize, bananas, manioc, and other crops. We can envision an indigenous Amazonian swidden-fallow agroforestry system, then, which would have a fruit orchard core, or series of cores, embraced by areas of regenerated forest. The forest, in turn, would be enriched by a variety of useful analog species able to compete in the viney subcanopy, or later on as a canopy species (fruit, timber) in high-forest fallow. Timber species would be appropriate late fallow-enrichment trees. Over a large area, swidden-fallow agroforestry would resemble Gordon's image. It would be more a thicket and less a plantation. Furthermore, the growth rate of managed successions may be as fast or faster than natural successions (Uhl 1983: 79).

The Bora process of swidden-fallow management involves the conversion of a short-term cropping system into a long-term agroforestry system. By examining swidden fallows of different ages, we have attempted to demonstrate how the useful component of the vegetation changes as fallows become older.

Fallowing is multipurpose. The secondary forest serves not only for nutrient storage for future cropping, but also as an important niche for secondary crops and useful non-managed plants. We identified 118 different useful species in Bora fallows. The term 'enriched fallow' could be used to describe the early stages of these fallows. In a subsequent 'forest-fallow' stage, economic plants are still present but are more dispersed, fewer in number, and are less managed or are not managed. However, over several cycles of clearing, burning, farming, and fallowing, while overall species diversity in the forest fallows declines, the relative number of useful trees increases (Peters 2000).

Viewed properly, a swidden site is never completely abandoned as a resource zone. Secondary harvests of fruits, other plants, and even animals continue until the forest is again removed for further cropping. There exists an identifiable sequence from original forest with some economic (useful) plants present, to a swidden with numerous individual economic plants present, to an orchard fallow or agroforestry phase combining managed economic plants and natural vegetation, to a mature forest (unless recleared) in which economic plants are fewer but still present in greater numbers than in the original forest (where plants useful for indigenous people are nevertheless numerous). Likewise, there is a corresponding sequence in the proportions of biomass which are cultivated or managed, spontaneous economic, and spontaneous non-economic.

An agroforestry system drawing on indigenous management methods could be an ecologically appropriate and potentially economically viable alternative to destructive short-fallow shifting cultivation in tropical areas. The ideal model would provide food crops during the swidden stage and cash crops and other products during the fallow stage. The cash crop perennials should be rapidly maturing species which can be harvested after ten to twenty years so that the cycle can be reinstated as soon thereafter as possible, given that renewal of soil fertility for new swiddens seems to take about twenty years. Such a system would help fulfill the need for sustained production of food and other needed products, as well as cash crops, and simultaneously reduce damage to a fragile environment. The Bora demonstrate that this is feasible.

The Shipibo: Multiple Biotope Use

The Indians of the major floodplains of the Amazon, such as the Tapajó and Omagua, were rapidly destroyed by European contact and few survive today. One exception is the Shipibo along the central Río Ucayali in Peru; they number over 15,000 and have much of their culture still intact. The economy of the village of Panaillo provides some idea of aboriginal floodplain subsistence, since the sustaining area lies entirely within the floodplain of the Ucayali. The cultural ecology of the village was studied in 1971–2 by a research team led by my student, Roland Bergman, and I spent a brief period with them in July 1971. The information here is drawn mainly from Bergman (1980; also see Myers 1990: 25–36).

Panaillo in 1971 was a small village located near the juncture of the Río Panaillo and the Río Ucayali, one of the major tributaries of the western Amazon in Peru (Figs 5.8, 5.9). The Shipibo there had regular contact with the river port city of Pucallpa, spoke Spanish, and had a resident Peruvian schoolteacher. Their subsistence system remains largely traditional, being based on plantain cultivation and fishing.

Typical of Amazonian meander-type rivers, the central Ucayali forms a floodplain that is some 30 km wide, an unstable biotope (microhabitat)

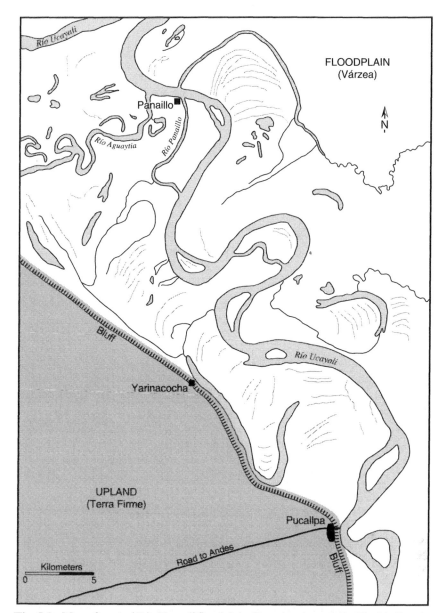

Fig. 5.8. Map of central Río Ucayali floodplain, Peruvian Amazonia. The bluff to the west separates the floodplain (*várzea*) from the non-flooded upland (*terra firme*). Adapted from Bergman (1980: 13). Panaillo and Yarinacocha are Shipibo Villages

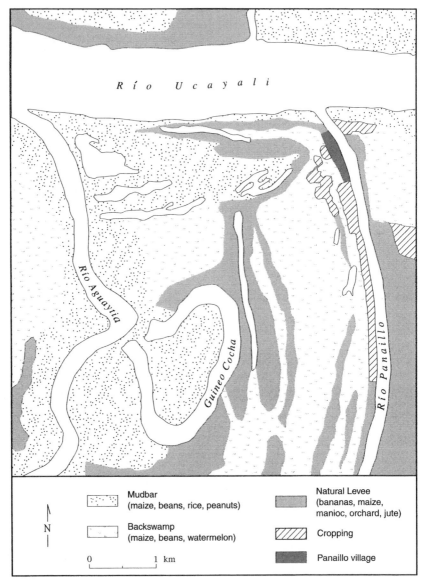

Fig. 5.9. Map of land types and major crops at Panaillo (Shipibo) on the Río Ucayali, Peruvian Amazonia. Most of the area shown is also used for hunting and fishing. Adapted from Bergman (1980: 90)

network of natural levees, side channels, backswamps, sandbars and mudbars (*barriales*), islands, and lakes (Lathrap 1968*a*). During low water, extensive *playas*, which may be wider than the river itself, are exposed. Tributary streams, such as the Panaillo and the Aguaytia, add their own levees and backswamps to the complex floodplain landscape. The village is situated along the top of the levee (*restinga*) of the Río Panaillo a few hundred meters from the Ucayali. Facing the river (foreslope), the levee breaks sharply down to the water. On the other side (backslope) the levee grades down more gently into a large backswamp (*tehuampa*), which is bounded by a former levee of the Río Aguaytia, and behind that there is a permanent lake (*cocha*). Continuing westward, the area to the Río Aguaytia consists of lakes, backswamps, levee remnants, and *playas*. To the north lie the large *playas* and levees of the Ucayali (Fig. 5.8).

The Shipibo farmer is thus faced with a varied environment, but with repetitive sequences of biotopes and an annual sequence of rise and fall of water level. This zonation is matched by a horizontal zonation of crops. However, the size of individual biotopes and distances between biotopes vary considerably. Major advantages are diversity and good soil. The major disadvantages are periodic early or unusually high floods. Success in subsistence is dependent on knowledge of changing river levels and the related availability of biotopes for different lengths of time and hence for different specific crops.

Bergman (1980: 60, 90, 93) mapped and diagramed the land use of Panaillo, recording houses, individual fields, and crops (Fig. 5.10). The crops associated with the principal biotopes there are as follows:

Mudbar In the Río Panaillo; flooded annually, silt loam soil; short growing season crops, including maize, beans, rice, peanuts, sweet potato, watermelon; potentially farmed every year, but only small portions are above water long enough to be cropped.

Levee Foreslope Silt loam soil; sugar cane.

Levee Top, Riverside Sandy soil; orchard and garden crops, including star apple, guava, cotton, shapaja, mammee, soursop, hog plum, and the Old World crops mango, lemon, grapefruit, sugar cane, and tangerine.

Levee Top, Center Houses, dooryard plants, similar to the riverside biotope above, but only a few scattered crops.

Levee Top, Backswamp Side Dominated by permanent bananas; a few short-fallow swiddens with maize and manioc; scattered guava, sapote, star apple, genipa, cacao, pejibaye, breadfruit, shapaja palm, tangerine; some commercial jute.

Levee Backslope Silt loam soil; sugar cane.

Higher Backswamp, Lower Levee Silt loam soils; swamp forest, cecropia, wild cane; maize, beans, watermelon, manioc, rice; potentially farmed every year.

Lower Backswamp Swamp forest; no cropping.

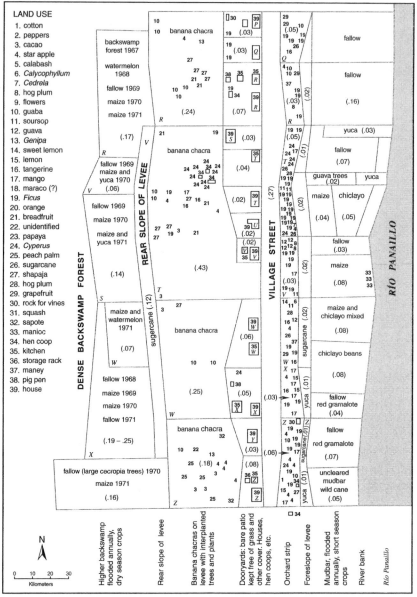

Fig. 5.10. Map of fields and crops at Panaillo, Río Ucayali, Peruvian Amazonia. According to Bergman: 'The trees and plants shown are economic, whether deliberately planted, as most are, spontaneously seeded, or wild but cared for. In the banana *chacras* [fields] useful species grow surrounded by dense stands of bananas. An orchard strip runs the length of the village between the street and the front edge of the levee. Only a few scattered banana plants are grown here amongst the fruit trees. In the orchard strip of [household] X manioc grows on the steep foreslope of the levee. On the rear slope of the levee [household] V grows sugarcane in a long narrow strip that is too low for reliable banana growing. Numbers in parentheses indicate areas in hectares. Capital letters indicate households. Crops are for 1971 unless otherwise indicated.' Adapted from Bergman (1980: 93)

Playas of the Río Ucayali Enormous sand- and mudbars; exposed as long as six months; flooded annually; beans, maize, rice, and peanuts (Fig. 5.11).

Thus, the Panaillo Shipibo make use of nine biotopes for several distinctive forms of agriculture—permanent, annual, and short fallow—each with a different group of crops.

Plantains (including bananas) were the principal crop, with manioc, maize, beans, and rice all secondary. These were supplemented by a variety of orchard and garden crops. Fish was the main source of protein except during the highest water (February–March), when fishing is poor. At that time, game, which

Fig. 5.11. Shipibo woman planting maize and rice on *playa* of the Río Ucayali at Panaillo, Peruvian Amazon. Photograph by Roland Bergman, 1971

tends to be trapped and concentrated on the levees, is important. Of the food crops, there were 10.75 ha of bananas, all on the levee. There were only 0.31 ha of manioc and 1.17 ha of maize on the levee, but 7.29 ha of jute for market. The low water *playa* and backswamp crops of maize (4.15 ha), watermelon (1.87 ha), beans (1.28 ha), and peanuts (0.10 ha) totaled only 7.40 ha, even though there was considerable unused backswamp land available. The low water crops, however, cannot provide year-round food, as can levee crops, hence the traditional underutilization of the low water biotopes despite their potential for high levels of annual production. The Panaillo Shipibo do not have much mudbar land on the *playas* of the Río Panaillo that are exposed long enough to plant rice for commercial production. Most of the large nearby *playas* of the Ucayali, where there is good rice land, were controlled by non-Indians.

There is also an ecological zonation of fish and game resources, based on types of water body, vegetation, and seasonality. The patterns of actual catches reflect the ecological zonation in combination with distance from the village. These patterns are mapped and described by Bergman (1980: 135–66).

Biotope agricultural patterning at Panaillo is representative of the kind of zonation that occurs throughout the floodplains of the Amazon Basin. However, it is not necessarily typical, given the wide variation in the environmental factors discussed earlier, in crop orientation, in accessibility to land or in land ownership, in population pressure, and in availability of commercial outlets. Variation from region to region is considerable, but there tends to be local consistency in agricultural zonation, as we see at Panaillo.

Most years the levees stand above flood levels. The highest local levees flood about every ten years, but the slightly lower Panaillo levee floods every five to seven years (Bergman 1980: 53–60). Houses are on raised platforms and people move about in canoes, but crops are vulnerable. In the 1971 flood, the river rose 9 m above the dry season low water level, although this was only a meter or so above normal floods. Staple crops, especially plantains, were mostly destroyed. Where the water covering the plants exceeded 60 cm for over thirty days, the crop loss was total. At lower durations more plantains recovered. Prehistoric crops of maize or manioc on the levees would have been more vulnerable than the post-conquest plantain. *Playas* are exposed for up to six months; however unusually high water levels during the dry season can result in destruction of *playa* and also backswamp crops (maize, beans, watermelon, peanuts, and rice today). When crop loss is severe, the Shipibo travel to distant swiddens located on high levees and in upland forest to harvest plantains and manioc.

Plantains, the staple, are a European introduction that replaced manioc on the better drained soils. Fish is the main source of protein except during brief high water when maize and game are more important (Blank 1981). This system of permanent (levee top plantains), short-fallow (levee top manioc), and seasonal (*playa* and backswamp maize, beans, etc.) cropping supported a population of 107 people in a permanent village (over 25 years old in 1971); the resource area exploited had a population density of about 4 per km^2 (Bergman 1980: 203).

The present Shipibo village of Panaillo is similar to other Shipibo villages in utilizing multiple microhabitats for cultivation. However, many Shipibo villages, both past and present, were located on bluff tops, in contrast to Panaillo, and thus also had direct access to the high-forest habitat where the poor weathered oxisols of the *terra firme* could be used for long-fallow shifting cultivation emphasizing manioc. This is true of the present village of San Francisco de Yarinacocha (see Fig. 5.8), as well as prehistoric occupations on the same site (Lathrap 1968*a*: 74).

The Karinya: Multiple Biome Use

Indian farmers commonly utilize multiple microhabitats (biotopes), as we have seen with the Shipibo and also the Kayapó. Most, however, are still thought of as specialized to a particular macrohabitat (biome), such as rainforest, savanna, or floodplain, but some cultivate in more than one biome. In Chapter 6, I argue that prehistoric riverine people farmed both the floodplains and the adjacent bluff forests. For both multiple biome and multiple biotope exploitations, the objectives are to maximize and diversify production and to minimize the risk faced if only one habitat is utilized, at the cost of increased travel time.

The Karinya of the Orinoco Llanos are a particularly good example of a society with multiple resource strategies. Farmers in the same community farm four biomes—the Orinoco floodplain, forest (*monte*), savanna, and palm swamp (*morichal*), comprising at least nine different biotopes.

The Karinya were studied by anthropologist Karl Schwerin (1966) in the early 1960s. When I was examining pre-European raised fields in the Venezuelan Llanos at Caño Ventosidad in 1972 (Chapter 11), I took the opportunity to also observe the Karinya ditched fields, assisted by Roland Bergman (Denevan and Bergman 1975). Schwerin and I later collaborated on a broader treatment of Karinya adaptive strategies (Denevan and Schwerin 1978).

The Karinya are a Carib-speaking group widely dispersed throughout the eastern Llanos west of the Orinoco delta in some thirty communities, including a few villages south of the Orinoco (Fig. 5.12). The total population in 1962 was 3,828, most Karinya having regular contact with regional gas and oil industries and with urban centers. Our study focused on the communities of Cachama and Mamo.

The Orinoco Llanos, which I consider part of Greater Amazonia, cover a vast area north of the Orinoco in Venezuela and Colombia. Much of this plain is low lying and poorly drained; however, there are extensive tablelands of Tertiary age north of the river forming what are often called the Llanos Altos.

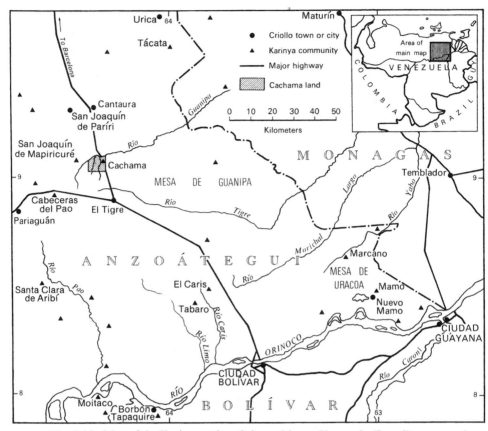

Fig. 5.12. Map of the Karinya region, Orinoco Llanos, Venezuela. From Denevan and Schwerin (1978: 6)

Despite a long dry season, the streams of the eastern Llanos flow year round, dissecting the terrain into large *mesas*. The *mesas* are covered by savanna— open grassland with a few trees, with very infertile, sandy soils. The stream channels, however, are lined with *moriche* palm (*Mauritia flexuosa*) gallery forests, and swamps (*morichales*), with fertile alluvial soils. The intervening region between the *mesas* and the Orinoco is a low-lying floodplain. During the rainy season the great river overflows its banks, flooding much of this area and replenishing the many lakes, lagoons, and backwaters which are found there, often at considerable distance from the main stream. At the same time there are high places in the floodplain which are rarely or never flooded, including natural levees, river terraces, and islands within the Orinoco channel. This region of sharply contrasting savanna, *morichales*, and floodplain comprises the Karinya territory (Fig. 5.12).

The Karinya have adapted their agricultural techniques to the exploitation of a range of macro- and microenvironments. Seven types of cultivation are

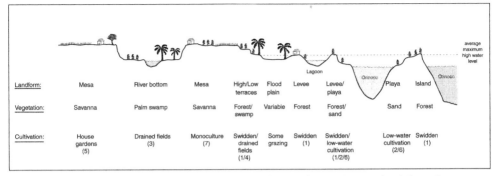

Landform:	Mesa	River bottom	Mesa	High/Low terraces	Flood plain	Levee	Levee/ playa		Playa	Island	
Vegetation:	Savanna	Palm swamp	Savanna	Forest/ swamp	Variable	Forest	Forest/ sand		Sand	Forest	
Cultivation:	House gardens (5)	Drained fields (3)	Monoculture (7)	Swidden/ drained fields (1/4)	Some grazing	Swidden (1)	Swidden/ low-water cultivation (1/2/6)		Low-water cultivation (2/6)	Swidden (1)	

Fig. 5.13. Biotopes and Karinya cultivation, Orinoco Llanos, Venezuela. Adapted from Denevan and Schwerin (1978: 14). See Table 5.2

practiced, some occurring in several different habitats, while some habitats are utilized for several different types of cultivation. The forms of agriculture are as follows: swidden mixed cropping, *playa* mixed cropping, river bottom drained fields, floodplain drained fields, house gardens, *playa* monoculture, and savanna monoculture. The locations of these systems are indicated in the profile shown in Fig. 5.13 and each is briefly described in Table 5.2; for more detail, see Denevan and Schwerin (1978). Drainage of the palm swamp (*morichal*) fields was undertaken by means of a network of ditches, which is described in Chapter 14.

The Karinya not only exploit a range of ecological situations for cultivation, but raise domesticated animals (chickens, pigs, cattle, goats, burros), and engage in fishing in small streams (Cachama) and in the Orinoco (Mamo), as well as some hunting (rabbits, iguana, deer, birds, capybara). Most men also participate to some degree in wage labor in the petroleum industry, ranching, as field hands, or as workers in the nearby cities.

The Karinya have adapted not only to a varied physical environment but also to a changing socio-economic milieu (Denevan and Schwerin 1978: 59–66). The result is a flexible and opportunistic economic system, one that is now being recognized as common for rural Indians and peasants in the Andes, Amazonia, and elsewhere.

The five case studies presented here have illustrated several aspects of Amazonian Indian field practices. It is difficult to reconstruct Indian agriculture from colonial documents, but there is information available (Mojo). Swidden cultivation sequences of activity may be quite variable even for a single household (Campa). Swidden abandonment to fallow may be a gradual process, rather than abrupt, with management of the early fallow stages for useful plants (Bora). The utilization of multiple microhabitats (biotopes) (Shipibo) and macrohabitats (biomes) (Karinya) is common. In addition, the limited protein availability explanation for dispersed settlement was examined for the Campa.

Table 5.2. Systems of Karinya cultivation

System[a]	Biotope	Types of crops	Tools used	Technological improvements	Communities
1. Swidden mixed cropping (short to long fallow)	Primary and secondary forests of river levees, high terraces, and islands	Subsistence; esp. tubers	Axe, machete, *garabato*	Minimal	Mamo, others
2. Playa mixed cropping (annual)	Levees, beaches, lake margins, islands	Subsistence; esp. grains, legumes	Axe, machete, *garabato, chicora*	Minimal	Mamo, Tapaquire, others
3. River-bottom swamp (*morichal*) drainage (continuous or short fallow)	Gallery swamps along rivers draining the *mesas*	Subsistence; esp. tubers, bananas	Axe, machete, *garabato, chicora,* shovel	Swamp drainage, some irrigation	Cabeceras del Pao, Tabaro, Cachama, Tácata, S. Joaquin, Santa Clara, Morichal Largo, Marcano, others
4. Floodplain swamp (*morichal*) drainage (continuous or short fallow)	Low terraces of the Orinoco floodplain	Subsistence; esp. tubers, bananas	Axe, machete, *garabato, chicora,* shovel	Very long drainage channels, efficient drainage, local crop specialization	Mamo, others
5. House gardens (continuous)	Savanna and elsewhere next to houses	Supplementary crops: fruit, herbs, medicinals, vegetables	Machete, *garabato, chicora*	Individualized care	All
6. Playa monoculture (annual)	Levees, beaches, lake margins, islands	Cash crops; esp. grains, fibers, legumes	Axe, machete, hoe, *chicora*	Minimal, but with crop specialization, large fields	Mamo, others
7. Savanna monoculture (annual)	Savanna	Cash crops; esp. grains, fibers, legumes	Tractor-drawn machinery	Mechanization, fertilizer, large fields, crop specialization	Cachama

[a] Numbered cultivation systems are keyed to Fig. 5.13. Crops are listed in Denevan and Schwerin (1978: 42–5).

Notes

[1] It has been incorrectly claimed that the Yanomami, with about 20,000 people (Chagnon 1992: 1), is the largest Indian group today in Amazonia.

[2] Studies also have been made of the agriculture of the nearby and linguistically related Machiguenga (A. Johnson 1983) and Amuesha (Salick 1989; Salick and Lundberg 1990).

[3] A minor crop of interest is the Amazonian potato, *Solanum hygrothermicum*, which I encountered in a few Campa villages at elevations of around 1,200 m (Ochoa 1984).

[4] The Gran Pajonal Campa cultivate at least forty-nine crops; however, fields are dominated by *manioc*.

[5] Bodley (1970: 36) documented four long distance shifts of Campa houses of from 20 to 80 km; in each case the reason given for moving was depleted game or fish resources.

[6] However, the Amahuaca of eastern Peru, one of the few Indian groups in the Amazon Basin for whom maize is the staple rather than root crops, are still semi-nomadic despite the apparently relatively high protein content of their vegetable diet (Carneiro 1964).

[7] Other early, but brief mentions of swidden-fallow management and utilization in Amazonia are for the Kayapó (Posey 1982: 21–2), Kalapalo (Basso 1973: 34–5), Yanomami (Smole 1976: 152–6), and the Andoke and Witoto (Eden 1980). More recent descriptions are for the Runa (Irvine 1989), Amuesha (Salick 1989), and Ka'apor (Balée 1994: 135–8).

[8] Another example of fallows dominated by single species is the Kalapalo piqui (*Cayocar brasilensis*) orchards (Basso 1973: 34–5). Other groves of useful trees may occupy disturbed areas around past or present villages; for example, species of *Orbignya, Astrocaryum, Elaeis, Mauritia, Acrocomia, Grias*, and *Platonia* (Balée 1992; Peters 2000).

6

Pre-European Riverine Cultivation

> The tropical forest farmer living on the bluff of old alluvium adjacent to
> the active flood plain, could simultaneously farm the limited but excellent
> recent alluvial soils . . . and the poor but essentially unlimited soils of the
> old [upland] alluvial deposits.
>
> (Donald Lathrap 1970: 44)

The *Várzea/Terra Firme* Dichotomy

In the *Handbook of South American Indians*, Julian Steward (1948: 886) stated
that: 'the important ecological differences were those between water-front and
hinterland peoples . . . The differences were in resources, and these partly
determined population density and community size.' Meggers' 1971 (also
1996) *Amazonia* book is organized on the basis of *várzea* and *terra firme* dis-
tinctions. The same emphasis is made by Lathrap (1968*a*; 1968*b*; 1970), Roosevelt (1980), Denevan (1966*b*; 1976), D. J. Wilson (1999: 63–5, 169–71, 176),
and others. A riverine concentration of settlement is supported by archaeology
and ethnohistory, although both are meager.

A continuing debate exists as to the cause or causes of the apparent dramatic
differentiation between floodplains and upland forests in terms of population
density, settlement size and stability, productivity, and general cultural complexity. There are several explanations.

Impoverished Soils of the *terra firme* (Meggers 1957; Roosevelt 1980: 79; Gross
1983: 445–6). Others have pointed out, however, that even with long-fallow
shifting cultivation much larger populations could be supported than at present (Carneiro 1961; Frank 1987). A variation on Meggers' deterministic thesis
is that poor soils (and other ecological factors) will not support a maize-based
diet which would have less dependence on animal protein than a manioc-based
diet (Gross 1975: 534). However, maize is a staple for some groups (Carneiro
1964).

Depopulation A second scenario is that the low, dispersed *terra firme* populations of today simply reflect post-European depopulation without recovery
due to continuing external pressures, rather than environmental limitations
(Beckerman 1979).

Warfare Chagnon and Hames (1979) argue that it is internal friction and warfare, involving competition for women and land, causing village fissioning, that effectively limits population concentration in the *terra firme*.

Protein Scarcity (limited game) in the *terra firme* compared with the *várzea* (rich aquatic wildlife) (see Chapter 5).

It is important to emphasize that the *várzea/terra firme* dichotomy is based on ethnographic analogies which are not necessarily valid for the past. Nevertheless, the distinction exists now (even for non-Indian populations) and probably existed earlier, but possibly for different reasons. On the other hand, the distinction between *várzea* and *terra firme* settlement in pre-European times may not everywhere have been as extreme as it has been in historic times.

The indigenous *terra firme* system we find in Amazonia today is a post-conquest adaptation: long-fallow shifting cultivation, manioc or plantain monoculture, a strong emphasis on hunting reflecting actual or perceived meat (protein) needs, and a pattern of mostly very small, frequently shifting settlments with low population densities (under 0.5/km^2). In contrast, pre-European patterns were more variable, with both sparse and dense populations, both large and small villages, both intensive and extensive cultivation, and different combinations of reliance on wild resources, forest management, and agriculture.

The general belief has been that riverine settlement and resource use were concentrated *in* the floodplain. Roosevelt (1980: 252), for example, repeatedly refers to agriculture and dense populations 'in' the floodplains. Hemming (1978: 190) says: 'In the sixteenth century the native population was very dense in the floodplains.' However, the floodplain is a high risk habitat because of regular annual flooding plus periodic extreme flooding of even the highest terrain, in addition to daily tidal flooding in the delta region. Extreme floods not only flood the highest ground but their duration is extended so that *playas* which most years would be exposed during the low water period will remain under water for additional months, thus preventing or reducing cultivation. Flooding is what floodplains do. What does this mean for riverine settlement? On levees within the floodplain, houses can be built on stilts and can be readily relocated or rebuilt, but loss of crops can be fatal, even with short-term storage. (Floodplain settlers today generally have access to means of subsistence in addition to floodplain crops and thus can survive the big floods.)

The floodplains were assumed (by this writer, among others) to have been zones of relatively spatially continuous settlement, an assumption not supported by archaeology, ethnohistory, or ethnographic analogy. This assumption of spatially continuous settlement is based in part on superior soils, rich aquatic resources, ease of transport, relative ease of clearing floodplain vegetation with stone axes, and availability of unvegetated seasonal *playas*. However, these advantages are countered by periodic severe flooding and by difficult

Fig. 6.1. Bluff-edge cultivation by unidentified Indians, *c.* 1780, Orinoco region, Venezuela. The person at the right is carrying cuttings, probably manioc, to plant in small mounds in front of him. The person at the left is harvesting tubers, possibly sweet potatoes, using a digging stick. Other crops include pineapple in the lower right, plantains in the center, and probably papaya at the left. From Gilij (1965/1: 186–7)

access to many portions of floodplains. I have proposed a bluff model, arguing that most prehistoric 'riverine' settlements were not located in the floodplain but rather on the valley-side bluff tops adjacent to active river channels where there was easy movement into the fluvial system (see Figs 4.2 and 5.8) (Denevan 1996). People had access to both *terra firme* and *várzea* resources—a complementary system (Fig. 6.1). Subsistence was a multiple strategy which involved the utilization of floodplain *playas* and levee soils and wildlife seasonally, in combination with more stable bluff zone gardens and fruit tree-based agroforestry, which both extended inland for 5–10 km or even more.[1] In addition, this was a patch pattern with dense settlement separated by relatively empty areas. The evidence is archaeological, historical, and agricultural.

Archaeological Evidence: *Terra Preta*

Striking evidence for pre-European bluff settlement is *terra preta*, or *terra preta do indio* (Indian black earth). This is an anthropic soil[2] of prehistoric origin, black or dark brown, rich in organic material, and often laden with cultural debris (ceramics, bones, carbon).[3] Such soil has been reported along the Amazon, Orinoco, Negro, Trombetas, Guaporé, Tocantins, Tapajós, Xingu, Napo, Ucayali, Vaupés, Caquetá, Corentyne (Guyana), other rivers, Marajó Island, and also in the Colombian Llanos[4] (Nimuendajú 1952; N. Smith 1980; Eden *et al.* 1984; Balée 1989: 10–14; Herrera *et al.* 1992; Falesi 1974: 214; Rodrigues 1993; Katzer 1944; Meggers 1993–5: 98; Andrade 1986: 22–3; Woods and McCann 1999; McCann *et al.* 2001). Dates are as old as 100–450 BC (Eden *et al.* 1984: 126). Many *terra preta* sites are on bluff edges. Some are enormous, such as the one underlying much of the city of Santarém. Roosevelt (1989*b*: 45) believes that some of these sites were large nucleated towns of chiefdom status,[5] with a permanence of several hundred years. 'Santarém habitation sites extend almost continuously along the river for hundreds of miles' (Roosevelt 1989*a*: 82).

Terra preta bluff sites are linear, paralleling the rivers. The river edge sites examined by N. Smith (1980: 563) range in size from about 1 to nearly 100 ha and average 21.2 ha. A site on the lower Rio Xingu near Altamira is 1.8 km long and 500 m wide, covering 90 ha.; one at Manacapuru on the Amazon near Manaus is 4 km long and extends 200 m inland, totaling 80 ha according to N. Smith (2 km by 400 m according to Myers 1973: 240). The *terra preta* at Juriti west of Santarém covers 350 ha and is 3.5 km long and 1.0 km wide (N. Smith 1999: 26). Roosevelt (1987: 157) says that the *terra preta* underlying Santarém covers 5 km² (500 ha). The Tapajós site near Belterra has about 200 ha of black *terra preta* and 1,000 ha of *terra mulata* (Fig. 6.2). The site at Tefé is 6 km long (*terra preta*?) (Myers 1973: 240). Although some *terra preta* sites have been excavated, only a few have been mapped (Sombroek 1966: 175; Heckenberger *et al.* 1999). Nimuendajú (1952; Meggers 1984: 642) visited and located sixty-five *terra preta* sites, most on bluffs, in the Tapajós region, and N. Smith (1980) located seventeen bluff sites in the central Brazilian Amazon. Heckenberger *et al.* (1999) excavated three adjacent *terra preta* sites at Açutuba on the lower Rio Negro.

An early description of *terra preta* soils was provided by the English traveler Herbert Smith (1879: 238), who found American Confederate families farming them near Santarém: 'It [tobacco] is cultivated on the rich black lands along the edge of these bluffs . . . All along this side of the Tapajós . . . [which] must have been lined with these villages, for the black land is almost continuous, and at many points pottery and stone implements cover the ground like shells on a surf-washed beach.'

Terra pretas are apparently former settlement and cultivation sites whose dark color is mainly due to organic waste, residue from domestic fires for cooking and warmth, and carbon and ash from agricultural fires; soil carbon is high. N. Smith (1980: 561–2) found that phosphorus levels are high, the result of ash; fish, game, and human bones; feces and urine; and shells. Bones also account for a high calcium content. The pH levels are also higher than for adjacent soils, and aluminum levels are moderately low. The soil fertility of *terra preta* is significantly higher than for most *terra firme* soils. In the middle Rio Xingu

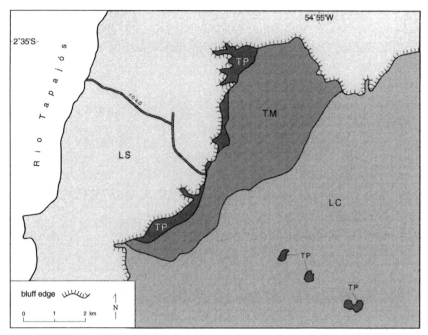

Fig. 6.2. Map of bluff *terra preta* (*c.* 200 ha) along the lower Rio Tapajós at Belterra, Brazilian Amazonia, with *terra mulatta* (*c.* 1,000 ha) inland. Note the small patches of *terra preta* within the *terra firme* latosols. TP, *terra preta*; TM, *terra mulatta*; LS, latosolic sand (sloping, eroded plateau edge); LC, latosolic clay (*terra firme*, plateau). Adapted from Sombroek (1966: 175)

region, Rodrigues (1993: 122–8) obtained detailed soil analyses of *terra preta* soils. Compared to surrounding red and yellow soils, the *terra preta* soils there are higher in organic matter, phosphorus, calcium, magnesium, cation exchange capacity, and base saturation, but are less acidic (pH of 6.0–6.7), and are lower in aluminum saturation and iron oxides. Both the black and the nearby red or yellow soils have the same type of clay minerals which suggests identical parent materials. Similar results have been obtained for *terra preta* soils elsewhere (Eden *et al.* 1984; Woods and McCann 1999; McCann *et al.* 2001; Glaser *et al.* n.d.).[6]

Currently, both Indians and non-Indians seek out the fertile *terra preta* soils for their fields, and ownership rights are established. People know of unused *terra pretas* deep in the forest that they may want to utilize some day (J. McCann pers. comm. 1999). Short-fallow shifting cultivation is practiced on *terra preta*, not because of declining fertility, but because of pest and weed invasion.

Terra preta on bluffs occurs on a variety of *terra firme* soils, including oxisols, ultisols, eutrophic soils (*terra roxa*), and spodosols (N. Smith 1980: 557). Most of these soils are of low natural fertility, the *terra roxa* being the main exception. Soil darkness varies considerably and there is no agreement on color criteria for a *terra preta*.

The Araracuara Project researchers on the Río Caquetá in the Colombian Amazon found brown soils adjacent to or surrounding pockets of black soil. They believe that the brown soils are not settlement zones but rather permanent or semi-intensive agricultural sectors that were maintained by organic additives which produced the brownish color. The brown soils differ from the black soils in color, less depth, less phosphorus, and having few cultural remains (Andrade 1986: 53–4; Mora *et al.* 1991: 75–7). Sombroek (1966: 175) points out the occurrence of brown soil, which he called *terra mulata* in the Belterra area east of the Rio Tapajós (Fig. 6.2). This is a soil lighter than *terra preta*, without artifacts, occurring in bands around *terra preta* on *terra firme*. He believed that: 'this soil has obtained its specific properties from long-lasting cultivation.' Undoubtedly, different kinds of *terra preta* originated and evolved through different pathways and on different parent soils (W. Woods pers. comm.). People in the Santarém region generally do not make a verbal distinction between *terra preta* and *terra mulata*, calling both *terra preta*, but they are aware of the greater fertility of the darker form (J. McCann pers. comm. 1999).[7]

McCann *et al.* (2001) found that in the Santarém region both calcium and phosphorus levels were high in *terra preta* where ceramics (middens) were abundant, but were relatively low for adjacent and more widespread *terra mulata* soils. These latter soils do have a higher organic content, and thus greater fertility than non-*terra preta* soils. McCann *et al.* believe that the addition of ash creates the dark color and raises pH which increases microbiological activity and fertility status, a process that may be self perpetuating long after burning has ceased.

McCann *et al.* (2001) suggest that agricultural activities involving burning and organic inputs were responsible for the majority of the dark earths they examined. They say that: 'Once infused with the self-perpetuating life force of an active soil biota and an adequate nutrient retention capacity, under proper management additional inputs may not be necessary to maintain a reasonable fertility.' Thus, 'agricultural systems more intensive than shifting cultivation seem possible' (W. Woods pers. comm. 1998). On the other hand, some earlier studies of *terra preta* reject the possibility that *terra preta* can be created by cultivation practices (N. Smith 1980: 555; Eden *et al.* 1984: 137). If *terra preta*-type soils indeed can be created by intensive, long-term cultivation in Amazonia, the implications for development are dramatic. Sombroek (1966: 261) commented that: 'Theoretically, it would be possible to attain gradually a level of soil organic matter which is comparable to that of the Terra Preta soil'; however, he goes on to say that: 'whether it will be economically justifiable is questionable.' The possibility must be investigated.

Darkness and depth of *terra preta* are probably indicative of length of occupation, whether continuous or periodic. The riverine sites examined by N.

Smith (1980: 563–4) had depths of up to 2 m with an average of 0.73 m. He suggests an accumulation rate of about 1 cm per ten years of occupation; thus 2 m depth would mean 2,000 years of settlement, but this seems unlikely. Ceramics, which are common in black earths, vary in style, indicating different cultural phases and hence, possibly, discontinuous settlement. Meggers (1992*a*) demonstrates this on the basis of seriation sequence analysis of ceramics in sites along the Rio Xingu, Rio Tocantins, and other tributaries. She believes that sites with large surface areas 'represent multiple reoccupations rather than large single villages' (Meggers *et al.* 1988: 291); and that: 'Hence, the surface extent of archaeological sites cannot be used to infer village size, as has often been assumed' (Meggers 1995: 29). Roosevelt (1989*b*: 45–6), on the other hand, believes that many large sites were of long duration, as do Mora *et al.* (1991: 39, 61, 77) and Herrera *et al.* (1992: 110) for Araracuara. DeBoer *et al.* (1996: 276) evaluate ceramic seriation and conclude that it does not support the argument for site reoccupation and 'the more general position that large sites did not occur in prehistoric Amazonia'. Reoccupation was probably a factor, but it remains to be demonstrated how long and how large most specific occupations were, a critical issue for estimating site populations.

An alternative view of large sites representing repeated abandonments and reoccupations, would be that some, at least, were permanent, accounting for the development of deep *terra preta*; however, there could still have been periodic abandonment of houses and even village sectors as people shifted residence location within the overall site because of house deterioration, high pest levels, deaths, and other reasons.[8] What portion of such sites was occupied at one time, on average, remains to be determined. In addition, as pointed out above, large portions of sites identified as *terra preta* may not be settlement middens but rather agricultural zones. Many of the bluff *terra pretas*, however, are continuous middens.

There have been a few *terra preta* sites reported within floodplains on high levees or on river terraces (Sternberg 1960: 417, 419; O. Coomes pers. comm. 1997). However, most riverine sites are on *terra firme* bluffs. N. Smith (1980: 562) found that even the twelve interfluve sites he examined were located either along a small river or within a few hundred meters of a perennial stream, indicating the importance of navigation, potable water, aquatic resources, and relatively easily cleared riparian forest. Bluff sites are often just above falls or rapids or where bluffs jut into rivers or at tributary junctions (N. Smith 1980: 562–3; Myers 1990: 19), suggesting strategic considerations. Sombroek (1966: 175) observed that *terra preta* sites in eastern Brazil 'are especially frequent at outer bends of the rivers, where no floodplains occur between the water and the upland, and where the waterway can be scanned freely'.

Certainly, not all bluff sites are *terra preta*. For example, the large Finca Rivera site near Leticia (Colombia) (1.5 km long, 45–60 m wide, 16 ha) is up to 30 cm deep, with no indication of black earth, though there is a brown midden layer (Bolian 1975: 22, 27).

Colombian scholars have studied the *terra preta* bluff soils at Araracuara on the Río Caquetá in the Colombian Amazon (Mora *et al.* 1991; Herrera *et al.* 1992; and Cavelier *et al.* 1990; also Eden *et al.* 1984). Analyses of soils, pollen, phytoliths, plant remains and ceramics, plus radio carbon dating provide systematic evidence for the nature of prehistoric settlement and land use.

At Site 2 at Araracuara there was continuous occupation from AD 385 to AD 1175, over 800 years, covering an area of 6 ha of brown anthropic soil. The dominant crop was maize, with some manioc. Palms and other fruit trees included *Iriartea, Oenocarpus, Jessenia, Mauritia, Astrocaryum, Bactris, Attalea*, and *Lepidocaryum*, all-important house garden and swidden-fallow trees today. This suggests that agroforestry systems had been established. By AD 800, agriculture had become intensified and was nearly continuous, with additions of chili peppers, caimito, and varieties of manioc. The maximum percentages of palms coincide with maximums of cultivated crops. Algae and silt occur in the soil, possibly from swamps in the floodplain, suggesting transport of alluvial silt and organic matter up the bluff (140 m high) to the fields to improve fertility and to reduce erosion.

At Site 3 at Araracuara, the anthropic soils span from AD 1 to AD 1800. Initial fields were small with long fallows. After AD 800 the site experienced intensified agriculture (long cropping, short fallowing), assisted by soil additives including domestic waste, dead leaves, wood, and weeds, plus silt and algae. These additives occur in greater quantities than at Site 2. The site is larger than Site 2, with 14.5 ha of brown and black soil, extending for about 1 km along the bluff. The project soil scientists believe that black soil probably originated at house sites and the brown soil at intensive field and garden sites. The Araracuara project provides evidence for lengthy, permanent bluff settlement and large populations; permanent agriculture involving artificial soil fertility maintenance, supplemented by forms of agroforestry; and interaction with an adjacent floodplain.

Present-day *terra firme* Indian settlements seldom produce *terra preta*, undoubtedly because they are of short duration and are supported by shifting cultivation (Pabst 1993: 142). Permanent villages would require some form of stable, sustainable agriculture. Exceptions might occur with some of the older mission villages, and these should be examined for black earth formation. Black earth is still being formed today in the backyards of Amazonian towns in Brazil visited by N. Smith (1980: 555–6). He notes that the rate of formation is probably slower now than in prehistory due to pig and chicken scavaging and the current practice of building fires on above-ground platforms.

This discussion of *terra preta* is an argument for large prehistoric settlement sites, especially on bluffs, which were occupied for long periods of time, even though an entire site may not have been fully occupied at one time. Such settlement had to have been supported by relatively intensive forms of agriculture (permanent or semi-permanent), rather than by long-fallow shifting cultivation as is common today with most surviving Indians, as well as non-indigenous settlers.

Most of the archaeological sites on bluffs, whether on *terra preta* or not, are located where active river channels impinge against the bluffs, as is true of village and city locations today. Prehistoric floodplain sites still exist but are less common than bluff sites, in part because most have been destroyed by river channel migration or have been buried under river sediment.

Historical Evidence

The accounts of early travelers on the Amazon provide some indication of the topographical location, size, and distribution of indigenous settlement. These include Carvajal (1934*a*; 1934*b*) in 1542, Salinas de Loyola (1965) in 1557, the men of Pedro de Ursúa (Orsúa)/Lope de Aguirre in 1561 (Vázquez de Espinosa 1948: 381–97; Mampel and Escandell Tur 1981), Laureano de la Cruz (1885) in 1647, Acuña (1942) in 1641, Heriarte (1952) in 1661, and Fritz (1922) in 1689–91. Examinations of these accounts are provided by Meggers (1971; 1996), Myers (1974*b*; 1992*b*), Porro (1994), and Denevan (1996).

Villages are variously described as being on hills, high spots, high land, and high banks. These are likely bluffs, but in some cases could be high natural levees. There is a description, for example, from the Ursúa expedition in 1561 of a village on a 'barranca muy alta' reached by a staircase of over 100 steps cut into an obvious bluff (Vázquez de Espinosa 1948: 385). Carvajal 'never mentions gardens on the várzea islands or other portions of the floodplain' (Meggers 1993–5: 96). By the seventeenth century, however, there were many descriptions for the upper Amazon of villages within floodplains, especially on islands, which Myers (1992*b*) believes reflect a locational shift for defensive purposes. Villages were initially described as large, numerous, linear, and continuous for as many as 2 to 5 leagues (12–30 km) (Carvajal 1934*a*: 198–212), each containing as many as 7,000–10,000 Indians (Porro 1994: 82),[9] but usually between 800 and 3,000. There were mentions of large uninhabited sectors between hostile groups, believed to have been buffer zones (Myers 1976; DeBoer 1981*b*). Such lands could also have been long sectors observed where the traveled river channels did not impinge against bluffs with villages.

The bluff location of villages is an indication of interaction with adjacent floodplain people and the use of floodplain resources. That there was also interaction between bluff settlements and the interior *terra firme* is indicated by reports of roads extending inland from the bluffs by Carvajal (1934*a*: 200–209) and the men of Ursúa (Vázquez de Espinosa 1948: 385–6). Fortresses were mentioned, and there were interior population centers as well as use of forest resources. However, roads also could have led to other bluff settlements.

Agricultural Evidence

Prehistoric riverine settlement in Amazonia has previously been associated with intensive floodplain cultivation on *playas* and natural levees. I believe that bluff cultivation was also very important.

The floodplains have clear advantages for agriculture. Soils, while variable in texture and nutrients, are mostly of relatively high fertility, renewed by annual flood deposits, with adequate moisture availability which is capillary and not fully dependent on rainfall. Forest clearance often is not a significant problem. The *playas* and low levees are cleared of vegetation by the annual floods. The high levees are forested with early successional growth which is relatively easily cleared. Some of the swamp-forest trees, however, are difficult to fell. The need for weeding varies with site. Cropping can be annual, although *playas* shift in location frequently. Cultivation reflects a sophisticated management of microhabitats based on differences in flood depth and duration and in soil (Chapter 4).

The disadvantage of floodplains is, again, the irregularity and variability of flooding. On the tributaries, destructive floods can occur without warning, even during low water seasons. The rise and fall of the main Amazon is more regular, but periodic extreme floods occur, filling entire sections of the floodplains, topping the natural levees, and destroying most crops. Thus floodplains are a rich but high risk habitat.

This risk was countered, in part, by food storage techniques: manioc tubers in covered pits, maize in raised cribs, and turtles in pens (Chapter 4). Manioc cuttings for new plantings are not preservable, however. Long-term flooding, or high floods for several years in succession, would have been devastating, and there are early reports of this. Thus, non-floodplain sources of food on *terra firme* would have been required for the feeding of large numbers of people.

The bluffs are part of the well-drained, low fertility *terra firme*, where prehistoric agriculture has been portrayed as long-fallow shifting cultivation comparable to that of surviving Indians today (Meggers 1971: 42ff.; Roosevelt 1980: 87; 1989a: 32). However, there are current indigenous examples of more permanent production, based on soil protection and maintenance, such as the swiddens and house gardens of the Kayapó, Waika, Siona-Secoya, and Amuesha, as well as agroforestry systems based on perennials such as those of the Bora.

An important consideration in reconstructing pre-European bluff agriculture is the use of inefficient stone axes to clear forest (Chapter 7). The bluff edge presents a narrow zone of relatively easily cleared vegetation, including small trees and softwoods, which would have been easier to clear with stone axes than the interior forest. This fringe, however, could have been gradually expanded inland for a short distance.[10] A second consideration is that agroforestry systems were probably integrated with permanent gardens and swiddens. Such agroforestry would have included swidden-fallow management and fruit orchards and forms of forest manipulation. The cultivation of fruit trees is mentioned in the early accounts (Carvajal 1934*a*: 210, 217; Heriarte 1952: 17; Acuña 1942: 37; Vázquez de Espinosa 1948: 384). The high concentration of fruit tree pollen at the Araracuara site on the Río Caquetá is indicative of agroforestry systems, such as orchards, house gardens, and managed fallows (Mora *et al*. 1991: 43).

Terra preta soil is indirect evidence of intensive bluff cultivation, in that permanent or semi-permanent settlement creating *terra preta* was probably associated with permanent or semi-permanent fields. The fertile black soil itself was undoubtedly cultivated, as it is today. In addition, brown anthropic soils (*terra mulatta*) adjacent to black soils (see Fig. 6.2) probably were agricultural soils enriched over time by ash and organic additives (see above).

Analogs of prehistoric bluff agriculture and agroforestry exist today but have been little studied. One example is that of river settlers along the Amazon in the Iquitos region. The village of Tamshiyacu is located on the east bluff of the floodplain, adjacent to the main channel (Hiraoka 1986). A small sector of levees and sandbars is planted in rice, maize, beans, and vegetables at low water. Cultivation extends inland from the bluff edge for between 2 and 9 km. The immediate bluff zone is used for short-fallow swiddens, with some agroforestry. A middle zone is used for short-fallow swiddens and especially agroforestry. The most interior zone is used for scattered long-fallow shifting cultivation. Beyond that is mature forest.

In the short-fallow fields, two crops of manioc are obtained over three years, followed by three to six years of fallow. No fertilizers are needed, even though the soils are very acid and low in nutrients. However, other crops do poorly. Several weedings are necessary, and this is the main problem rather than declining manioc yields. The low-protein manioc staple is supplemented with fishing and hunting and floodplain seed crops. The agroforestry zone consists of young managed fallow with pineapple and fruit trees. The older sectors are dominated by orchards of umarí (*Poraqueiba sericea*), which last twenty to thirty years. These agroforestry systems are derived from Indian antecedents. Most families also have a home garden.

Projecting to prehistory, the long-fallow zone at Tamshiyacu probably did not exist when stone axes were used. The short-fallow zone adjacent to the bluff edges probably consisted of a mosaic of rotating semi-permanent fields, orchards, house gardens, and managed secondary forest. This zone was probably not very wide, 1,000 m or less, given the difficulty of clearing primary forest, except along tributary streams. Beyond was a probable zone of manipulated forest, agroforestry patches, small gardens located at tree falls and fallowed field sites, forest extraction, and hunting. The total radius of a village sustaining area from bluff edge inland was probably up to about 10 km, given present patterns.

Other examples of combined *várzea* and bluff cultivation by Indians and river settlers (*ribereños, ribeirinhos, varzeiros, caboclos*) today are at Santa Rosa (Padoch and de Jong 1992), along the Río Tahuayo (Coomes 1992*b*: 163–7) near Iquitos, at Yarinacocha and Sarayacu on the Río Ucayali (Myers 1990: 33, 57), at Nuevo Nazareth and Puerto Nuevo on the Ucayali (Tournon 1988), Ticuna villages near Leticia (Bolian 1971; 1975: 20), at Coari on the middle Amazon (Parker *et al.* 1983: 181–3), and at Ituqui east of Santarém (Winkler-Prins 1999: 265–6).

Complementarity

I have briefly examined historical, archaeological, and agricultural evidence to support the thesis that prehistoric riverine settlement in Amazonia was primarily located in fringing bluff zones rather than in the floodplains, based on a dual strategy or complementarity of *várzea* and *terra firme* resource use. The critical archaeological evidence is *terra preta*, anthropic black soil created by prehistoric settlement on the bluffs. The evidence for intensive agriculture is mainly inferential. Large, permanent villages require productive, stable agriculture. Complementary bluff/floodplain production systems today are indicative of what was possible. Bluff/floodplain interaction is indicated by sixteenth-century locational information, and by reports of roads leading from river edge into the interior, of landing places 'down' on the river, of fruit trees being harvested in the interior, and of fish being traded inland.

There may well have been a form of transhumance or shifting of residence from temporary huts on the *playas* during low water to the bluffs during high water. Or people living in more permanent villages on natural levees may have had recourse to living temporarily with kin on the bluffs during high floods. Or, during high floods there may have been an exchange of fish from *várzea* villages for crops from bluff villages, as is done today between Shipibo villages on the Ucayali (Tournon 1988: 43). There probably also was a seasonal shift from *várzea* fishing during low water to bluff zone hunting during high water. Additional temporary residences may have consisted of hunting camps inland from the bluffs. Subsistence strategy varied depending on environment, season, demography, and distance. Bluff cultivation and forest resource use provided a safety valve when *várzea* fields were destroyed by high floods. In addition, settlements may also have been located on bluffs for defensive purposes.

The concept that prehistoric riverine food production involved the integration of bluff and floodplain habitats has been presented previously by several scholars, including Peter Hilbert, Lathrap, Meggers, Myers, Carneiro, and N. Smith (Denevan 1996: 674). However, the implications were not adequately acknowledged for understanding settlement patterns, demography, and subsistence. Nor was the patchy nature of bluff settlement and agriculture recognized. The large bluff villages apparently were separated by long stretches of sparse settlement where bluffs were isolated from river channels, possibly 80 per cent of the riverine zone (Denevan 1996: 673). When a river channel shifted away from a bluff, the bluff villages probably were rapidly abandoned and shifted. Lathrap (1968a: 64) recognized this locational pattern: 'A community ought to be within easy walking distance of a canoe-launching site either on the active channel of the river or on a large body of open water which in turn connects to the main channel of the river, which always functions as the major avenue of communication between communities. Present Shipibo-Conibo settlements conform to these expectations' (see Fig. 5.8). Sombroek (1966: 175) also pointed out that riverine patches of *terra preta*, indicating former settlement, 'are usually located at sites where the upland is near the navigable waterways'. However, this observation was not pursued by either Lathrap or Sombroek.

Notes

[1] In terms of settlement, a riverine (floodplain and fringing upland bluff zone)—interfluve (non-riverine) distinction is more appropriate than a *varzea* (floodplain)—*terra firme* (upland) distinction.

[2] Or 'anthropogenic', the result of human activity.

[3] Falesi (1974: 210–14) discusses possible natural origins of *terra preta*, none of which seem plausible.

[4] *Terra preta* is seldom reported in western Amazonia, which may just be a lack of awareness. Yarinacocha, a noted site of long duration on a bluff near Pucallpa on the Río Ucayali (Fig. 5.8), was not described by Lathrap as having dark anthropic soils. However, color photos I took there in 1956 adjacent to his excavations clearly show dark brown to black surface soil in an area of otherwise red and yellow latosols.

[5] A 'chiefdom' is a regional, socially stratified political unit comprised of several or many villages with a paramount ruler (chief), in contrast to independent villages on the one hand and large states on the other.

[6] Black carbon, which is 'chemically and microbiologically inert and persists in the environment over thousands of years', is converted into stable soil organic matter ('SOM enrichment'). 'Hence, black carbon is a sustainable and effective cation exchanger and probably responsible for the high nutrient storage capacity of Terra Preta compared to the Oxisols they originated from . . . these soils maintain their high fertility even after their abandonment centuries ago' (Glaser *et al.* n.d.).

[7] Woods and McCann (1999: 7) designate both *terra preta* and *terra mulata* as 'anthropogenic . . . dark earths'.

[8] Bororo villages abandoned today may be entirely relocated within 100 m of the previous village (Wüst 1994: 325). Even Meggers (1994: 411) notes that houses may be moved as little as a few meters.

[9] Meggers (1992a: 203) believes that: 'The conclusion that early eyewitness accounts exaggerate the indigenous population density seems inescapable.' A similar statement appears in the new 'Epilogue' of her revised edition of *Amazonia* (1996: 187). Careful examination of the early Amazon reports suggests considerable credibility, in my opinion (Denevan 1996: 661–4).

[10] Along the Rio Arapiuns, there are zones of vine forest up to 7 km deep which appear to be anthropogenic (J. McCann, pers. comm. 1998).

7

Pre-European Forest Cultivation

> The population density and level of cultural complexity achieved by protohistoric Amazonians is one of the most controversial topics in American archaeology.
>
> (Betty Meggers 1993–5: 91)

The forms of Amazonian Indian forest farming examined in Chapters 4 and 5 are mostly recent. It is tempting to believe that pre-European field practices were similar; however, we do not know this. We can only speculate as to what fields may have been like, based on limited evidence.

A common assumption is that shifting cultivation was the dominant form of prehistoric cultivation, with forms of intensive (permanent or semipermanent) agriculture being absent or rare (Meggers 1957; 1993–5; 1995; Steward and Faron 1959: 292; Willey 1971: 339; D. J. Wilson 1999: 63). However, there is no archaeological method that I know of that can conclusively demonstrate the presence of shifting cultivation.[1] Nor is there an 'archaeological signature' to indicate even relative cropping to fallow ratio (Killion 1990: 192). While ash can indicate field burning, and sediment transfer can indicate erosion following forest clearing, neither prove field shifting. Stone axe heads used for tree felling found in or near archaeological sites are probably associated with agriculture, but what kind of agriculture? The earliest historical account that I know of that clearly indicates shifting cultivation is for tropical forest in the lower Río Magdalena region of Colombia in 1578. Because of 'weak' soil, the Cabrillas Indians could not obtain two crops in succession on the same field and therefore had to make new clearings each year (Aguado 1931: 2: 325).

Ethnologists and ethnohistorians have generally portrayed surviving indigenous hunters and gatherers (foragers), shifting cultivators, and other traditional economies as representative of prehistoric food production systems. Even where such groups have clearly undergone considerable acculturation, it has been suggested that their food-obtaining ecologies (and settlement behavior and population densities) are essentially intact and of long (pre-European) standing, even with changes in crops and tools (Meggers 1995: 35). This perspective, however, is coming under increasing attack. Few groups anywhere in the world have been isolated completely from global economy and technology, directly or indirectly.

Recently, archaeologist Anna Roosevelt stated that in Amazonia 'theories about pre-Conquest subsistence cannot be tested with ethnographic data' and that 'present-day Indians' resource management modes may not be representative of prehistoric ones' (Roosevelt 1989*b*: 31). Colchester (1984: 311) provides further emphasis: 'it is time that we started examining Amazonian societies in terms of the recent radical transformations that have occurred and that are occurring in their technological, demographic, and economic bases.' Beckerman (1987: 88) points out that: 'the systems we see operating today are for the most part tiny remnants of what was once a much larger system of farmers and fields' in Amazonia. Furthermore, most surviving Indians are located in the *terra firme* high forests of the interfluves, where resource conditions (soils, game, and fish) are relatively poor, whereas numerous pre-European Indians were located in or adjacent to the resource-rich floodplains.

Our knowledge of indigenous adaptations to the *terra firme* is based primarily on present-day observations. Indian agriculture now is characterized by short cropping/long-fallow shifting cultivation, and low-protein crop staples (manioc, sweet potato, plantain), in association with small, temporary settlements (*c.* 10 to 100 people) and very low population densities (below 0.5 per km^2). There are, however, examples of productive agroecological techniques, which may extend cropping and reduce fallowing. Considerable debate exists over the reasons for the shifting cultivation now practiced by most *terra firme* Indians, particularly over whether soil poverty or game scarcity is the key limiting factor (see Chapter 5).

I believe that the short cropping/long-fallow shifting cultivation pattern is primarily a post-conquest development, reflecting the shift from forest clearing with stone axes to the much more efficient iron or steel axes, and that in prehistoric times forest clearing was too labor-intensive to be a common or frequent agricultural strategy. Hence, most agriculture was permanent or semipermanent. To support this conclusion, I will examine: (1) the role of stone axe efficiency for cutting forest, and (2) *terra preta* soils as indicators of sedentary *terra firme* settlement and cultivation.

Stone Axes and Forest Cultivation

> The mastery of forest by man requires no axe.
>
> (Carl O. Sauer 1958: 108)

Tree Cutting

I am more and more impressed by the evidence that clearing forest with a stone axe is a very laborious business, even with the assistance of girdling, tree falls, and fire (Denevan 1992*c*). There is some data available on the technology and the differential labor involved in clearing forest with stone and metal axes in Amazonia, in particular from Carneiro (1974; 1979*a*; 1979*b*), who conducted

Fig. 7.1. Héta Indian cutting a log with a stone axe, 1960, Paraná, Brazil. Photograph by Vladimír Kozák. From Carneiro (1974: 122)

experimental research with the Amahuaca in eastern Peru, the Kuikuru in the Brazilian Amazon, and the Yanomami in southern Venezuela. Felling trees with stone axes was clearly time-consuming, difficult work (Figs 3.1 and 7.1).

Carneiro (1979*b*: 69–70), for example, calculates that a 610 mm diameter tree of moderate hardness could take from 11.7 to 14.4 hours to fell with a stone axe versus 0.52 hour with a steel axe. Felling times, of course, vary with the type of axe, arm strength, cutting technique,[2] trunk thickness, and hardness of wood. The ratio of felling time, stone to steel, increases progressively with trunk size. The ratio is only 10:1 for a 152 mm diameter tree; it is 23:1 for a 610 mm diameter tree; and it is 32:1 for a 1,219 mm diameter tree, or 115 hours versus 2.4 hours. Differences in felling times between stone and metal axes are much greater for hardwoods than for softwoods: 'holding diameter constant, a tree twice as hard as another [density or specific gravity] will take twice as long to fell.' This is for a steel axe. The difference for a stone axe would be even greater (Carneiro 1979*b*: 62).

K. Hill and Kaplan (1989; Kaplan 1985), working with Ache Indians in Paraguay and the Yora in Peru, confirmed Carneiro's hypothesis that the rates of clearing times increase disproportionately with increasing tree diameter and tree hardness, with a significantly greater rate of increase for stone compared with steel axes. The hardness, however, had a much greater effect on stone axe felling times than did tree size. 'For hardwoods, the time cost for stone-axe clearance can be 60 times greater than for metal tools' (K. Hill and Kaplan 1989: 331). Overall, the average efficiency ratio was about 10 to 1, stone to steel.

For a family plot of 0.69 ha of trees of mixed size and hardness, Carneiro (1979*b*: 71) calculated a total of 1,229 labor hours for clearing with a stone axe versus only 64 hours with a steel axe, a ratio of 19 to 1. The former equals 246 five-hour work days, which is simply not tolerable. He asks how swidden clearing could be done then. His answer is that labor time was reduced with the assistance of trunk burning, girdling (cutting a ring through the cambium layer), tree falls to knock down additional trees (sometimes notched), and leaving the largest trees standing. He calculates an average efficiency ratio of only 7 or 8:1, stone to steel, using auxiliary techniques, and 10:1 if all trees over 610 mm in diameter are left standing.[3] Kaplan (1985), however, found that the multiple tree-fall technique did not reduce clearance time significantly. Killing and deleafing trees by girdling allows the trees to remain standing, but will bring in sunlight to some of the adjacent ground surface (Chagnon 1992: 64). Piaroa oral tradition indicates that tree felling by girdling with a stone axe, waiting for the tree to die and dry out, and then burning at the base of the trunk, took three to four months (Zent 1998: 268).

A good description of clearing with stone axes by the Antipas (Jivaro) is provided by Up de Graff (1923: 203–4; also quoted in Harner 1972: 197–8):

If you saw the one-handed stone axes which are the only tools these people have with which to fell the enormous trees, many of them three to five feet in diameter, to make their clearings (often five acres in extent), you would wonder how it were possible to accomplish this feat. It is a feat of patience rather than of skill. The wood is not cut, but reduced to pulp, six or eight men working round one tree at the same time.

The first step in making a *chacra* is to remove the undergrowth; the soft stems are cut with hard-wood machetes, what can be torn up by the roots is torn up, and the small saplings are snapped off by main force. Then the attention of the workers is turned to the larger trees. A ring is cut round the trunks of all the trees within a radius of, say, a hundred feet of some picked giant, enough to weaken them, and prepare them for the final strain which breaks them off. Finally the giant itself is attacked by a party with axes which works for days and weeks, until at last there comes a day when the great trunk has been eaten away sufficiently for it to crack and fall. But it does not fall alone, for it drags with it all the smaller trees in its vicinity which are bound to it and to one another by an unbreakable network of creepers among the upper branches. With a rending crash, a hole in the roof of the forest is made, and the sunlight pours in . . . I have examined the stumps of these fallen trees many a time; they resemble in every respect those of a clearing made by beavers.

In 1745, Padre Gumilla (1963: 429, 433) reported that the Orinoco tribes took two months to fell a single large tree with stone axes. Barandiaran (1967: 25) interviewed a Sanema-Yanoama (Yanomami) Indian in Venezuela who said his parents required several days to fell a single tree using a stone axe. Not knowing the number of people or hours involved, both of these reported lengths of time are nevertheless impressive.

An important consideration is the availability of stone for axe heads. It can take several days to make a stone axe (Kozák *et al.* 1979: 401–2); proper stone sources may be far away requiring long treks or trade (Denevan 1966*a*: 47–8, 97). Axe heads dull or break and shafts break.[4] Axes are lost or stolen. The rapidity by which Indians shifted to metal axes when available, their struggle to obtain them, and the major role of metal axes in trade are well known, reflecting their great saving of labor, as is related in 'The Revolution of the Ax' by Alfred Métraux (1959). Ferguson (1995: 17) describes the extent to which Amazon people will or did go to obtain Western tools: trade, war, long journeys, relocating villages, raiding Westerners, high payments, excessive labor— 'willing to kill and risk death to get them'.[5]

Agricultural Implications

The inefficiency of the stone axe has dramatic implications for prehistoric agriculture in Amazonia. Several anthropologists have suggested this (Colchester 1984; K. Hill and Kaplan 1989; Zent 1992: 412–18). Kaplan (1985), in an unpublished paper, presented the hypothesis that, given stone axes: 'aboriginal farmers, particularly in interfluvial regions, were highly selective regarding their choice of potential gardening sites and that as a result the distribution of forest types placed important constraints on settlement pattern and subsistence practices throughout the Amazon basin.'

Sites for fields would have been sought where the vegetation lacked large hardwood trees and was dominated by smaller softwood trees, essentially secondary vegetation (Lizot 1980: 8), such as fallow-field regrowth, or along streams, or sites disturbed by tree-falls and landslides. The Machiguenga say that when they had stone tools, settlement was concentrated along small streams[6] where clearing for gardens was easiest (K. Hill and Kaplan 1989: 332).

Even today, they often clear fields from thickets of giant bamboo (Baksh and Johnson 1990: 205).[7] The Yora's reliance on foraging apparently reflected limited availability of metal axes (K. Hill and Kaplan 1989). Allan Holmberg (1969: 270, 272) reported that the Sirionó gave much more attention to gardening and became more sedentary as soon as he provided them with steel hatchets.

The Yanomami use of stone axes for clearing and the impact of the introduction of steel axes in the twentieth century are described by Lizot (1980), Colchester (1984), and Ferguson (1998). Secondary vegetation and stands of soft-stemmed musaceous species were sought for swiddens because of greater ease of clearing with stone axes compared to mature forest, even though more labor is required for weeding in secondary vegetation. Plots were small, the larger trees were not felled, and trekking for game and wild plant foods was of major importance for subsistence. '[T]he Yanoama [Yanomami] of the seventeenth century were interfluve foragers, who supplemented their subsistence by the cultivation of small plots, widely dispersed about their foraging territory' (Colchester 1984: 308). Early explorer accounts support this pattern. Robert Schomburgk (1841: 221) described a wandering Yanomami group that cleared small plots in which they planted peppers and manioc and then returned later for harvest. With the introduction of metal axes, the Yanomami changed from a foraging economy supplemented by agriculture to an agricultural economy supplemented by foraging, with larger fields and villages and less mobility (Lizot 1980: 9; Colchester 1984: 310; Ferguson 1998: 291–7). Likewise, Machiguenga Indians said that in the past, when it was difficult to obtain steel axes, their gardens were much smaller and they relied more on forest products for food (A. Johnson 1977: 164). The Wõthihã (Piaroa) in Venezuela became 'more horticulturally oriented with the arrival of steel' (Zent 1992: 415).

There is another consideration to the problem of cultivating high forest without metal axes to facilitate clearing. This is the relatively frequent availability, spatially and temporally, of natural clearings as result of (1) blowdowns and (2) forest fires. Recent tree blowdowns resulting from strong convective wind storms were identified on Landsat photos in a sector between Venezuela and Rondônia. There were 330 patches ranging from 5 to 3,370 ha in size and totaling 90,000 ha (Nelson *et al.* 1994: 856). Also, forest fires are much more common in Amazonia than is generally believed (Nelson and Nascimento Irmãol 1998). Even in very moist forests, several successive unusually dry years will make it possible for fires to occur. Fire scars and possible fire-derived vegetation, identified on Landsat images (1987–92) for the forested Brazilian and Peruvian Amazon Basin, cover several hundred thousand square kilometers. Charcoal is common in Amazon soils. Samples in northern Amazonia mostly date to the past 6,000 years (Saldarriaga and West 1986: 364). Most forest fires are probably of human origin, including escaped field burns and other intentional or accidental ignitions by farmers, trekkers, and hunters. Lightning-set fires are infrequent because lightning is usually accompanied by heavy rainfall in Amazonia. Prehistoric farmers with stone axes probably took advantage of both blowdowns and forest-fire sites for their fields. They may well have started fires in dried-out forests in order to create openings. Extensive blowdowns and fire sites may have attracted large numbers of Indians who established large and/or numerous villages in the vicinity.

I do not argue that clearing of interfluve forest was rare in pre-metal axe times, but that it was probably more restricted and less frequent than with tropical forest Indians today, most of whom clear new fields every two or three years. Sites were undoubtedly carefully selected, based on ease of clearing; however, a field could be initiated at a tree-fall gap, which is random. Once a field was established it was probably maintained in cultivation as long as possible. It was likely that less labor was required to combat weeds (partly suppressed by controlled shade) and other pests, and to use soil maintenance techniques, than was required to establish new clearings. The Achuar today, even with availability of steel axes, meticulously weed and thereby extend the life of a swidden for a year or more and thus reduce the frequency of clearing new plots (Descola 1994: 186). The same is true for the Kuikuru (Carneiro 1983: 87). However, the time required for weeding may be greater than that for clearing with a metal axe, especially as a field becomes older (Carneiro 1961: 57). This is not the case with stone axes.

The argument here is hypothetical, as there is little direct evidence on the nature of prehistoric *terra firme* agriculture. There probably were pockets of fairly intensive farmers, in relatively large numbers, mainly along small streams. Overall populations were probably low, but possibly larger than scholars have believed, including myself (Meggers 1992a; Denevan 1992a: xxv–xxvii).

In contrast, on the floodplain and adjacent levees there was either no vegetation to clear or only easily cleared vegetation, and the stone axe was less a liability. Soils were fertile and wildlife resources rich. Fields did not 'shift' and populations in some places were dense. Also, those savannas where the soil fertility and drainage were either not severe, or could be managed, could have been attractive to permanent farmers given that there were few or no trees to clear (e.g., Denevan 1966a: 94–5; Posey 1985a: 140–4). Pohl *et al.* (1990: 235) believe that in prehistoric Belize wetlands were farmed before upland forests, in part because of the work required to fell forests.

We do know that hardwood forests were cleared for agriculture elsewhere by relatively dense populations using stone tools, as in Yucatán, in western Central America, and the Caribbean (but see C. Sauer 1966: 51). However, these were areas with soils (limestone, volcanic) much superior to those of Amazonia, so that fields could be cropped for numerous years and new fields did not have to be cleared frequently. The considerable labor involved in clearing with stone tools thus could be tolerated. A study by Doolittle (1992: 392–3) argues that prehistoric shifting cultivation in eastern North America was less common than permanent fields. Carl Sauer (1966: 52) did not find early colonial reports of shifting cultivation in the Caribbean.

The adoption of metal axes and machetes in the New World was generally very rapid where Europeans were present. According to Hans Staden (1928: 74, 90), who lived with the Tupinambá, iron tools were an important trade item on the Brazilian coast as early as 1554. Jean de Léry (1990: 101), who was on the Brazilian coast at the same time as Staden, said that 'goods' [particularly iron axes] from the Europeans let the Indians 'have big gardens'.[8]

Remote regions obtained metal tools indirectly through trade and raiding, probably on an irregular basis. Isolated tribes continued the use of stone axes well into the twentieth century, although few still do so. This raises a question. How was agriculture affected when people at times had metal axes and at other times still had to depend on stone axes, or when some farmers in a village had metal axes while others did not? There seems to have been little reporting on this. An 'overnight' adoption of metal axes as reported by Holmberg (1969: 268) and others, was probably unusual.

There are other questions that need to be pursued. How effective were the iron axes that were introduced in colonial times?[9] Iron axes must have been less effective than steel axes, but most of the experimental data from Carneiro and others is for steel axes. Also, to what extent were metal axes present in Upper Amazonia in pre-European times? Lathrap (1970: 178) found bronze axes on the Río Pachitea and Río Pisqui in eastern Peru, clearly traded from the Andes. These were small, however, and probably were not used to cut down large trees. Finally, what was the significance of the stone axe for agriculture elsewhere in tropical America? For a discussion of stone and metal clearing tools in Central America, see Gordon (1982: 57–61). He notes that: 'Clearing wet evergreen forest without metal cutting tools would clearly have been a slow and laborious process.'

Thus, shifting cultivation as an ancient, dominating practice in Amazonia may be a myth. There is little or no direct evidence for it. At best, it was uncommon, at least in short-cropping cycle form. It is not logical, given the stone axe. It is a relatively modern adaptation resulting from the introduction of the metal axe.[10] What is the significance of this, if valid? Certainly it tells us something about pre- and post-Columbian adaptation and the impact of European technology on agriculture. 'Production and social organization were altered as each settlement chose a particular method for acquiring manufactured goods, particularly iron tools' (Golob 1982: 269).[11] But stone-axe technology also tells us that the Amazonian forest can be farmed successfully and sustainably with minimal clearing by means other than shifting cultivation, which is one of the main instruments of forest destruction today.

The introduction of metal tools changed the situation dramatically. With forest much easier to clear, any type of forest vegetation could serve for swiddens. Fields could be abandoned readily if there were problems with soil decline, weed and pest invasion, game depletion, or shift in settlement, even after just a year or two since relatively little labor need be invested in new clearings. Short-cropping, long-fallow systems became efficient. And there may have developed a greater casualness towards site selection, soil maintenance, and weeding. Social factors could have primarily determined site selection. Also, short-cropping monocultural systems emphasizing food staples probably became favored over shady polycultural systems (house gardens, intercropped swiddens, agroforestry). The latter are more protective of soils and hence more sustainable. Long-fallow systems, however, can only support small villages and sparse populations. We have seen this post-Columbian pattern in several variations: Campa, Bora, Kayapó.

Terra Preta

Terra preta, which is primarily known from riverine bluffs, was discussed in Chapter 6. In addition, numerous *terra preta* sites have been reported in interfluve *terra firme* forests in Brazil, including a total of about 50,000 ha between the Rio Tapajós and Rio Curuá-Una (Katzer 1944: 35–8). Such sites are usually much smaller than most bluff sites; N. Smith (1980: 563) obtained an average size of 1.4 ha for twelve interfluve sites. Some are as small as 0.3–0.5 ha. Most are shallower than bluff sites, suggesting shorter periods of occupation, but sufficient to create black earth. They tend to be circular and probably represent a single, large communal house or a circle of smaller houses.

These small *terra preta* sites support arguments for very small, permanent, interfluve villages and possibly larger temporary villages which did not create *terra preta*. Recently, however, large interfluve *terra preta* and *terra mulata* sites have been reported. Michael Heckenberger (1996: 94–104; 1998) has done archaeological research on *terra preta* sites (AD 1000–1500) of 30–50 ha on *terra firme* in the upper Xingu Basin. These are ringed by multiple defensive ditches or moats. His logical explanation for a second or third ditch is that the entire area within the initial ring became fully occupied as population grew over time. Hence a new outer ditch was dug, and possibly others later. Concurrent populations of 1,000 to 2,500 people were supported, he believes, by surrounding intensive manioc gardens on improved poor, natural soils, plus protein from locally rich aquatic resources. Numerous other ringed villages, as many as six rings, possibly permanent with intensive agriculture, have been reported in southern Amazonia, including northeastern Mojos in Bolivia (Heckenberger 1996: 114–16; Wüst and Barreto 1999). There are other, even larger interfluve *terra preta* sites, such as Oitavo Bec (120 ha) south of Santarém (Woods and McCann 1999: 12) and at Comunidade Terra Preta (200 ha) between the lower Tapajós and the Rio Arapiuns (N. Smith 1999: 26).

Indians today are well aware of the locations and value of *terra preta* soils. The 'Arawete, Kuikuru, Mawé, Mundurucú, Xikrin-Kayapó, and many other

Indians prefer to plant nutrient-demanding crops in this most fertile cultural horizon' (Balée 1992: 42). Maize is often planted on *terra preta*, and this crop reduces dependence on animal protein.

Possible Forms of Pre-European Forest Cultivation

I am suggesting that shifting cultivation in prehistoric Amazonia was uncommon because of the inefficiency of the stone axe, especially in the mature, high, hardwood forests of the *terra firme*. *Terra preta* and *terra mulata* are probable evidence of more intensive cultivation than shifting cultivation. Indian shifting cultivation today has a short-cropping period, reflecting poor soil, pest invasion, game depletion, and social friction, but it is made possible by the steel axe which makes clearing new plots a relatively easy process—a matter of a few weeks to create a field large enough (0.5–2.0 ha) to feed a family. What then was the nature of pre-European high-forest agriculture? We do not know and may never know. However, there are several possibilities (Denevan 1998).

House Gardens: permanent plots of mixed annuals and perennials around houses, with careful weed control and soil management using household refuse for fertilizer. Lathrap (1977), in his classic article 'Our Father the Cayman, Our Mother the Gourd', maintained that the earliest agriculture in Amazonia was in such house gardens. He believed that the first gardens were along or near rivers, but they were undoubtedly also an important form of prehistoric crop production in the interfluve forests since they do not require frequent clearing. Today, house gardens are poorly developed in most forest villages given the frequency of village shifting.

Intensive Swiddens: located on sites where tree clearing was unnecessary or relatively easy, such as naturally disturbed places and old field plots with young secondary growth of softwoods, as well as at blowdown and forest-fire sites. Present-day examples of intensive swiddens are the highly diverse (polycultural) fields of the Waika and the Kayapó (Chapter 4). Such fields contrast with the monocultural swidden dominated by a single species, usually manioc, which is the common form of tropical forest Indian field today (Beckerman 1983*a*), even for the Yanomami (Hames 1983*a*: 18–19). Most current monocultural fields are only used for one to three years. Beckerman (1983*a*: 4–6) gives several reasons for the monocultural field, but he does not consider the role of the steel axe in making short-lived swiddens feasible. Prehistoric short-fallow, monocultural fields of manioc with protein supplied by locally rich aquatic resources were possible, as argued by Heckenberger (1996: 98–100; 1998) for the upper Xingu Basin.

Intensive swiddens (short-fallow or permanent) were feasible, based on known indigenous practices today. Weeds could have been controlled by shading and hand weeding; pests could have been reduced by crop diversification; soils could have been artificially maintained to varying degrees by short fallowing and external organic inputs; and rich soils could have been unintentionally created by settlement activity and by agriculture in the form of *terra preta* and *terra mulata*. Labor inputs often would have been high, but probably not as high as that required by long-fallow shifting cultivation with frequent field shifts necessitating considerable difficult tree clearing with stone axes. This assumes that time and energy are critical factors, and they usually are (K. Hill and Kaplan 1989: 331; and others).

Examples of traditional semi-permanent intensive cultivation elsewhere in the tropics are the composting systems of the Wola (Sillitoe 1996: 375–92), the Enga (Waddell 1972: 44–6), and other groups (Denevan 1966: 125–6) in highland New Guinea. Sweet-potato vines, grass, leaves, and burned vegetation are incorporated into small sweet-potato mounds or larger raised fields. Sweet potatoes have low phosphorus and nitrogen requirements, and the composting (especially grass) boosts available potassium. Yields for the Wola actually progressively increase with cultivation. The sweet potato is, or was, a staple for many Indian groups in eastern Brazil. Manioc also has low nutrient demands but depletes soil of potassium (Cock 1982: 757–8, 760) and thus responds to composting.

Slash/Mulch Cultivation (see Chapter 4): this technique minimizes nutrient loss, protects the soil, and can be of long duration.

Patch Cultivation: the planting of small natural clearings, such as tree falls, or easily cleared vegetation such as bamboo. This could have been done by bands of semi-nomadic foragers, such as the Sirionó. These people have been very mobile, returning periodically to small plantings. Households could also have been permanently settled at small clearings, as suggested by *terra preta* sites of 1 ha or less. In addition, people in large villages with surrounding fields could have also obtained production from small patches. The Kayapó villagers today plant both domesticates and semi-domesticates in natural clearings created by tree falls, as well as along trails and at camp sites (Posey 1984*b*: 117, 122). Another report of tree-fall planting is for the semi-nomadic Nambicuara in the Guaporé Valley in 1968: 'we came to a place where a huge tree had fallen, taking several smaller trees with it. Among the tangle of fallen limbs, tobacco plants were growing' (D. Price 1989: 127). A Waíwai Indian in Suriname told ethnobotanist Mark Plotkin (1993: 194) that: 'A well-planned garden should look like a hole in the forest opened up when a giant *ku-mah-kah* tree falls over.'

Tropical forest tree mortality is more frequent than generally realized, *c.* 1–2 per cent per year for all trees (Denslow and Hartshorn 1994: 124). At Barro Colorado Island in Panama, on 50 ha over *c.* three years, the death rate of large trees (over 64 cm diameter at breast height) was 105 out of 678 trees, or 5.1 per cent per year, or 0.7 trees per ha per year (Hubbell and Foster 1990: 531). Some gaps are as much as $1,000 \, m^2$ (0.1 ha) in area. This size natural opening compares with current indigenous swiddens which mostly range form 0.4 to 0.6 ha, but are as small as 0.02 ha (Beckerman 1987: 59, 69). Many prehistoric fields could have originated as tree-fall gaps, with gradual enlargement using stone axes, fire, and other techniques.

Agroforestry: forest manipulation via intentional and unintentional planting and management of perennial crops along trails, campsites, fallow swiddens, and other activity areas (Posey 1985*a*; Denevan and Padoch 1988). These 'food forests' contained domesticates, semi-domesticates, and spontaneous tree growth much of which was managed for useful species. Suggestions of this for prehistory can be seen in the present Kayapó 'forest fields' (Chapter 4) and the Bora 'forest orchards' (Chapter 5), as well as the anthropogenic forests of the Huastec in Mexico (Alcorn 1984), the Mayan *per kot* forests (Gómez-Pompa *et al.* 1987), and the 'tree gardens' in Bocas del Toro, Panama (Gordon 1982: 52–98). Gordon points out that the introduction of the machete was detrimental to such forest management. Clearing and weeding by hand better allows for decisions as to what plants are to survive and which are to be destroyed, whereas slashing by machete tends to be less selective. He believes that polycultural fields were an integrated component of anthropogenic forests. However, while agroforestry systems do not require frequent clearing, they are 'shade' systems which suppress weeds, but they are not conducive to production of staple annual crops. Hence, population densities would have remained low, unless associated with primary fields.

The cultivation of fruit trees is mentioned in the early accounts (Carvajal 1934*a*: 210, 217; Heriarte 1952: 17; Acuña 1942: 37; Vázquez de Espinosa 1948: 384). The high concentration of fruit tree pollen, seeds, and pits at archaeological sites is also indicative of agroforestry systems, such as orchards, house gardens, and managed fallows (Mora *et al.* 1991: 43; Roosevelt 1999: 380–3).

These five models of *terra firme* agriculture with a stone axe technology in reality were likely manifested by numerous transitional forms and combinations, varying with habitat, mobility, time, and demography. These activities, plus foraging, contributed to the creation of anthropogenic forests, or semi-managed forests, with a larger than natural number of useful plants present. The Amazon forest was not pristine in 1492, nor is it today. Probably all of these forms of agriculture and agroforestry were present in the *terra firme* in a mosaic of variable population densities that may have included sectors of sparse semi-nomadic foragers and small but permanently settled households and extended families. In some advantageous places large and permanent fields and associated villages developed, such as on Amazon bluffs, in the upper Xingu Basin, and in the Rio Arapiuns Basin, where there are large *terra preta* sites. These fields and villages could have originated in small clearings which were enlarged over a long period of time by gradually eliminating the trees at the periphery. By people coming back repeatedly to the same site and/or by shifting houses and fields within the same site, the interfluve *terra preta* soils could have been created.

Conclusions

Prehistoric Amazonian Agriculture Reconsidered: A Patch Pattern Model

The portrayal of prehistoric agriculture, settlement, and population presented here differs from the ideas of Meggers, Lathrap, Carneiro, Roosevelt, and D. J. Wilson, although various elements appear in their work. My interpretation is based on bluff zone-*várzea* complementarity, stone axe inefficiency, anthropogenic *terra preta* soil, uneven access to floodplain river channels, and available archaeology. This evidence suggests a patch pattern in which sectors of intensive agriculture and dense settlement, both along the rivers and in the interior forests, were separated by large, sparsely occupied sectors. Probably no more than 20 per cent of riverine land, mainly where bluffs impinged on river channels, might have been heavily settled. In the interior *terra firme* forest, in contrast, tree-fall openings and other feasible sites were converted into small fields and gardens whose locations were somewhat random, with very sparse settlement. However, some of these fields were enlarged and intensified and supported patches of large populations, as in the interior of Santarém and in the upper Xingu Basin. The factors often given to explain differences between indigenous *várzea* and *terra firme* settlement in historical and recent times— natural soil characteristics and/or protein availability from wildlife—were probably less important in prehistory. Fertile *terra preta* supporting protein-rich crops could be created almost anywhere. I do not mean to say that patterns of natural soils and of game and fish were insignificant for human settlement, but rather that they have been overemphasized. In central Brazil, '[n]o correlation was found between village size and village location in relation to the environment . . . contradicting hypotheses of environmental limitation for cultural development in the lowlands' (Wüst and Barreto 1999: 17, 19). The prehistoric indigenous people of Amazonia did have effective systems of cultivation, often very productive and sustainable, which were related to settlement patterns more than to specific habitats.

Given bluff living and cultivation, the dichotomy between *várzea* and *terra firme* settlement based on soil constraints, expressed by many Amazonianist scholars, breaks down or at least is not as distinct. If bluff soils could be made to support relatively large numbers of people, so could other *terra firme* soils. Interfluve *terra preta* sites confirm this. That such sites are sporadic (patchy) in the interior forests was largely the result of non-natural soil factors, such as the use of inefficient stone axes, the lack of demographic pressure, high labor requirements for intensive cultivation, village fissioning related to social conflict, and absence of resource circumscription. (On the latter, see Carneiro 1970*b*.)

The patch model, if valid, has significant implications for contemporary Amazonian development. First, contiguous settlement and cultivation of the major riverine zones is unlikely, given that floodplains are subject to periodic destructive floods and given the limited extent of unflooded bluffs which are also adjacent to year-round navigable river channels. The resulting locational pattern of patchy pre-European riverine settlement, continued with colonial missions and towns (Denevan 1996: 673), exists today (with some exceptions), and is likely to persist in the future. Second, we have seen that *terra firme* soils can be cultivated permanently by traditional (low-technology) methods. To be sure, high labor costs, polycultural cropping, a diversity of food production systems, and forest management and extraction, while appropriate for subsistence, may not be feasible for a strongly market-oriented economy. On the other hand, a complementary bluff zone/floodplain system of land use can maintain some degree of forest, ensure reliable security, and support a modest population density, such as apparently existed in the indigenous past.

Environmental Limitations in Amazonia: Fact or Fancy?

Do the environmental characteristics of Amazonia set limits to indigenous food production, hence to population size and density, and thus to cultural development? In 1954 and 1957, Meggers presented a theory of environmental determinism based on her archaeological research in Amazonia: 'The level to which a culture can develop is dependent on the agricultural potentiality of the environment it occupies' (Meggers 1954: 815). The Amazon was considered to have low agricultural 'potential', with only swidden cultivation possible, and thus limited cultural development. Advanced cultures in Amazonia, as on Marajó Island, were considered to have developed elsewhere and then deteriorated after moving into Amazonia. Meggers was widely criticized in the 1960s and 1970s (Carneiro 1995), but she still persists over forty years later (Meggers 1987: 152; 1992*b*: 38; 1996: 173),[12] and similar thinking continues to appear in the Amazonian literature (e.g., Lamb 1987; D. J. Wilson 1999: 63, 169). Roosevelt (1980: 79, 87), for the *terra firme*, said that: 'Meggers is correct . . . sustained-yield intensive cultivation . . . is not possible with aboriginal methods . . . agriculture in the tropical forest is limited to swidden cultivation . . . intensive cultivation is an impossibility.' Roosevelt (1980: 112–19; 1987; 1989*b*: 40) disagreed, however, with Meggers on the *várzea* and wet savannas, finding convincing evidence of large settlements, dense populations, and intensive agriculture, particularly with maize as a staple. Roosevelt (1980: 119), also made the general statement that: 'evaluation of subsistence potential . . . [requires] consideration of the ability of specific technologies to express that potential.' Later, Roosevelt (1991*b*; 1999) vigorously rejected ecological determinism in Amazonia, for both *várzea* and *terra firme*. Another recent critique of Meggers' determinism is by Myers (1992*a*), who points out the antecedents to her thesis by Kirchoff, Alfred Métraux, and Julian Steward.

Gross finds himself 'drawn to a position that seeks to recognize the broad limits to human population growth and settlement density imposed by environmental factors'. He says that: 'no study has yet suggested that upland horticulturalists can long exceed the limits imposed by Amazonian nutrient cycles' (Gross 1983: 445–6).

Hames and Vickers (1983*b*: 19) are open to the role of resource limitations on settlement but urge caution with 'monocausal determinism'. Hames (1983*b*: 405) states that: 'environmental constraints . . . and technological factors . . . are the variables that govern settlement pattern.' Vickers (1983*a*: 464–5) says that: 'The relative density of resources and the level of technology powerfully influence the density of the human population that can be supported in a given area . . . differences in settlement size and permanence, and in the levels of sociopolitical organization attained.' One can only agree as long as technology is included, for technology can override so-called 'constraints'.

The problem with the concept of agricultural (or environmental) 'potential' is that such potential is not something inherent in nature independent of culture. Agriculture is a cultural phenomenon, and thus agricultural potential is in part culturally determined. It involves an interaction between technology and environment. The fact is that there were agricultural systems that were more intensive than long-fallow shifting cultivation: semi-permanent polycultural systems, permanent house gardens, near permanent cultivation on *várzea playas* and levees, raised fields in wet savannas, as well as less intensive but long lasting and productive agroforestry systems. The interfluve forests apparently were farmed intensively in places in pre-European times, as indicated by the presence of *terra preta* and *terra mulata* soils. If so, the same should be possible today.

The real, practical limits are technological and economic (costs of technology and labor, availability of capital), not environmental. There are no fixed carrying capacities. This does not mean the Amazon Basin can become a world breadbasket; the costs at present are considerable.

The traditional view of the pre-European people of the upland Amazonian forests—sparse populations, small dispersed autonomous villages, long-fallow shifting cultivation—is based: (1) on a belief in technological and environmental (poor soils, limited animal protein) 'limitations', and (2) on ethnohistoric and ethnographic survivals. The view that is now emerging, however, is that in earlier times there was considerable social, demographic, and economic diversity, which in some areas included intensive agriculture, large permanent villages, and probably chiefdom-level societies (Heckenberger 1996: 206–8; Whitehead 1994: 38–41). Ceramics of the Tapajó chiefdom, generally considered to be riverine in location, are also found in *terra preta* soils on *terra firme* tens of kilometers inland from the Amazon, Tapajós, and Arapiuns rivers

(J. McCann pers. comm. 1999). This is one response to D. J. Wilson's (1999: 444) challenge in his important recent text: 'Go out in the *terra firme* and find a place where either chiefdoms or states existed in the recent or ancient past.' Such places did indeed exist. Soils were maintained by simple but effective techniques, and fertile anthropogenic soils were created. Maize and legumes were (and still are) grown on these improved soils, providing high-protein levels, in contrast to manioc, thereby reducing reliance on game and fish for adequate protein, thus repudiating the limited-protein explanation for perceived extreme contrasts in *várzea* and *terra firme* populations. In part, I maintain, intensive agriculture and improvement of soil fertility resulted from the inefficiency of stone axes for clearing forest, a situation that changed dramatically with the introduction of metal axes. A tool handicap in the past encouraged a positive soil strategy.

A New Synthesis

Viveiros de Castro (1996), a Brazilian ethnologist, recently published a perceptive analysis of scholarly 'images of nature and society' in Amazonia, in terms of an old synthesis ('Standard Model' or 'Tropical Forest Culture' model) and an emerging new synthesis. The old synthesis was articulated by Julian Steward in the *Handbook of South American Indians* (Steward 1948: 883–99; 1949*b*: 697–710; also Steward and Faron 1959: 284–318), and later by Gordon Willey (1971: 398–9). Willey viewed the Tropical Forest Culture as 'a lifeway based on the slash-and-burn cultivation of manioc', with a continuity from the past to the present.[13]

The typical Tropical Forest Culture was characterized by Steward as consisting of autonomous, egalitarian villages of limited size and duration, with a simple technology, in an unproductive environment, with a shifting cultivation system unable to produce an agricultural surplus. During the period from 1954 (Meggers) to 1980 (Roosevelt) the development of cultural ecological research in Amazonia diverged in several related conceptual directions, including ecosystemicism, adaptationism, materialism, and environmental determinism. However, most scholars continued to accept the basic Tropical Forest Culture model, which they attempted to explain by single-factor ecological (poor soils, protein deficiency), social (warfare, migration), or demographic causation.

The passing of the old synthesis has been a slow process, with roots in Lévi-Strauss (1963) in 1952 and in the dissertations by Carneiro (1957), Lathrap (1962), and Denevan (1963*a*), followed by various studies in history, anthropology, demography, ecology, pedology, and botany. Recent contributions toward the new synthesis have come from archaeologists Anna Roosevelt, Irmhild Wüst, Clark Erickson, Peter Roe, Warwick Bray, Michael Heckenberger, Thomas Myers, Eduardo Góes Neves, and others. Involved is the rejection of accepted concepts about cultural ecology, cultural history, historical evidence, and ethnographic projection.

Amazonianist anthropologists now acknowledge that there was and is considerable diversity in environment, agricultural form and intensity, population density, settlement pattern, and social organization. Emilio Moran (1995: 71) speaks of 'Disaggregating Amazonia'. He says that recent scholarship has 'begun to tear the veil that hid the reality of Amazonia: its enormous environmental and cultural diversity'. At certain times and places native populations were dense, agriculture-intensive, settlements large and permanent, and societies stratified and relatively complex politically (chiefdoms). There was a significant human impact on Amazonian nature, so that adaptive responses[14] were not to a pristine environment, but rather to a culturally transformed nature. Lathrap and then Roosevelt have maintained that prehistoric societies in Amazonia were not 'archaic' but rather were important contributors to the rise of civilization.

In addition, the perceived distinction between *várzea* cultures (large permanent villages) and *terra firme* cultures (small shifting villages) has lessened, given the 'bluff model' thesis, the discovery of enormous *terra pretas* in the uplands, and new archaeological evidence for complex societies in the uplands. Also, there is documentation of the interaction between people in different environments as well as control and use of different environments by the same people (Whitehead 1993).

There is also increasing recognition that contemporary surviving Indians do not necessarily represent those present when Europeans first arrived and drastically intervened, although those societies would have changed in various ways regardless. We have to rely more on evidence from the past than on projection to the past.

Viveiros de Castro (1996) has hopes for a new theoretical synthesis, still in formation, that reflects these perspectives, with multicausal explanations, as well as consideration of structure, political economy, and historical process (also see Crépeau 1990; Descola 1994). Cultural ecology will remain part of the story, and we must accept that how people make a living from the Amazonian environment was and is highly diverse, including forms of productive, sustainable agriculture as argued here. There was a cultural mosaic reflecting variation in productivity, social organization, previous history, settlement pattern, and demography, and not necessarily variation in environment. The relative importance and distribution of the variants that we now know once existed, with some survivals, remain to be demonstrated. This is the great challenge for future research on indigenous people and their agriculture in Amazonia.[15]

Notes

[1] For the Ecuadorian Amazon, Piperno and Pearsall (1998: 258–61, 312–20) identify maize pollen, *c.* 3300 BC, which they believe was cultivated by slash-and-burn methods. However, as they point out, ground and polished stone axes did not appear in the American tropics until after 1000 BC. In my opinion their evidence of forest clearing does not indicate whether agriculture was

shifting cultivation or more permanent gardens. They also report maize presence on seasonally receding lake margins by *c.* 3300 BC, and this would not have involved forest clearance.

[2] Hodder (1983: 79–80) questions the validity of Carneiro's (1979*a*; 1979*b*) results for Yanomami tree-cutting efficiency because the practitioner had no prior experience using a stone axe.

[3] Other studies have obtained lower stone axe to steel axe efficiency ratios (*c.* 3:1 to 6:1), but they generally do not take into consideration variability in tree diameter and hardness; for example, Saraydar and Shimada (1971; 1973); Townsend (1969: 203–4); Salisbury (1962: 220); Steensberg (1980: 38–9).

[4] See Carneiro (1979*a*: 41), Townsend (1969: 201), and Lewenstein (1987: 35–43). Lewenstein compares the time to make and sharpen, efficiency, and durability of ground stone versus chipped stone axes for the Maya. Carneiro (1974: 115) quotes an Amahuaca man on former use of stone axes: 'They say it was always breaking. They say it was always getting dull. That stone axe is no good!'

[5] There are many examples of the Indian obsession for metal axes in Amazonia; for example, DeBoer (1981*a*); Métraux (1959); Isaac (1977: 141); and Golob (1982: 115, 126–7, 153–4, 201–2).

[6] Wilk (1985: 55) believes that in the pre-European Maya lowlands riverbank recessional farming preceded long-fallow swidden farming in the uplands, in part because of inadequate land-clearing axes.

[7] Even with steel axes, Amazonian tribes today prefer secondary forest for new swiddens because of the 'relative ease of cutting early successional soft woods' (Beckerman 1987: 72).

[8] Harner (1972: 198), however, indicates that Jivaro swiddens were reduced in size by one-quarter to one-half following conversion from stone to steel axes after 1925, reportedly because of a shift from communal to individual clearing; food production did not increase. This may be an exception.

[9] The development of the iron axe was instrumental in the clearing of the forests of northwest Europe in the Middle Ages. Some iron axes had strips of steel welded to the head ('steeling'), but single piece, high-carbon steel axes did not become common until the twentieth century.

[10] Lathrap 'insisted that shifting slash-and-burn agriculture was a secondary, derived, and late *phenomenon* within the Amazon Basin' (Lathrap *et al.* 1985: 54–5). This is because he believed that initial settlement was riverine, involving permanent fields, with shifting cultivation in the upland forests coming after the floodplains were fully settled, not because of stone-axe inefficiency.

[11] For a treatment of the social and economic impact of the transition from stone to steel in highland New Guinea (the Siane), see Salisbury (1962), and for Cape York, Australia (the Yir Yoront) see Sharp (1952).

[12] The second edition (1996) of the well-known survey and interpretation by Betty Meggers, *Amazonia: Man and Culture in a Counterfeit Paradise*, is a 'revised edition'; however the text of the original 1971 edition is reproduced with only a few minor changes. A new 'Epilogue: Recent Developments', was added, and additions were made to the References.

[13] Lathrap had a somewhat different but ambiguous view of 'Tropical Forest Culture' (Raymond 1994: 176–9), also see Neves (1999).

[14] My own view of adaptation is that it can only be considered as 'selective adaptation' (Denevan 1983). People confronted with a particular environment, or a changing environment, always have options (technological, social) as to how they will cope with it.

[15] For a recent statement providing archaeological evidence in support of these conclusions, see Heckenberger *et al.* (1999).

Part III

Andean Irrigation and Terracing

The ancient Andean populations managed natural resources better than we manage them today.

(John Murra 1983: 5)

8

Irrigated Fields

> In the building of irrigation canals, I doubt that there has ever been a
> people or nation in the world who constructed and conducted them over
> such rough and difficult terrain.
>
> (Pedro de Cieza de León [1553] 1959: 176)

Andean Environmental Diversity

The Andean mountains are mostly arid, with high elevations and often poor
soils. The adjacent Pacific Coast from southern Ecuador to central Chile is one
of the driest deserts on earth. The eastern Andes north of Argentina have tor-
rential rainfall, montane tropical forest, and some of the most precipitous terrain in the hemisphere. Nevertheless, the Andean region in prehistoric times
witnessed intensive forms of cultivation that supported locally dense populations and complex states and civilizations. Much of this agriculture involved
irrigation and terracing or, on many mountain sides, a combination of both.
The forms of Andean irrigation are highly diverse, complex, and innovative,
reflecting sophisticated water-management skills (Ortloff 1993; 1995).

Generally, irrigation is necessary for cultivation in regions of the Andes with
under 500 mm annual precipitation (ONERN 1976: 52). The actual limit
of rainfall-dependent agriculture, however, is dependent on the field capacity
of local soils, seasonal distribution of rainfall, potential evapotranspira-
tion, type and length of the growing season of a crop,[1] level of demand, and
level of acceptable risk. Even areas with higher quantities of precipitation
(500–1,000 mm) may benefit from irrigation in order to increase production
and to ameliorate periodic drought. Figure 8.1 indicates the approximate areas
of the central and northern Andes with under 500 mm annual precipitation.

The least rainfall is experienced in parts of the Bolivian *altiplano* and especially on the coast of Peru and northern Chile; in Peru precipitation declines
from about 250 mm/yr near Tumbes to 31 mm/yr near Chiclayo. Most of the
central and southern Peruvian coast receives less than 30 mm/yr. Some hill
slopes, however, near the ocean receive over 125 mm/yr and support seasonal
vegetation (*lomas*) (ONERN 1976).

The Peruvian coastal desert is crossed by numerous rivers originating in the
wetter, higher elevations of the Andes. These vary from dry washes which carry

Fig. 8.1. Map of precipitation under 500 mm/yr, western South America. By Gregory Knapp

water only after periodic *El Niño*-associated rains to a few perennial rivers supplied by mountain rainfall or by glaciers and lakes. As a result, there is little agreement on the number of 'rivers' on the Peruvian coast. Romero (1961: 13: 147–75) says there are fifty-two valleys, and he describes forty-nine of them. Robinson (1964: 165–6) maps fifty-seven rivers, most with headwaters where there is regular rainfall; the amount of land irrigated in each valley is indicated. Dobyns and Doughty (1976: 6) state there are only thirty-six rivers which are used for irrigation. Only ten rivers consistently reach the coast throughout the year (Sherbondy 1969: 116).

Most of the flow of the coastal rivers occurs between January and May, reflecting the rainy season in the highlands. There is considerable fluctuation within this period, however, in volume and duration of peak flows. For the Virú Valley, and others, there are long-term cycles of greater or lesser abundance lasting five years or more (M. West 1981: 53). In the low water period, some water is available in the dry river beds from springs or from groundwater tapped by sunken fields (see below).

The Andean region, including the narrow Pacific coastal plain, is one of tremendous climatic and topographic diversity, often over short distances. On steep mountain slopes there are narrow horizontal climate-vegetation bands based on temperature decrease with altitude. Within each of these bands, however, there is a patchy variability based on degree of slope, rainfall related in part to slope orientation, soils, and drainage, and thus vegetation (Zimmerer 1999). It is not surprising, then, that there has been a corresponding mosaic of adaptative strategies for producing food, particularly strategies for managing water. In addition, environmental change is frequent and at times violent, bringing potential disaster to farmers: earthquakes, landslides, floods, volcanic eruptions, drought, hail, and frost. Major droughts might last for several years. Periodic El Niño events bring destructive floods to the north coast of Peru and drought to the mountains. Thus farmers not only had to adapt to marginal and spatially highly variable natural conditions, but they also had to be sufficiently flexible to adapt to temporally changing, often hazardous conditions, in contrast to a predictable regularity. Crop failure was frequent, but was countered, in part, by food storage mechanisms. Polo de Ondegardo in 1571 observed that crops failed one year out of three in the highlands (D'Altroy 2000).

Canal Irrigation

Irrigation has received more attention than any other form of prehistoric agricultural technology in South America. This is because irrigation works in arid lands, especially coastal Peru, are readily visible, often are well preserved, and are impressive in their number and size. Although some major canal systems were abandoned by Inca times, others were functioning when Pizarro arrived in Peru and were described by the early chroniclers. Some are still in use; others have been replaced by modern systems. In several valleys the extent of prehistoric irrigation exceeded that of today (Schaedel 1986: 320). Here, I will examine the archaeological and historical evidence for the distribution and characteristics of indigenous irrigation.

Irrigation, the artificial transfer of water from a source to agricultural fields, involves canals and diversion and collection features. I will focus first on large surface canals, and then on lesser features. The Chicama-Moche Intervalley Canal system, the largest in the prehistoric New World, is discussed as an example of the problems of attempting to interpret prehistoric agricultural landscape features. Irrigated terracing, mostly relying on small local canals, is discussed with terraces in Chapters 9 and 10.

Irrigation Studies

The classic study of prehistoric irrigation in western South America is *Life, Land and Water in Ancient Peru* by historian Paul Kosok (1965). The volume is primarily a general survey of ancient sites, but it is notable for the use of aerial photographs and the mapping of irrigated areas. Illustrated with hundreds of ground and air photos, the study is based on field research in Peru in 1940–1 and 1948–9. Explorations were made by plane, by vehicle, and on foot, guided by historical accounts initially and later more by air photos taken during World War II. Aerial survey had earlier been used for archaeological discovery in Peru by Shippee and Johnson in 1929 and 1931 (Denevan 1993), but Kosok was the first to combine systematically air photography and ground survey in archaeological research in South America. The Peruvian coast is particularly apt for this, given complete aridity and the starkness by which ancient ruins and canals are etched in the desert.

Kosok's volume has been a stimulus for more analytical studies of prehistoric irrigation in Peru. Some of the subsequent studies were of the Virú Valley (Willey 1953), the Jequetepeque Valley (Eling 1986), and particularly the Chicama-Moche Intervalley system. Studies of highland irrigation sites exist for Cuzco (Sherbondy 1982), Arequipa (Bernedo Málaga 1949) and Cusichaca (Farrington 1983*b*). Surveys of Andean irrigation include Sherbondy (1969), Regal (1970), B. Price (1971: 24–33), Farrington (1980*a*), Horkheimer (1990: 133–44), Ravines (1978*a*: 21–32), Zegarra (1978), Park (1983), and Netherly (1984). Sherbondy (1969) lists and maps the location of 303 sites in Peru having prehistoric irrigation, with references for each. Regal's (1970) book on Inca hydraulic works gives brief descriptions of prehistoric irrigation for numerous valleys on the coast and in the Andes.

The Distribution of Irrigation

There are four dry environments that have been irrigated by Indians in South America, all in the Andean region: (1) coastal valleys between southern Ecuador and central Chile, (2) scattered intermontane valleys from northern Ecuador to northern Argentina, (3) the high plain (*altiplano*) of southern Peru and northern Bolivia, and (4) a few valleys near the Caribbean coasts of Colombia and Venezuela. No Indian irrigation has been reported in the Chaco or anywhere in Brazil. Areas of coastal irrigation in Peru are shown on Fig. 8.2, and Fig. 8.3 shows areas of early reported irrigation in Ecuador, Colombia, and Venezuela. For detailed mapping of prehistoric irrigation in northern Ecuador, see Gondard and López 1983.

Peruvian Coast The best evidence of canal irrigation, both archaeological and ethnohistorical, is from the coastal Peruvian desert. Sherbondy (1969) shows twenty-five to thirty coastal valleys where there were pre-colonial irrigation canals and other hydraulic works, based on reports by Spaniards in the six-teenth- and early-seventeenth centuries and on studies in more recent times by historians and archaeologists.

The largest prehistoric canal systems and areas irrigated, which date from *c.* AD 1000, were on the north coast of Peru from the Río Motupe (north of the Río Chicama) to the Río Casma (Fig. 8.2). The zones (twelve valleys) irrigated in this sector, plus the Río Chillón and Río Rimac valleys near Lima, were mapped by Kosok (1965: 24, 86, 146, 180). The valleys on the central and south coasts were less extensively irrigated. Kosok worked on measuring the extent of prehistoric irrigation, but the results and methodology were never published. He did determine that the area irrigated on the Peruvian coast *c.* 1940 was between 500,000 and 600,000 ha, and that the prehistoric maximum exceeded this; the greatest difference was on the north coast (Kosok 1965: 34). In 1960, the area irrigated in thirty-six major valleys was 627,000 ha (Robinson 1964: 166–7).

T. Pozorski (1987: 119–20) believes that the maximum extent of prehis-toric agriculture in the Chimú region of the north coast was reached about AD 1300. The total non-marginal area was comparable to that in 1940. In AD 1300 a major El Niño flood damaged parts of the canal system, and the marginal areas were abandoned and remained so through the Inca period and beyond.

At the time of Spanish conquest, most of the Peruvian coastal valleys that had been cultivated in pre-Inca periods were still irrigated. In some cases, such as the Chicama, Moche, and Virú valleys, irrigation had declined, but else-where the Incas had expanded cultivation. The ethnohistoric record on this is unreliable, however, due to land abandonment in the sixteenth century. Early writers could not be certain whether abandonment observed had been pre-Inca or post-Inca.

Northern South America Some of the drier areas near the Caribbean Coast were apparently irrigated in prehistoric times (Fig. 8.3). There are remains of irrigation canals, and associated reservoirs which were utilized by the Tairona of northern Colombia during the late pre-Columbian period (Donkin 1979: 90). The Timotean tribes of the Venezuelan Andes historically irrigated from reservoirs called *guimpúes* (Donkin 1979: 84; Métraux and Kirchhoff 1948: 356). The Caquetío Indians near Barquisimeto in Venezuela had an irrigation

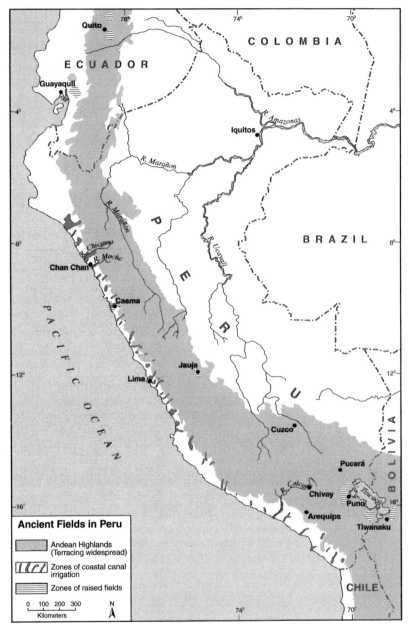

Fig. 8.2. Map of the central Pacific Coast of South America showing Andean highlands, coastal irrigation, and raised fields. By W. M. Denevan

Fig. 8.3. Map of early colonial indigenous irrigation in northern South America. By Gregory Knapp

system (Hernández de Alba 1948: 470). The Cumanagoto villages of the Araya Peninsula of Venezuela irrigated coca (Kirchhoff 1948: 481–2). Kroeber (1946: 898; no source) reported that the Guane in the northern Chibcha region practiced irrigation. G. Reichel-Dolmatoff (1961: 96) gives several early mentions for the Guane, Tairona, Muzo, Colorado, and other Indians. I am unaware of archaeological proof of prehistoric irrigation in these areas, however.

Inter-Andean Valleys Irrigation has been practiced in many inter-Andean valleys. Frequently, it has been associated with terracing. Andean valley bottoms are usually narrow, so that flat, irrigable land is limited. Highland irrigation systems are thus on a much smaller scale than those on the coast. Also there is rainfall in the highlands, so that irrigation is a supplement rather than the sole source of water (Farrington 1980*a*: 298).

Some of the highland valleys and towns in Peru with reported prehistoric irrigation include Cuzco, Arequipa, Cajamarca, Colca, Apurimac, Ollantaytambo, Yucay, Anta, and Pisac (Sherbondy 1969). The canals of Cuzco are described by Sherbondy (1982), who found that irrigation districts corresponded to Inca political subdivisions.

The Quishuarpata Canal in the Cusichaca Valley north of Cuzco is described by Farrington (1980*a*: 298–302). This canal is lined with cut granite blocks and is up to 80 cm wide, 30 cm deep, and is 6 km long. It is fed by a river at 3,500 m elevation and by three smaller streams along its course. The gradient is 13 per cent overall, with steep chutes of from 43 to 64 per cent slope. The canal was designed to carefully control water velocity from becoming excessive.

For northern Ecuador, Mothes (1986) describes earthen irrigation canals in the semi-arid, intermontane valleys as being persistent, successful, and efficient. One prehistoric canal, the Pinampiro, is located along the fringes of the Chota River. It draws water from a canal intake on a stream with fluctuating flow levels at 3,040 m elevation, and then flows smoothly downslope for 20 km, until its waters are diverted to fields for irrigation on often steep slopes extending for over 1,000 m. A series of 'vertical jumps' along the canal disrupt an otherwise gentle flow.

Myers (1974*a*) observed prehistoric irrigation in the highlands of northern Ecuador between the Mira and Guayllabamba Rivers. Canals have been discovered beneath *tolas*, or large earthen mounds. Myers postulates that these canals were intentionally filled in after water became scarce, and the *tolas* were erected over them for burial and ceremonial purposes.

Highland irrigation in Ecuador has also occurred on unterraced slopes through use of *canterón* or serpentine-furrow irrigation. In this technique, observed for example in the Ambato and Riobamba-Alausi dry valleys of central Ecuador, water is led downhill through S-shaped furrows to irrigate crops (Knapp 1991: 70). *Canterones* are a means of irrigating steep slopes without terracing. This technique may be prehistoric in the highlands; furrow form does resemble relic furrows of the Peruvian coast.

Systematic irrigation was practiced in northwestern Argentina in pre-Inca times in highland valleys such as the Calchaquí and Tafí (Núñez Regueiro 1978).

Altiplano Irrigation canals are common on the *altiplano* or high plain north and south of Lake Titicaca. Many are undoubtedly prehistoric, but there has been little archaeological research on them. The lake plain itself is poorly drained, and much of the prehistoric agriculture was by means of raised fields (see Chapter 13). It has been suggested that these fields were also irrigated by water standing in ditches during the dry periods (C. Smith *et al*. 1968: 362; Lennon 1982: 226). Canals associated with raised fields could have served for irrigation by bringing water from the lake or local streams and ponds, rather than being drainage canals. Prehistoric canals and aqueducts on the Bolivian *altiplano* are described by Soria Lens (1954: 89–90).

Elsewhere On the central coast of Chile the Mapuche (Araucanians) were irrigating north of the Río Maule in the sixteenth century. On the north coast of Chile, the Atacameño were irrigation farmers (Latcham 1936: 278–81). In semi-arid Córdoba in central Argentina there is no reported archaeological evidence of irrigation of any significance, possibly some small-scale canals associated with terracing (Aparicio 1946: 676–7).

Chronology

Frédéric Engel has proposed that Peruvian coastal agricultural history can be differentiated into a long period of 'archaic' agriculture, and a short late period of extensive and intensive agriculture. During the earlier period, prior to AD 1000, floodwater farming took place in the large river valleys, and only the lower portions of the smaller coastal stream valleys were irrigated, with the use of diversion dams. Large canals and major irrigation works were constructed after about AD 1000. A much greater area was thus made available for cultivation, often with double or triple cropping; for example, in the Cañete Valley there was an increase from 1,500 to 150,000 ha with irrigation. However, a lower concentration of riverine silt in irrigation water meant that soil fertility maintenance was more of a problem under irrigation than under floodwater farming (Engel 1976: 164, 175–7, 181, 285–8).

Irrigation on the north coast of Peru may have been as early as 3000–4000 BC in the Zaña Valley (small ditches) (Dillehay *et al.* 1997: 54). The location of the Caballo Muerto complex, 1400–800 BC, in the Moche Valley is associated with the intake of the Vichansao Canal, suggesting a minimum antiquity for that canal (Moseley and Deeds 1982: 28). The intakes and initial sections of other canals of the Moche Valley were related to Salinar times (about 350–500 BC) or even earlier sites. Willey (1953: 362–3) in the Virú Valley identified the Gallinazo Period (AD 300–700) for which there was certainly extensive irrigation. T. Pozorski (1987: 111) says there is no physical evidence on the coast of canals and irrigated fields before 200 BC, but that irrigation as early as 1800–900 BC can be inferred from the change then in settlement pattern from a coastal concentration to inland valley sites.

There were major canals on the north coast by the start of the Moche period (*c.* AD 100). 'Almost all techniques of canal construction were in existence prior to the Chimú period [AD 1000]—the Chimú refined techniques but did not invent new building methods' (Kus 1972: 78). Radiocarbon dates for the great Chicama-Moche Intervalley Canal are between AD 1040 and 1310 (Pozorski and Pozorski 1982: 860).

In the highlands there has been little early dating of irrigation canals. Irrigated bench terraces at Huarpa in Ayacucho date to 200 BC–AD 600 (Brooks 1998: 121). Given that agricultural features, such as canals and terraces, experience ongoing deterioration and repair, dating is very tenuous. However, associated features can be dated. Recently, Zimmerer (1995: 481) dated charcoal in sediments apparently derived from 'hybrid floodwater-canal irrigation' to *c.* 1500 BC near Tarata in the Cochabamba region of the Bolivian Andes. At

La Galgada in the western Andes at 1,100 m elevation east of Chimbote, irrigation canals may date to 2400–3000 BC based on presence of crops where both rainfed farming and floodwater farming are not possible (Grieder *et al.* 1988: 2, 138, 144).

Early Colonial Descriptions of Irrigation Canals

The exceptionally rapid depopulation of the Peruvian coastal valleys in the sixteenth century has deprived us of good first-hand observations of functioning indigenous irrigation. Probably the most valuable account is that of Cieza de León, written in 1553, on the basis of a reconnaissance in 1547. The following passages are representative:

I may say that all the land of the valleys which is not sand is among the most fertile and productive to be found in the world . . . such water as they have comes by irrigation from the rivers that flow from the sierras . . . they built irrigation ditches at intervals, and, strange though it seems, both in upland and low-lying regions and on the sides of the hills and the foothills descending to the valleys, and these were connected to others, running in different directions . . . The Indians took and still take great care in bringing the water through these ditches . . . For, as the rivers never dry up, the Indians can conduct the water where they will   (Cieza de León 1959: 316, 318).

He described several individual valleys on the Peruvian coast where irrigation canals and fields had been abandoned, apparently since the Spaniards arrived but possibly earlier: 'This valley of Tumbes was once thickly settled and cultivated, covered with fine, cool irrigation canals, channeled from the river, with which they watered their crops abundantly and harvested much corn' (Cieza de León 1959: 300). 'So the sights of this valley [Santa] are the graves of the dead, and the fields they cultivated when they were alive. They used to dig great irrigation ditches from the river, with which they watered most of the valley . . . But now that there are so few Indians, as I have told, most of the fields are untended and grown up to thickets and brush' (Cieza de León 1959: 326). 'In the time of its prosperity, before it was conquered by the Spaniards, when it enjoyed the rule of the Incas, in addition to the ditches with which they watered the valley [Ica], they had ones larger than all the others which had been artfully brought from the heights of the sierras, so that it made no drain on the river. Now, when they have a shortage of water, the big irrigation canal having been destroyed, they dig cisterns along the river and the water collects in them, and they use it for drinking and dig little ditches out of them to water their crops' (Cieza de León 1959: 348).

Garcilaso de la Vega, born in Cuzco in 1539, made several interesting remarks about Inca irrigation: 'When the Inca had conquered any kingdom or province . . . he ordered that the agricultural land should be extended. . . . For this purpose irrigation engineers were brought: some of these were extremely skilled, as is clearly demonstrated by their works' (Garcilaso 1966: 1: 241). He described the diversion of water from an eastward flowing river to the west to the Ica Valley by means of a large canal: 'But now, with the aid of the canal, which was bigger than the river, they more than doubled the extent of their cultivable land' (Garcilaso 1966: 1: 350).

Garcilaso, however, had a tendency to glorify the Incas by exaggerating the size of irrigation works (Sherbondy 1969: 114). For example, he reported canals in the highlands which were 120 and 150 leagues long and 3 to 4 m deep. 'These works are certainly so great and wonderful that they exceed all the description and praise that one can devote to them.' But he was correct in noting in the same passage that the Spaniards have let the canals 'go to rack and ruin either deliberately, or what is more probable, through complete indifference' (Garcilaso 1966: 1: 296).

An astute observer of the natural environment and post-conquest agricultural technology was Bernabé Cobo in 1653, who wrote well after Cieza and Garcilaso. He described the skill with which canals were constructed across difficult terrain (Cobo 1990: 213).

Figure 8.4 shows a copy of a 1567 map of the Taymi Canal leading from the Río Lambayeque to a series of named secondary canals.

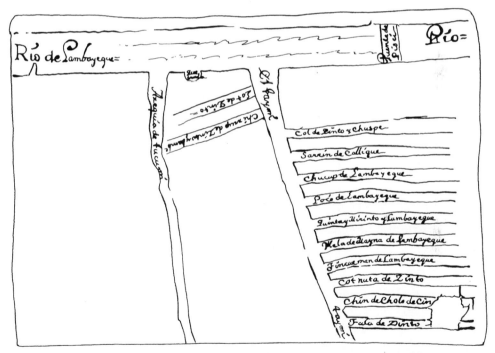

Fig. 8.4. Map of the Taymi canal (right center) in 1567, Lambayeque, coastal Peru. Secondary canals (right) are named. The Río Lambayeque (top) is also called the Río Chancay. From Netherly (1984: 241)

Canal Characteristics

From surviving prehistoric canals in Peru it is possible to understand their design, functions, and organization and hence the intentions and engineering skills of the Indians who built them. We have considerable knowledge now of relic canals as the result of field mapping, site surveying, and excavating by archaeologists, geographers, and engineers, especially for the canals in the Chicama and Moche valleys. The pre-Inca and Inca canals of the Pacific Coast demonstrate an impressive knowledge of engineering principles and environmental management.

Functions The main feeder canal (*acequia madre*) was designed to collect water from a river flowing from the Andes, the water then moving continuously by gravity to secondary distribution canals and from them into field channels and furrows. The system had to maintain a sufficient gradient and capacity in order to: (1) deliver the necessary water when, where, and in amounts needed, and (2) to do so without either excessive scouring or sediment deposition and with a minimum of maintenance. Efficiency of water preservation was critical. Canals were mostly exposed at the surface and usually lacked an impermeable lining; some were stone-lined but were still permeable unless sealed by clay sediment. Hence, the longer and wider a canal was, the more water that was lost from evaporation and seepage. Measurements of water loss today from mud-lined or stone-lined canals in the Moche and Chicama valleys range from 2 to 5 per cent per kilometer, a serious problem for long canals (Kus 1984: 413–14).

The prehistoric irrigation systems had no storage capacity of any significance other than the canals themselves (see Reservoirs below). The volume of water entering the system had to be controlled by sluice gates (*boca tomas*) at the canal intakes from stream sources. For abandoned canals, such gates have not survived but presumably consisted of boulders and wood as is common today. These devices, while crude, have been quite effective in regulating intake water. They are adaptable to the extreme seasonal fluctuations in Andean river flow ranging from raging torrents to small trickles. Traditional *boca tomas* are described for the present Jequetepeque Valley by Eling (1987*b*). In several ways they are more effective than modern cement intakes.

Velocity of moving water, which affects volume moved and also scouring (which can lead to collapse) and sedimentation (which reduces capacity), was controlled by gradient, sinuosity, and canal width, depth, and profile, which the Indian engineers seemed to understand (Park 1983: 161). Flow velocity on steep slopes was slowed in some canals by zig-zag courses (Ravines 1978*a*: 28; Fig. 8.5). Large canals in the Moche Valley range up to 2.7 m in width and 2 m in depth, with gradients of less than 1 : 500 (for profiles, see Fig. 8.6). For further details on canal hydraulics, including flow conditions (velocity, discharge) see Farrington (1980*a*), Farrington and Park (1978), and Park (1983).

Fig. 8.5. Canal cut in rock, with zig-zags to slow flow, near Cajamarca, Peruvian Andes. Photograph by James Kus

Construction 'A high degree of civil engineering still was necessary to construct and maintain such complex systems [Chimú canals]; knowledge of surveying and of open channel flow hydraulics was paramount' (Ortloff 1995: 55).

Canals follow contours as much as possible, and construction varied accordingly as terrain changed. Many canals were dug into the earth, at times stone-lined for stability to prevent collapse from undercutting. Some canals were cut into solid rock, as for example the Cumbemayo Canal in Cajamarca (Zegarra 1978: 112; Fig. 8.5) and the Huarancante Canal in the Colca Valley (Brooks 1998: 202). Actual tunnels, short in length, are rare. One example is the Pinampiro Canal in northern Ecuador, which has tunnels as much as 200 m long cut into a volcanic duripan to bypass unstable areas (Mothes 1986: 104–7, 119–27). Other canal sections are elevated in order to maintain a gradient, and on steep slopes the outer side of a canal may be embanked. Aqueducts, or elevated canals constructed of rock and/or earth, were built across large low places such as streambed crossings. Some are quite large. The Mampuesto Aqueduct on the Moche Canal is 9 m high and 600 m long (Park 1983: 162). The Ascope Aqueduct in the Chicama Valley is 15 m high and 1,400 m long (Kosok 1965: vii, 106; Sherbondy 1969: 118). An aqueduct in the Pampa de Zaña is about 3 km long and up to 8 m high in places (Kosok 1965: 137). There is a prehistoric aqueduct for irrigating terraces in the Colca Valley (Chapter 10), and small aqueducts occur elsewhere in the highlands. At Lake Titicaca, some aqueducts functioned not for irrigation but rather for drainage (Kolata 1993: 227–9).

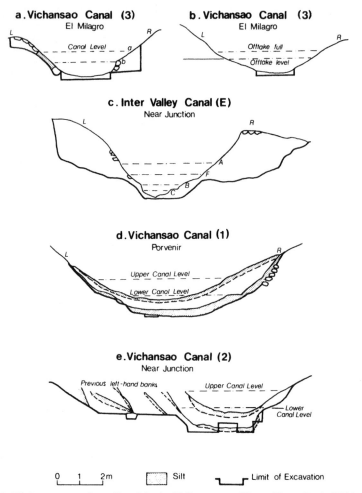

Fig. 8.6. Vichansao canal profiles, Moche Valley, coastal Peru. From Park (1983: 160)

Canals were not only regularly maintained to clean out silt and blowing sand and to repair collapses, but canal form was remodeled to improve efficiency (Park 1983: 163). Parallel channels are indicative of attempted construction of more efficient sections or routes, or abandonment of damaged sections.

Early canal systems were probably constructed piece by piece via gradual expansion, with considerable trial and error. Later, conscious planning of large systems is apparent, with objectives of supplying given amounts of water to specific field sites (Park 1983: 165). Water would have been introduced into canal segments during construction to check on gradient and other characteristics (flow testing) (Kus 1984: 409). Thus uphill gradients should have been readily detected.

A possible means by which Chimú engineers measured slope and controlled canal gradient is a ceramic bowl filled with water with a sighting tube. Such an apparent device was found in a museum in Huaraz in Peru and is described by Ortloff (1988: 104; 1995: 64–5).

Lower canals were built first. When it became necessary to irrigate more of the land above the first canal, a second canal, or several over time, was built at a higher level. The result, as in the Moche Valley, was a series of nested Vs open toward the valley mouths (Moseley and Deeds 1982: 31). The highest canals are referred to as 'maximum elevation canals'. Figure 8.7 shows the canal system in the Moche Valley.

Intervalley Canals These are canals which extend from one valley to another, thereby expanding cultivation in the recipient valley. They run along the slopes of bordering hills within a few hundred meters of sea level. The valleys on the Peruvian coast united in prehistoric times by such canals were the Rimac and Chillón, the Jequetepeque-Lambayeque-Motupe five-valley megasystem, and the Chicama and Moche described below (Netherly 1984: 238–9).

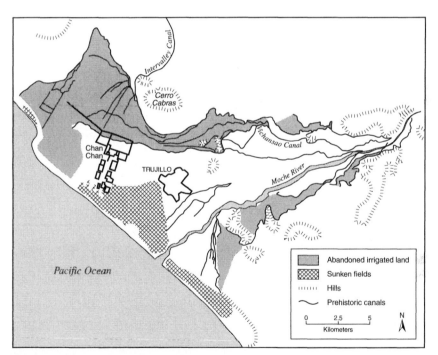

Fig. 8.7. Map of the Moche Valley canal system, coastal Peru, showing canals potentially fed by the Intervalley Canal (northwest) and canals fed by the Moche River (center). Portions of the white areas between canals are under modern irrigation. Also, sections of the sunken fields are still cultivated. Adapted from Ortloff *et al.* (1985: 79)

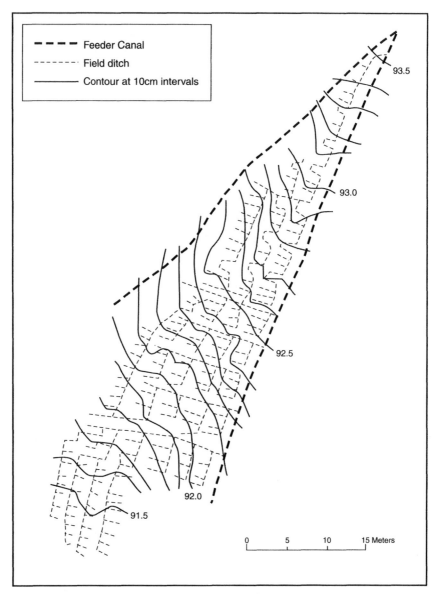

Fig. 8.8. Map of Chimú field with 'E'-type irrigation furrows, near Intervalley Canal, coastal Peru. Adapted from Kus (1972: 175)

Field Systems Water was distributed to individual crop plants on the coast either through furrow systems or by means of irrigation basins. Kus (1972: 167–85) mapped a number of sites in the Chicama and Moche valleys where feeder canals fed prehistoric furrow systems. He identified six patterns which he calls 'straight', 'interrupted straight', 'E' (Fig. 8.8), 'serpentine' (Fig. 8.9),

Fig. 8.9. Furrowed fields (often serpentine) and canals, Pisco Valley, coastal Peru. Photograph by Robert Shippee and George Johnson, 1931. American Museum of Natural History

'stone pile', and combined 'E-serpentine'. Furrow patterns seem to have varied with slope, soil, and type of crop. The E and serpentine furrows served to pond water, and also down cutting was minimized by the many turns. In the stone-pile fields, clusters of stones are spaced at 1–2 m intervals within straight furrows or in a few instances where there were no furrows. Their purpose is unclear.

Relic furrowed fields occur mostly on the north coast of Peru. One of the few sites on the south coast is in the Pisco Valley, shown on a photo taken by the 1931 Shippee-Johnson air-photo expedition (Bennett 1946: pl. 23) (Fig. 8.9). These have recently been destroyed (J. Kus, pers. comm.). Serpentine furrows (*caracoles*) have been described in the Azapa Valley of northern Chile (Rivera Díaz 1987: 232, 249). These are present-day furrows, but they were reported in nearby Arica in 1713 (Frezier, 1717: 152), and they are undoubtedly an old practice in the region. Antúnez de Mayolo (1986a: 176) mentions Inca zig-zag ditches (*wecu wecu*) which allowed water to settle in the soil.

Canal Collapse and Abandonment

There is a considerable and controversial literature on the causes of abandonment of prehistoric irrigation canals in Peru. Abandonment of Inca canal systems was related to the disruption by the Spanish Conquest, including collapse of hydraulic management and Indian population decline. Some canals continued to function into the colonial period and even to the present, in modified form. However, there was also pre-Inca abandonment. From their inception, canals were re-routed, enlarged, restored, and abandoned permanently.

Reasons given for abandonment include: tectonic uplift; El Niño mega floods destroying canal segments and causing canal down cutting and erosion; earthquakes and landslides; climatic change, particularly extended periods of drought; dune encroachment; re-routing to obtain greater efficiency or shorter routes; clogging with sediment; possibly salinization; and down-cutting which isolated the canal intakes (Nials *et al.* 1979; Moseley 1983; Moseley *et al.* 1983; Ortloff *et al.* 1985). Social breakdowns and warfare were also disruptive, but are difficult to document and correlate with canal abandonment.

The prehistoric canal systems of the Peruvian north coast are indeed impressive. In several valleys the extent of pre-Hispanic irrigation exceeded that of today, possibly by 15–20 per cent for the entire coast (Kosok 1965: 34). The 80 km Talambo Canal system in the Jequetepeque Valley once irrigated over 30,000 ha, but today the canal is one-third as long and only irrigates 3,477 ha (Eling 1987*b*: 178–9). In the Moche Valley only 12,833 (64 per cent) of the 20,139 ha irrigated in Chimú times was shown as cultivated on 1942 air photos (Moseley and Deeds 1982: 32–3, 52–3); however, by 1960, a very comparable 20,026 ha were in cultivation (Robinson 1964: 165). Differences between modern and pre-Hispanic areas of irrigation can be explained in part by a change in primary crops from less water-demanding maize and cotton in prehistoric times to more water-demanding sugar cane and rice today (Kosok 1965: 16; Netherly 1984: 236). However, not all prehistoric irrigation systems were in use at the same time. The modern calculations include land irrigated by well pumps, which inflates those figures. Also, considerable water now may be diverted from the water-supplying rivers for non-irrigation purposes. Nevertheless, the collapse of prehistoric water-supply systems that have yet to be restored or replaced must also be taken into consideration.

The Chicama-Moche Intervalley Canal

[The] Intervalley Canal system is a prodigious monument to indigenous engineering capabilities [. . . but, was it] built with uphill segments that never worked?

(Ortloff *et al.* 1983: 375)

The most spectacular and most studied prehistoric irrigation systems in South America are on the north coast of Peru. The largest (in length) of all is the Chicama-Moche (La Cumbre) Intervalley Canal, which was constructed by the Chimú state after AD 1000 in order to carry water from the Chimu Valley to the Moche Valley just south in order to expand irrigation there (Figs 8.7, 8.10). This canal is of particular interest, and worthy of brief, separate treatment here, because of its great size and because of the controversies over its functions, labor inputs, whether it was ever used, whether sections run uphill, and cause of abandonment. Was the canal an example of sophisticated engineering, a technological disaster, or both? The debate involves a group of scholars who were all associated with the Chan Chan-Moche Valley Project directed by Michael Moseley from 1969 to 1974 (Moseley and Day 1982) and the Programa Riego Antiguo Projecto directed by Moseley and Thomas and Shelia Pozorski from 1976 to 1979: archaeologists Moseley, Robert Feldman, Kent Day, and the Pozorskis; engineer Charles Ortloff; and geographers Ian Farrington and James Kus.

The basic reference on the Chicama-Moche Canal prior to the 1970s was Larco Hoyle (1945: 4), who referred to the '113 km long' La Cumbre Canal,

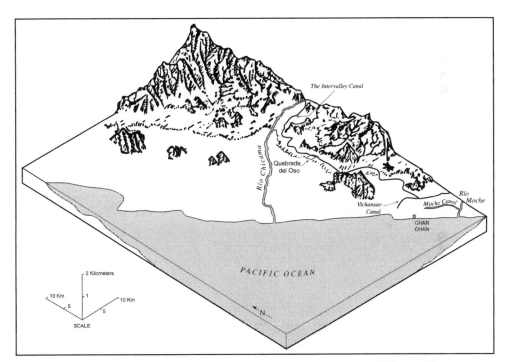

Fig. 8.10. Map of route of the Intervalley Canal, coastal Peru, from the Chicama River to the Vichansao Canal in the Moche Valley. The canal apparently functioned at least as far as the Quebrada del Oso but not to Chan Chan. Adapted from Kus (1984: 409)

which carried water from the Chicama Valley south to the pre-Inca city of Chan Chan in the Moche Valley. Kosok (1965: 94) traversed and mapped the route of the canal on foot, observing that the canal 'at times completely disappeared'. Much of the dissertation by Kus (1972), based on field work in 1969–70, is a detailed examination of the canal.

Ortloff *et al.* (1985: 78) and Ortloff (1988) have described the Intervalley Canal as a product of irrigation technology which evolved over a long time in a changing environment. The earliest canals were constructed in the form of unlined 'great trenches' cut straight through a sand dune landscape on the desert coast (Ortloff 1988: 101). These trenches were fed by inlets from the Chicama and Moche rivers. Irrigated 'breadbaskets' were thus created near Chan Chan, the Chimú capital, as well as the coastal region around the former Mochica capital of Moche.

These irrigation systems were eventually replaced by the construction of contour canals located upstream at the base of the Andes (Ortloff 1988: 103). The canals required more precise engineering and surveying skills as they hugged the contour lines of the landscape. Various associated structures had to be built, such as aqueducts across the *quebrada* (narrow dry channel) bottoms, and supportive terraces along the steep, rough Andean terrain in order to ensure even flow (Fig. 8.11).

Ortloff *et al.* (1985: 78) have categorized the major original trench canals as Class 1, having 'unlined, parabolic channel cross sections typical of the

Fig. 8.11. Section of the Intervalley Canal, coastal Peru. The terraces in the center foreground apparently served to shore up the canal on a steep slope, but they may also have been irrigated from the canal. Photograph by James Kus, *c.* 1970

equilibrium shapes produced by flow-induced, side-wall erosion'. However, after flooding destroyed a large proportion of these channel networks around AD 1100, large-scale reconstruction was undertaken by the Chimú, involving more sophisticated hydraulic engineering, channeling water with greater efficiency from the Moche River through the Vichansao canal (see Fig. 8.7). These have been designated as Class 2 canals, being 'masonry lined with cobbles set in adobe mortar and generally [having] a trapezoidal cross section' (the most efficient flow form).

The construction of the Intervalley Canal was undertaken by the Chimú so that water could be channeled from the Chicama River to the system of contour canals in the northern portion of the Moche Valley, in an effort to reclaim and expand land under cultivation in that valley. Construction was mainly between *c.* AD 1000 and 1300 (T. Pozorski 1987: 113, 116). The total length is 74 km to the juncture with the Vichansao Canal, not the 100 km or more claimed by earlier observers (Ortloff 1988: 103).

The Pozorskis (1982: 866–7) calculated that the Intervalley Canal required the excavation of $3,727,600 m^3$ of earth and stone and $148,000 m^3$ of masonry for lining canals and walling bank terraces. They estimated that full construction with a work force of 1,000 men working 300 days per year would have only taken twenty years or less. Ortloff *et al.* (1985: 96) calculated that it would have taken 5,000 men working 183 days per year 100 years to construct the canal. Kus (1984: 413), however, believes that 'we are not dealing with a single construction effort but rather with a discontinuous series of construction episodes . . . a cyclical project taking place over a period of several centuries', mainly during periods of severe drought.

Because of apparent coastal uplift, measurement of the canal by Kus (1972: 9) 'yielded impossible results: some sections of the canal appeared to go uphill!' Farrington (1983*a*: 373), however, found no sections running uphill, and he suggested faulty surveying of the canal gradient by the other scholars involved. He did measure channel cross-sections and calculated capacities and velocities and concluded that the canal was a 'technological disaster' capable of carrying only a small portion of its intended discharge to the Moche Valley (Farrington and Park 1978: 255). Ortloff (1988: 104), on the other hand, found that Moche Valley cross-sections and gradients approximate 'the shape modern engineering shows is the most hydraulically efficient'.

Ortloff *et al.* (1982) and the Pozorskis (1982: 857–8) both found that segments of the Intervalley Canal indeed do run uphill (one stretch gains 70 m) in both the northern (or upper) Chicama section and the southern (or lower) Moche section. Ortloff *et al.* (1982) explained this as the result of tectonic uplift subsequent to canal construction, resulting in abandonment of the canal. The Pozorskis, however, point out that in the two valleys 'only canals which never functioned have uphill slopes. Moreover, all canals that functioned in the past at present have downhill slopes' (Pozorski and Pozorski 1982: 857). They attribute the uphill segments to engineering error due to hasty

construction, with some later attempts at correction, and not because of tectonic uplift. Ortloff *et al.* (1983) subsequently defended their original position.

Was the Chicama-Moche canal never used? If the canal once carried water,
then this should be indicated by the presence of sediment and soil color
changes. From these criteria the Pozorskis (1982: 885) concluded that none of
the Intervalley Canal ever functioned. This seems certain for the lower canal,
south of the Quebrada del Oso (see Fig. 8.10), which is unlined and thus apparently never completed. Also, Kus (1984: 414) points out that if water loss due
to seepage and evaporation was as much as 3 per cent per kilometer, as it is
today, then 'less than one-tenth of the water entering the canal would have
reached the fields near Chan Chan'. Kus (1984: 409–10), in contrast to the
Pozorskis, believes that the upper canal, north of the Quebrada del Oso, was
utilized because of the presence of multiple channels and stone-lining; evidence of canal cleaning and thus functioning; the presence of irrigation-
deposited silt layers (attributed, however, by the Pozorskis to canal lining
washed off by rains); and agricultural fields along the upper canal that exhibit
elaboration of the furrow patterns which only occurred in fields that were actually irrigated. Kus (1984: 410) concludes that since there are upslope segments
in the upper canal which may have been in use, then there may have been some
post-construction tectonic uplift.

The evidence for use of the northern canal may be inconclusive, however,
and it is questionable whether the necessary magnitude of uplift could have
occurred in the short period of time involved or even over the 700 years following AD 1300. More research is needed on the northern canal, including the
points of argument by Kus. The recovery of organic evidence of the cultivation
of associated fields would be a positive indicator of canal use.

If the upper canal were in fact used, the uphill sections have to be the result
of subsequent tectonic uplift. It is curious, however, that both the upper and
lower canal systematically go uphill entering *quebradas* and downhill leaving
quebradas. What tectonic action could have resulted in this regularity? The
Pozorskis (1982: 858) believe that tectonic forces could not have created this
pattern, and that therefore it is the result of consistent survey error by the
Chimú. They feel that the ancient surveyors used *quebrada* bottoms as survey
reference points. Canals entering *quebradas* seem to go downhill when they
actually are going uphill because the reference *quebrada* floors themselves were
sloping uphill. On the other hand, Moche and Chimú irrigation engineers
demonstrated an ability to successfully design canals across *quebradas* elsewhere (Ortloff *et al.* 1983: 386).

Kus (1984) believes that the southern Intervalley Canal was never completed
because it was essentially a 'public relations' measure or 'political response'
during times of drought:

[I]t seems as if the Chimu elite tried to show that the Intervalley Canal would supply a
large amount of water to the north side of the Moche Valley by simply constructing a

large canal on the Moche Valley side of the divide—as if wishing for a large flow of water could make it so . . . a severe drought would have reduced the need for agricultural labor . . . the Intervalley Canal thus made sense in terms of absorbing surplus labor . . . stored agricultural supplies could have been 'paid' to the workers on the canal project, effectively distributing state goods to those in need . . . a year or two later, at the end of the drought, that same labor force was probably better used on state fields. . . . Decades, or even centuries, later during the next severe drought, the project was revived but again abandoned before completion.   (Kus 1984: 414)

In this interpretation, the failure of the lower canal was perhaps due to neither tectonics nor survey error; it was not really intended to succeed agronomically (and the engineers may have been aware of the problems of water loss involved), but rather to demonstrate state concern and action.

One of the last statements by one of the Chan Chan Project protagonists is that of Ortloff (1988: 103, 107), who says that: 'Although the Intervalley Canal certainly functioned as far south as Quebrada del Oso, [tectonic] distortions along its length led to its being abandoned before hookup with the Vichansao Canal could be accomplished.' He otherwise attributes abandonment of canals, amounting to a significant agrarian collapse in the Moche Valley, to down-cutting of the Moche River which cut off the upriver canal systems.[2]

To summarize: the southern portion of the Intervalley Canal never functioned, perhaps because it was meant as a 'make work' public project. It is still unclear whether the northern canal carried water or not, but it was elaborately developed with associated furrowed fields. If the northern canal did function, subsequent tectonic uplift has caused some stretches of it to run more than 10 m uphill. If it did not function, it may have been because of errors in slope measurement during construction. Tectonic uplift, poor engineering, or a fake canal? The Chicama-Moche Intervalley Canal, the greatest aboriginal irrigation project in the Americas, remains an enigma.

In this volume I examine prehistoric and contemporary indigenous cultivation systems that were ecologically sound and sustainable for long periods of time. Much of the Chicama-Moche Canal was a failure for reasons that were human or environmental or both. There were other failures of field systems in prehistory, but we know very little about them.

Other Water Management Features

Reservoirs

In Peru, reservoirs (*represas, pozas, estanques*) were not a major means of storing water and regulating water flow. Nevertheless, prehistoric reservoirs have been reported in many of the north coast valleys.[3] Reservoirs were excavated, embanked, or a combination of both. The never-used reservoir at the junction of the Chicama-Moche Canal and the Vichansao Canal in the Moche Valley was excavated 9 m into sand dunes, with stepped perimeter walls of cobble set in mud mortar. It is rectangular in shape, 138 by 40 m (Park 1983: 158). Most relic reservoirs are much smaller, especially those associated with terracing in the highlands. Reservoirs were fed by streams, canals, or directed runoff.

Some reservoirs were created by one or two earthen walls aligned with natural features such as *quebradas*. Natural ponds, including glacial lakes, were also used as reservoirs in the highlands, some in part dammed. Several are described by Ravines (1978*a*: 28–30). Tacto, for example, located in the headwaters of the Río Casta above Lima, is only 32,000 m^2 in size, with an earthen dam 1.5 m high and 29 m long.

Coursed, stone-lined Inca reservoirs that fed irrigated terraces at Raqay-Raqayniyog, Tipón, and Callachaca near Cuzco are described by Sherbondy (1982) and Niles (1987: 149–62). Guamán Poma depicts a stone-walled reservoir with irrigation water flowing to fields (Fig. 8.12). Brooks (1998: 197–8) describes small pre-Hispanic stone-lined reservoirs in the Colca Valley.

Diversion Embankments

Earthen walls designed to guide floodwaters or runoff into canals, reservoirs, or field plots are more common in the Southwest United States than in South America. They do occur, however, in some of the Peruvian coastal valleys, such as the Virú (M. West 1981: 65).

In the Chilca Valley on the south coast of Peru, raised and depressed water management features (used from *c*. AD 800–1150 to the present) have been interpreted in the past as sunken fields which tapped subsurface moisture (see below). Knapp (1982), however, interprets most of the irregular walls there as being flood diversion embankments for irrigation rather than spoil heaps from excavated depressions. Evidence for flood diversion includes embankment morphology and topographic position, indications of recent frequent flooding, and the presence of floodwater farming practices in the region today.

The Chilca site consists of *c*. 5.0 km^2 of embankments and basins currently accessible to periodic flooding by the Río Chilca. There are about 100 linear and complex embankments 50 to 1,000 m long, 10 to 30 m wide, and 5 m high (Fig. 8.13). They often create semi-enclosed basins which could have trapped floodwaters. The floors of the basins are at the same level as surrounding terrain, suggesting that they are not sunken fields. The amount of earth in the embankments could have come from only 60 cm of excavation which has since been leveled by flood sediments. The linear nature of many of the ridges parallel to the beach suggests that the embankments (full of cultural garbage) could have been built on top of pre-existing natural beach ridges.

This interpretation has proved controversial (R. Smith 1983; Moseley 1983; Moseley and Feldman 1984), but Knapp (1983) remains convinced of its

Fig. 8.12. Bordered gardens irrigated by a stone-lined reservoir, *c.* 1615, coastal Peru. An alternative explanation is that the double line grids are irrigation canals. From Guamán Poma de Ayala (1980: 1059)

correctness. Moseley (1983) has argued that tectonic uplift rendered earlier sunken-field systems inoperable and led to the current dominance of flood-water farming. Tectonic uplift would not, however, have altered the morphology of the embankments (which clearly reflect flood use rather than water

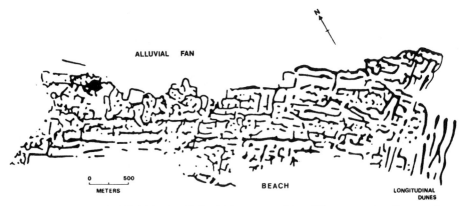

Fig. 8.13. Embanked fields in the Chilca Valley, coastal Peru. These served to divert and trap floodwaters and/or as sunken (water-table) fields. Adapted from Knapp (1982: 146)

exclusion), the flood frequency of the Chilca *quebrada*, or the position of the Chilca embankments at the foot of the *quebrada*. If significant uplift had occurred, of course, the embankment system would now be located at a higher elevation and significantly inland from the beach. Such is not the case.

Prehistoric embankments also occur in highland basins, such as Titicaca (C. Smith *et al.* 1968: 357) and Cayambe in Ecuador (Batchelor 1980: 678–82) where they are associated with raised fields and probably had both drainage and irrigation functions. On the Bolivian *altiplano*, the Aymara constructed long dikes to divert floodwaters around areas to be cultivated (Soria Lens 1954: 91).

Albarradas

A unique form of embankment is the *albarrada*, found on the arid south coast of Ecuador and north coast of Peru. These are crescent or U-shaped, hand-constructed ridges which trap rainwater and runoff water which is used today for crops, pasture, and domestic use. Those on the Santa Elena Peninsula of Ecuador (Fig. 8.14), numbering about 200 used and abandoned ridges, have been described by McDougle (1967), Stothert (1995), and Bogin (1979: 25). These *albarradas* are 10–30 m long, 5–20 m wide, and up to 5–6 m deep. They slope downward from the mouth of the U. Excavation of cultural material from the ridges indicates that they date to as early as 900–100 BC. Antúnez de Mayolo (1986*a*: 176) reports similar 'half-moon' features from Lambayeque to Tacna in Peru, some as long as 100 m.

Protective Dikes

Another form of embankment in Peru consists of walls designed to protect canals and fields from flash floods and slope runoff. Horkheimer (1990: 138–9) describes such a wall at Mazo in the Huara Valley and others in the Chancay

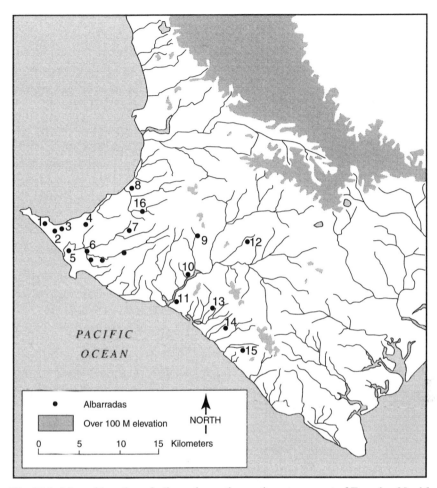

Fig. 8.14. Map of location of *albarradas* on the southwestern coast of Ecuador. No. 16 is near the town of Santa Elena; No. 3 is near La Libertad; No. 1 is near Salinas. Adapted from Stothert (1995: 136)

Valley. Others have been reported on the north bank of the Río La Leche (Shimada 1982: 177) and on the Cañete River (Regal 1970: 79, quoting Antonio Raimondi in 1862). Squier (1877: 203) noted parallel stone walls on slopes of the Nepeña Valley which may have protected canals and fields from landslides. Other walls along coastal valleys, such as the Virú, may have had defensive functions (Willey 1953: 363–4).

Filtration Galleries

These are underground canals (*puquios*), mainly found in northern Chile and southern Peru, similar to the *qanats* of the dry lands of the Old World, including Spain. They are nearly horizontal tunnels dug underground from a higher groundwater table, or other water source, to a reservoir or to the fields to be irrigated. Most include vertical shafts to facilitate construction and repair and to provide ventilation. They may or may not be rock-lined or cut in rock. Their purpose seems to have been to reduce evaporation and to provide a means of controlling gradient. They occur in at least twenty-four coastal and Andean sites, including the Nasca, Moquegua, Ica, Pisco, Rimac, and Santa valleys, the Pampa de Tamarugal in Chile, Paucartambo near Cuzco, and Potosí in Bolivia (for mine drainage) (Barnes and Fleming 1991: 51).

Carl Troll (1963) believed that the subsurface canals in South America were indigenous in origin, as do Schreiber and Lancho Rojas (1995) more recently for Nasca. However, following a careful literature and field study, Barnes and Fleming (1991) argue that most or all are of Spanish origin, the concept being brought to Peru in the mid sixteenth century. Lack of early mention in the colonial literature, however, does not necessarily mean that filtration galleries were not prehistoric. Raised fields and certain other clearly prehistoric agricultural features generally were not mentioned in the early accounts. Clarkson and Dorn (1995) obtained controversial dates of AD 560 to AD 600 from rock varnish (accumulations of oxides) on stone lintels of *puquios* at Nazca.

Bordered Gardens

Rectangular cultivation plots (bordered gardens or irrigation pens or basins) occur on the Peruvian coast. A drawing by Guamán Poma apparently depicts such plots (see Fig. 8.12). In the north, abandoned bordered plots in the Virú Valley vary in size from 2.5 by 3.5 m to 30 by 40 m (Willey 1953: 364–7, pl. 54). They are edged by ridges of rock and apparently were irrigated by canals by 'turning water in from the top and letting it gradually fill the numerous rectangles'. It was suggested to Willey by Richard Schaedel that some of these in the Santa Valley may be the remains of colonial rice paddies, but Willey (1953: 367) believed those in the Virú Valley were probably prehistoric. Kosok (1965: 208–9, 214) shows photos of similar features in the Santa and Nepeña valleys.

Hatch (1976: 65–8) describes current small-farm irrigation using earth-bordered (*bordos*) basins (*posas*) in the Motupe Valley (Lambayeque). *Posa* irrigation requires less frequent water applications than furrow irrigation and cuts water use in half. The soil in a *posa* remains humid for two or three weeks after a deep soaking (10–15 cm depth). Gates (*bocas*) allow water entry. In the Azapa Valley of northern Chile, bordered gardens called *melgas* are fed water via canal (Rivera Díaz 1987).

On the south coast, squares with earth and rock borders, called *canteros*, are abandoned in the Quebrada de San Juan (Fig. 8.15), but are still cultivated (prickly pear) in the Chilca Valley (Fig. 8.16) (Núñez Jiménez 1987: 2: 471–6). There are some 1,500 stone-lined pits in *lomas*, or fog oases (see below) at

Fig. 8.15. Abandoned bordered gardens (*canteros*) in the Quebrada de San Juan, coastal Peru. The rock borders may have served as lithic mulches to conserve moisture. From Núñez Jiménez (1987: 2: 474)

Chilca, dating from 100 BC to AD 200; stone strips, mounds, and ridges in north-western Argentina date to AD 300–700 (Lightfoot 1994: 173–4). Some of these may have served as lithic (rock) mulches for conserving moisture.

Sunken Fields: *Hoyas*

Sunken fields or gardens might be considered the opposite of raised fields. Instead of artificially raised surfaces to facilitate drainage, depressions are dug to collect water for crops. There are two main forms. The coastal sunken fields (*hoyas*) are in normally dry river beds where large pits have been dug to the ground water level to tap moisture for cultivation. The highland sunken fields or pond fields (*qochas*), in contrast, are artificial depressions dug to trap rainwater.

Hoyas have also been called *hoyadas, mahamaes, wachaques, chacras hundidas, puquios* or *pukios*, and *canchones*. A good descriptive term is water-table farming (Soldi 1982: 19). They date to prehistoric times, at least AD 800 at Chan Chan (see Fig. 8.7) (Moseley and Feldman 1984: 406). They have also been found in the Chicama, Pisco, Moche, Ica, Virú, and other valleys of coastal Peru. In northern Chile they occur in the Pampa de Tamarugal (Soldi 1982:

Fig. 8.16. Bordered gardens in cultivation (prickly pear) in 1977 in the Chilca Valley, coastal Peru. These are watered by occasional floods in the wet season and by container irrigation via burro in the dry season. From Núñez Jiménez (1987: 2: 476)

67–71), and in the northern highlands of Ecuador they occur near Laguna de Colta (Knapp 1991: 69). M. West (1979: 139) dates the Virú fields to about 100 BC.

Several Peruvian chroniclers described *hoyas* (Rowe 1969), and they have been well known ever since. They have received attention from archaeologists J. R. Parsons (1968), Rowe (1969), and Moseley and Feldman (1984); from geographers Dagodag and Klee (1973), Psuty (Parsons and Psuty 1975), R. Smith (1979; 1983), Knapp (1982; 1983); and from historian Ana María Soldi (1982) in a comprehensive monograph.

One of the first descriptions of *hoyas* is by Cieza de León (1959: 337) in 1553: 'It is an amazing thing to hear what they do in this valley [Chilca]. To provide the needed humidity, the Indians dig wide, deep holes [*hoyas*] in which they sow and plant the things I have described, and with the dew and dampness God makes them to grow.'

Hoyas vary considerably in size. Those at the mouth of the Virú Valley measure from 30 by 30 m to 50 by 100 m. They are about 1 m below the natural ground surface, with the removed soil forming embankments 2–4 m high (Willey 1953: 16–17). The *wachaques* at Chan Chan are up to 8 m or more deep (Moseley and Feldman 1984: 404). Salinization apparently was a problem in prehistoric sunken fields (Rowe 1969: 322–3).

Sunken fields still in use on the Peruvian coast include: (1) pits for reed cultivation (Knapp 1979: 85–7), (2) cultivation using deep trenches, especially for annuals (Knapp 1979: 81–7), and (3) cultivation in small pits (*c.* 1 m^2 by 1–2 m deep) (*huecos*), especially for perennials such as figs (Knapp 1979: 71–81). Occasionally natural topographic lows (interdune areas) have been expanded for fig cultivation in *huecos* (Knapp 1979: 65–72). Very large sunken fields in the interior of the Pisco and Ica Valleys are apparently not prehistoric, but rather date to later times when they were used for vineyards and date palm production by *haciendas* (Parsons and Psuty 1975: 264; Rowe 1969: 323).

Another form of sunken field is the trench-shaped *canchón* of northern Chile (Field 1966: 157–60), which is 1–2 m deep and 20 m or so long. The cropping surfaces are 3–4 m above groundwater, but capillary action provides moisture to the crop roots. Abandoned *canchones* (a probable result of salinization) are common, but their age is unknown.

The best known *hoya* site is in the Chilca Valley. Knapp (1982; 1983), however, argues that the Chilca features are actually water-control embankments rather than artificial depressions (see 'Diversion embankments' above). Possibly both embankments and *hoyas* are present.

Sunken Fields: *Qochas* (*Cochas*)

Qochas or pond fields are excavated pits which fill with rainwater. They were reported by anthropology students at the Universidad Nacional in Cuzco, and have since been described by Jorge Flores Ochoa (1987). They are only known on or near the Lake Titicaca plain, at elevations around 3,850 m. The largest numbers are located to the north in the region of Pucara. Others have been reported at Ayaviri west of Pucara, at Sillustani west of Puno, and on the Copacabana Peninsula on the Bolivian side of the lake.

The sides and bottoms of *qochas* are progressively cultivated as the water in them recedes. Some are systematically fed by water via canals. Many are lined with *huachos* (narrow beds) which provide drainage for waterlogged soil. Potatoes are the most common crop; other crops include ulluco, oca, cañihua, quinoa, añu, and barley. *Qochas* are also utilized by livestock for pasture and as waterholes, and by local people for domestic purposes.

Qochas vary in shape from circular to oblong to rectangular. The average depth is 2 m, with some up to 6 m. They average 90–150 m in diameter, with some over 200 m. At Pucara, abandoned *qochas* occur within a sector of 128 km^2 and cultivated *qochas* occur within a sector of 256 km^2 (Flores Ochoa 1987: 283). Densities for both sectors range from 80 to 100 per km^2. Using the lower density for both gives a total of about 31,000 *qochas*.

When viewed from the air, *qochas* form a striking, pocked landscape (Fig. 8.17). While many are cultivated today, little is known about their construction, ecology, and history. Their origins are probably local and prehistoric.

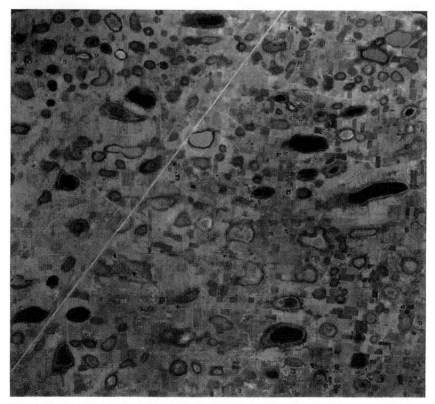

Fig. 8.17. Air photo of *cochas* (sunken fields) near Taraco, northern Lake Titicaca, Peru. Servicio Aerofotográfico Nacional, Lima, 1986

In the province of Espinar in southern Cuzco, there is a form of mini-*qocha* (*qochawiña*) (Orlove 1977: 95–6). These are only about 7–10 m² in area and 1–2 cm deep and are easily dug. They retain moisture from rainfall longer than adjacent soil and thus protect crops and pasture from short dry spells and extend moisture for a week or two after rain.

Seepage

Hoyas are one form of seepage system, although designed to utilize subsurface water. In the Andes surface seepage (springs) on slopes may be directed to canals and to terraces, as for example in the Colca Valley (Chapter 9). Molina (1968: 67) in 1542 said that (in Peru): 'And where there is no river, they have springs with which they irrigate their lands and gardens.' There is a prehistoric site at Ramaditas, Tarapaca, in the Pampa Tamarugal in the Atacama Desert of northern Chile, dating to *c.* 300 BC, in which there is a canal and ditch system with associated diversion walls and wells, which Shea and Rivera (1995–6) call 'reticulate irrigation'. The ditched fields are 30–40 cm deep and are positioned on gentle slopes. The sources of water are seepage from aquifers and canals from *quebradas* on a 2 per cent slope.

Floodwater Farming

Floodwater farming ('self-watering') refers to the cultivation of land watered by periodic river overflow. This is not true irrigation in that water is not artificially transferred; however, overflow water may be diverted or trapped by embankments, as in the Chilca and Virú valleys (Knapp 1982; M. West 1981: 65). The land to be cultivated may require clearing of riparian vegetation. Thus floodwater systems may involve some special preparations. Ravines (1978b: 94) believes that on the coast during the pre-canal period, 3000–1800 BC, floodwater farming was the earliest form of cultivation (also M. West 1981: 64–5). It undoubtedly persisted along with or independent of canal irrigation, and it continues today, as in the Virú Valley (M. West 1981: 65). A single flood may be sufficient to water one crop, but there is a risk of crop loss from unexpected subsequent floods after planting (Netherly 1984: 235–6).

Flood-farmed sectors are called *bañados* in the Andes. Early descriptions for the coast seem to be lacking, but there are some sixteenth-century mentions for northwestern Argentina: 'They [Indians of Córdoba] are great farmers, for there is no remote place with water or flooded land [*tierra bañada*] which they do not plant' (Luis de Cabrera 1965: 388). 'The rivers of this province, especially the Esteco and Santiago del Estero, are in the winter like the Nile: they leave their beds and extend themselves across the plains watering lands which are called *bañados* . . . they plow and sow in them' (Lizárraga 1968: 185). [Tucumán] 'is a land very abundant in food, because they get [it] from *temporal* [rainfall fields], irrigation, and *bañados*' (Sotelo Narváez 1965: 393).

The term *temporales* is used today to refer to coastal valleys or *quebradas* which seldom carry water, but when they do people divert the water on to land which can be cultivated (Kosok 1965: 118). Antúnez de Mayolo (1986b: 176) describes ditches dug in the shape of half-moon depressions which fill with runoff from occasional rains and are then planted on the north coast of Peru.

Complex Systems

In Bolivia near Desaguadero at Lake Titicaca, there is a complex water management system involving canals, drainage ditches, tributaries, and the Río Desaguadero for the varied purposes of irrigation, flood reduction, controlled siltation, and reduction of salinity by diverting saline tributaries (Soria Lens 1954: 88–9). This is also an area of raised fields. Such complex, multifunctional systems are probably not unique and need to be identified, described, and understood.

Lomas Cultivation (Fog Oases)

The heavy fog (*garúa*) which hangs over the arid central and southern coast of Peru during the winter is capable of producing moisture from condensation, especially where impinging on the slopes of outlying hills of the Andes at *c.* 200–800 m elevation (Engel 1973). The resulting vegetation, referred to as *lomas*, consists of grasses, scrub, and even low forest. This vegetation has been of economic importance over time for hunting, pasture, wood, and agriculture. Extent and location have changed with climatic change and coastal uplift. Grassy *lomas* trap up to 150 mm of condensation, and tree-covered *lomas* can capture up to 900 mm (Benfer *et al.* 1987: 198). Today, most of the *lomas* have been considerably degraded by fuel-wood cutting and overgrazing. There is physical evidence, however, of prehistoric farming of *lomas*, utilizing canals, terraces, diversion walls, and bordered gardens.

Vázquez de Espinosa (1942: 495) in 1628 and Bernabé Cobo (1956: 1: 88–9) in 1653 described *lomas* cultivation at Chala, Atico, and Atiquipa on the south coast of Peru—gardens, orchards, and pastures. Cultivation, now mostly abandoned, was based directly on *garúa* moisture as well as on water from wells and springs.

Benfer and colleagues (1987) and Engel (1973) describe early *lomas* cultivation in the middle Chilca Valley. *Melgas*, a type of bordered garden, are small, rectangular plots with raised rock borders which trap and retain water condensed from *garúa*. Some are still in use. (These may be identical to the *canteros* described earlier.) *Maceteros* are stone-lined pits, about 1 m deep, often in low spots where moisture collects. The stones may have served to collect condensation as well as to reduce evaporation. Stone-wall terraces are also present and probably functioned to slow runoff and sediment from infrequent slope wash, possibly at a time when precipitation was greater than the present 25 mm/yr average.

In the desert *lomas* of Atiquipa and Chala on the coast in southern Peru, there are abandoned terraces and canals which apparently were fed from *quebradas* in which *lomas* moisture accumulated. In associated deposits there have been found remains of maize, cotton, squash, and quiñoa along with *guano* and llama and guinea pig manure (Canziani 1995: 116–17, 129; Engel 1973: 278).

In conclusion, pre-Hispanic people developed complex, diverse, and successful systems for efficiently managing limited amounts of water in the arid regions of western South America. Today, we may have superior canals, associated structures, and machinery for moving water; however, much more is involved in irrigation and water conservation than physical features. There is also the knowledge of where to locate the canals, appropriate dimensions and gradients, the control of velocity, volume, sediment accumulation, salinization, evaporation, and the reliability of delivery, and how to adjust to both environmental change and population growth. These types of know-ledge rivaled or surpassed those of today's irrigation engineers (Schaedel 1986: 223).

Notes

1. Since maturation time is shorter at lower elevations, rain-watered crops can often be obtained in lowland areas with somewhat less than 500 mm/year precipitation.

2. For elaboration of this thesis, see Ortloff *et al.* 1985; Moseley (1983), and Moseley *et al.* (1983). Another possible factor contributing to canal abandonment was El Niño flooding (especially *c.* AD 1100) which destroyed canal systems and also increased down-cutting thus stranding canal intakes (Moseley *et al.* 1983; Nials *et al.* 1979).

3. Moche (Park 1983: 158); Casma (Regal 1970: 84, quoting Julio Tello); Chancay (Horkheimer 1990: 139); Rimac (Regal 1970: 71, quoting Villar Córdova); Chira, Piura, Nepeña, Pacasmayu, Huarmay (Latcham 1936: 276); Chillón (Rostworoski 1972: 256); Jequetepeque (Eling 1986: 141).

9

Terraced Fields

> They also leveled the land by making terraces that they call *pata* on the hillsides . . . the Indians planted all the hills, even the highest and steepest which would have been too rough to cultivate without terracing. As we see these hills today they look as if they are covered with flights of stairs.
>
> (Bernabé Cobo [1653] 1990: 212–13)

In contrast to irrigation, which has received considerable research but no comprehensive synthesis, terracing in the Andes has been little studied until recently. Much of what is known is surveyed in a magnificent volume on New World terracing by Cambridge geographer Robin Donkin (1979). Donkin provides information on regional distributions of terracing, dating, classification, and functions, with numerous maps and photos and an impressive bibliography. Thus a lengthy review is not needed here.

Terraces have been a highly visible, defining characteristic of Andean cultural landscapes since well back into pre-Inca times. There may be a connection between the Spanish term '*andenes*' for bench terraces and 'Andes'[1]. The Inca improved and expanded terracing, in part to increase production of maize, which was of ceremonial significance to them (Guillet 1987*b*: 409). Most of the early chroniclers were impressed by the terracing they saw and so commented, as have many travelers, tourists, and scholars subsequently. Most of this information, however, has been superficial and descriptive. There were few examinations of terrace morphology, construction, chronology, functions, productivity, ecology, and origins until recently. Few terraces had been excavated; few had been mapped. When we began our study of the Colca Valley terraces in southern Peru in 1984, I was surprised to learn that more was known about raised fields, which were only recently discovered, than terraces which had been observed, photographed, and admired for so long. We were ignorant not only about prehistoric terraces but also about the terraces being farmed today which continue to be an important component of Andean agriculture.

Early Descriptions

Apparently the first mention of Peruvian terraces was in 1535 by Pedro Sancho de la Hoz (1917: 149): 'all the mountain fields are made like stairways of stone.'

Other early reports are by Cieza de León in 1550, Pedro Pizarro in 1571, Ortiz de Zúñiga in 1562, Sarmiento de Gamboa in 1572, Viceroy Francisco de Toledo in 1575, Ulloa Mogollón in 1588, Baltasar Ramírez in 1597, and Garcilaso de la Vega in 1604. Their observations and others are cited and quoted by Donkin (1979: 20–1). Terraces elsewhere in the Andes received less early attention.

An example of an early description, and an indication of expansion of terracing under Inca rule, comes from Sarmiento de Gamboa (1907: 98). He said that Pachacuti Inca Yupanqui (ruled *c*. 1438–71) 'supplied by art what was wanting in nature. Along the skirts of the hills near villages, and also in other parts, he constructed very long terraces of 200 paces or less, and 20 to 30 wide, faced with masonry, and filled with earth, much of it brought from a distance. We call these terraces *andenes*, the native name being *sucres*. He ordered that they should be sown, and in this way he made a vast increase in the cultivated land' (trans. by Donkin 1979: 20).

Later travelers often mentioned terraces, but the person who especially made the general public aware of the drama of the Peruvian terraces was the botanist O. F. Cook in a 60-page, well-illustrated article in *National Geographic* in 1916 entitled 'Staircase Farms of the Ancients'. Cook had been a member of Hiram Bingham's 1915 expedition to southern Peru. In particular, Cook described the terraces in the vicinity of Cuzco, Machu Picchu, Pisac, Ollantaytambo, and the Vilcanota and Urubamba rivers. Fifteen years later, in 1931, the Robert Shippee-George Johnson aerial survey of Peru provided even more spectacular photography of terraces near Cuzco, Arequipa, the Colca Valley, and elsewhere (Denevan 1993). The results were written up in widely read articles in *National Geographic* (Shippee 1933; 1934) and *The Geographical Review* (Shippee 1932*a*; 1932*b*). Shippee toured the United States giving slide lectures and showing his film *Wings Over the Andes*. The photography is excellent, and the black and white aerial prints continue to be used and published by scholars.

Pre-Donkin monographs and research articles on Andean terraces are few. They include the doctoral dissertation by Chris Field (1966) on the terraces of northern Chile and northwestern Argentina. This is based on a detailed field survey of numerous abandoned and still cultivated terrace sites. These terraces are generally of lesser technical accomplishment and extent than those in southern Peru. The study is descriptive, with no excavations, but it is nevertheless very useful and merits greater attention than it has received. Terraces were mapped in the Machu Picchu region by the Vilcabamba Expedition (Fejos 1944). Bonavia (1967–8) is one of the few Peruvian archaeologists to excavate terraces (Mantaro Valley). Some excavation was done earlier (1945) of terraces in the Rimac Valley by Maldonado and Gamarra Dulanto (1978). Ravines and Solar la Cruz (1980) describe terraces at Pisac near Cuzco. A. Wright (1963) studied terrace soils in northern Chile. Broadbent (1964) briefly describes abandoned terraces in Chibcha regions of highland Colombia. Swanson (1955) published a brief survey of Peruvian terraces, with few details.

Recent Studies

Donkin's 1979 monograph certainly was a stimulus to terrace research and for more systematic local study. At the same time, the growth of an ecological perspective in archaeology, ethnology, and cultural geography and a greatly increased interest in prehistoric cultivation contributed to a boom in terrace studies. Major microregional studies have been undertaken at Moray near Cuzco (Earls 1989) and in the Colca Valley at Coporaque (Treacy 1989*a*; 1994*b*) and the Río Japo Basin (Brooks 1998). The Cusichaca Project in the middle Urubamba Valley mapped, excavated, and restored terraces and associated canals (Keeley 1984; 1985; Farrington 1980*b*; Kendall 1992). Keeley (1988) and Keeley and Meddens (1993) also studied pre-Hispanic terrace soils in the Río Salado Basin in northern Chile and in the Chicha-Soras Valley in southern Peru. Earls and Silverblatt (1981) discuss experimental temperature control as a function of the circular Inca terraces at Moray. The Moquegua Valley Project included an examination of prehistoric terraces and canals (Stanish 1987). Sherbondy (1982; 1987) describes terraces in association with Inca irrigation at Cuzco, Niles (1999: 208–31) provides a detailed description of the terraces of Inca Huayna Capac at Yucay (Fig. 14.5). Studies of still functioning indigenous terraces are mostly brief and descriptive. The Colca Valley Project, which was specifically on abandoned terraces (Denevan 1987*a*), is discussed in Chapter 10. Guillet (1987*b*; 1992) also worked on Colca terracing and irrigation.

Chronology

All ancient agricultural earthworks are difficult to date as to time of construction, period of use, and abandonment, given disturbance by both erosion and deposition, by the process of agriculture, and by frequent rebuilding. This is especially true of terraces. The fill behind terrace walls may contain datable material (ceramics, bone, carbon) which could be either younger or older than the terrace walls. Cultural material in a terrace wall can be dated, but terrace walls collapse frequently and are rebuilt. Thus a terrace wall can have segments of different ages. Type of terrace and form of construction can provide a relative age. Dating can be done when a terrace is in close association with dated ruins. For example, at Coporaque in the Colca Valley we found terrace walls forming one side of a house wall, indicating contemporaneous construction. Also, soils (organic A horizons) beneath terrace walls can be dated, providing a possible date of construction or at least a maximum date.

Willey (1971: 131) believed that major Andean terracing began *c*. 200 BC–AD 600. Terraces at Tafí del Valle in northwestern Argentina have been dated at AD 300–400 (González and Núñez Regueiro 1962: 493, 495). However, Brooks (1998: 270–1, 282–3, 286) has more recently dated a soil sample from the fill (at

34 cm below the surface) of an unirrigated bench terrace to 2480–2320 BC in the Río Japo Basin in the Colca Valley. Also, Grieder *et al.* (1988: 2, 138, 144) have identified irrigated terraces dating from sometime between 2400 and 1395 BC at La Galgada in the western, central Peruvian Andes. The Río Japo Basin and La Galgada terraces may be the oldest known terraces in South America. However, unirrigated sloping field and cross-channel terraces are probably the earliest terraces (Donkin 1979: 32; A. Wright 1963: 66, 68; Keeley 1988: 195; Brooks 1998: 379).

Distribution

Donkin (1979: 24) has mapped the general distribution of terraces (in use and abandoned) in South America (Fig. 9.1). In addition, he has ten regional maps which show specific sites where terraces were observed by him, were reported in the literature, or appear on air photos. In the Andes, terracing extends discontinuously from Tuñame (near Trujillo) in Venezuela to Salamanca (north of Valparaiso) in Chile and Ancasti (near Catamarca) in northwestern Argentina. Terraces are heavily concentrated in the hills and mountains within or bordering the *altiplano* of southern Peru and northern Bolivia, in the Cuzco region, and in many of the western valleys of central and southern Peru. They are much more localized south of the *altiplano* and from northern Peru on north. Brooks (1998: 128–9) gives general locations and elevations for different terrace types in Peru.

On the eastern slopes (*montaña*) of the Peruvian Andes there are large sectors of rainfed terraces, usually abandoned. Most are under forest and not readily visible; they are unmapped and poorly known. These regions include the tributaries of the upper Río Inambari north of Lake Titicaca, the Chanchamayo Basin, the Río Abiseo-Gran Pajatén region, the Huanuco Basin, the Río Mantaro, some of the Bolivian *yungas*, and apparently sites in southern Ecuador (Isbell 1968; Lathrap 1970: 171–9; Lennon *et al.* 1989; Bonavia 1967–8; and Donkin 1979: 122–5).

As Donkin (1979: 22) emphasizes, most New World terraces are located in arid and semi-arid regions and hence are irrigated. He estimates that 95 per cent of known terracing occurs in drylands. Some of these terraces are entirely irrigated, especially on the western Andean flanks, whereas for many others irrigation supplements natural rainfall, as in the Colca Valley (Chapter 10). Terraces on the eastern slopes are mostly rainfed. Since these are extensive, a higher portion of central Andean terraces are rainfed than elsewhere.

In Peru the Oficina Nacional de Evaluación de Recursos Naturales (ONERN, 1987) has partially measured the extent of terracing using air photos. Available totals include 74,750 ha for the southern departments of Arequipa, Puno, Tacna, and Moquegua. For all of Peru, ecologist Luis Masson

Fig. 9.1. Map of terracing in South America, with southern boundary of agriculture, *c.* AD 1500. Adapted from Donkin (1979: 23)

(1986: 208) originally estimated a total of one million ha of terracing, but he later reduced this to between 500,000 and 600,000 ha, which is more realistic (Masson, pers. comm. 1988); however, he did not take into consideration large expanses of forest-covered terraces on the eastern slopes.

Actual detailed mapping of terrace distribution is rare anywhere. The best example is from the Ifugao terrace project in Luzon in the Philippines, which produced magnificent color maps at a 1 : 10,000 scale matched to 4 m contour interval maps (Conklin 1980). For the Colca Valley, we mapped cultivated and abandoned terraces from air photos at an original scale of 1 : 17,000 (Denevan 1988).

Types of Terraces

Terraces have been classified by J. Spencer and Hale (1961), Treacy and Denevan (1994: 96–101), and others, based primarily on field morphology (form). The four main types are check dams and cross-channel terraces, sloping-field terraces, bench terraces, and broad-field terraces (Fig. 9.2). The first two involve natural or controlled runoff (water harvesting), and the

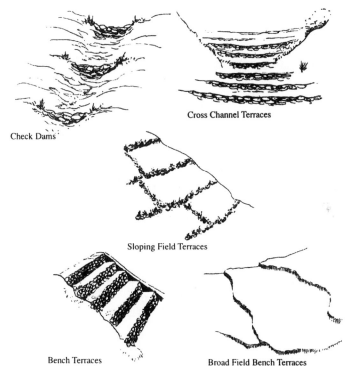

Check Dams

Cross Channel Terraces

Sloping Field Terraces

Bench Terraces

Broad Field Bench Terraces

Fig. 9.2. Examples of terrace types. From Brooks (1998: 2)

second two usually involve irrigation. Water harvesting requires a constant state of alert in order to manage runoff and thus favors dispersed settlement with houses near fields.

Cross-channel Terraces and Check Dams

These have retainer walls (*c.* 0.5–2.0 m high), usually of stone, built across (at right angles to) narrow valleys or *quebradas* with intermittent streams. The walls are backed by soil fills with cropping surfaces which have been partially or fully flattened from the accumulation of sediment behind the walls. The primary functions are: (1) erosion control and soil accumulation, and (2) water catchment, settling, and spreading. Water drains into the terrace area from the surrounding slopes as well as from the original intermittent stream, which essentially has been destroyed by a series of cross-channel terraces. Where water flow is infrequent, the primary function of the retainer wall is to prevent flood damage and erosion and/or to accumulate soil and flatten the slope, rather than floodwater farming, and the fields may be dry farmed most of the time. Or, fields may be farmed only after an infrequent flash flood.

Check dams are the most common form of floodwater control. They are very similar to cross-channel terraces, although often shorter; the term 'check dam' emphasizes the management and conservation of water functions, whereas 'cross-channel terrace' (silt-trap terrace) emphasizes the soil management functions. The two terms are often interchanged. With a new wall, there may be little or no soil accumulation or slope flattening initially, but neither may be necessary for cultivation. Water is concentrated from large catchment areas on to small fields in both types.[2]

Cross-channel terrace walls and check dams of stone usually do not contain water drop structures. When built of earth, however, they may have step-like stone drops so the water can pass to the next lower field without eroding the wall. After abandonment, the earth wall will likely be destroyed, but the stone drop may survive as an indicator of the wall. Also, stone walls may mostly wash out over time, but the sides built into the embankments may survive as evidence of the former walls.

Cross-channel terraces and check dams are not common in the central Andes. We observed some that were abandoned on the higher slopes above the Río Colca. Regal (1970: 40) mentions dams (*tajamares*), unbroken after 400 years, in *quebradas* of the Río Chiva Basin on the north coast of Peru. Donkin (1979: 92, 126, 130) describes others, mostly abandoned, in western Ecuador, northern Chile, and northwestern Argentina.

Sloping-field Terraces

These are probably the most common form of terrace. They are contoured, or nearly so, and are located on valley sides rather than in the channels of intermittent streams. The cropping surface is sloping, but usually part of the natural slope has been reduced by soil accumulation behind a retainer wall, which is of stone, earth, or vegetation. These terraces function to reduce erosion, build up soil depth, conserve moisture, and control runoff. Soil accumulation may be the primary function on rocky slopes with very thin topsoil. Usually these slopes are steep, but lesser slopes may also be terraced for this reason. The rate of soil accumulation may have been accelerated by the removal of vegetation, surface stones, or a surface hard pan. Sloping-field terraces may be segmented into short units rather than being continuous along a contour (Fig. 9.3).

Some sloping-field terraces are irrigated from above via horizontal canals; however, most occur at higher elevations (over 3,500–3,700 m) where there is adequate rainfall or runoff for crops (Guillet 1987*d*: 81–2). In semi-arid regions the entire slope between walls may not be cultivated, only the lower sections where runoff is trapped behind the walls and moisture thus accumulates. During wet years, a larger portion of the slope would be cultivated (Treacy 1989*a*: 127–33; Brooks 1998: 133, 190, 382). Side walls may serve as catchment deflectors. Sloping-field terraces may also receive water from springs and slope seepage. Sloping-field terraces throughout the Andes are described and illustrated by Donkin (1979), and in Chile and Argentina by Field (1966).

Fig. 9.3. Segmented, sloping-field terraces, near Tarma, Peruvian Andes. Photograph by W. M. Denevan, 1956

Bench Terraces (Fig. 9.4)

The most spectacular and best preserved terraces are the large, stone-wall bench or staircase terraces with horizontal planting surfaces (*andenes, patas, bancales, takhanes*). They provide leveled terrain and deep soil on often steep slopes, but their primary function is to facilitate an even distribution of irrigation water over the cultivation surface.

Bench terraces are characterized by: (1) high $(1-5\,m)^3$ retaining walls of stacked, interlocking stones, (2) level platform planting surfaces, (3) valley-side positions following slope contours, (4) arrangements in vertical serial rows, (5) cut-and-fill construction, (6) inward sloping walls, and (7) built-in irrigation and other devices, including.

The stone in the retaining walls may be cut or shaped and fitted, with the riser or face of the wall inclined back for greater support. The large, wide walls may consist of two rows of stone separated by and often underlain by rock rubble. The outer face has to be high enough and strong enough to support a deep fill that may become saturated with water. Usually, the walls extend below the surface for anchoring. In Peru, terrace walls are often 100 m or more long. Bench terraces may include side walls, niches for storage, water drops, drains, service stairways, and steps in the walls (Fig. 9.5). Some bench terraces are cut entirely

Fig. 9.4. Inca bench terraces still in cultivation, Pisac, Urubamba Valley, Peruvian Andes. Photograph by Robert Shippee and George Johnson, 1931. American Museum of Natural History

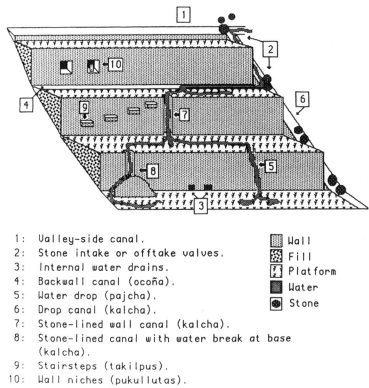

1: Valley-side canal.
2: Stone intake or offtake valves.
3: Internal water drains.
4: Backwall canal (ocoña).
5: Water drop (pajcha).
6: Drop canal (kalcha).
7: Stone-lined wall canal (kalcha).
8: Stone-lined canal with water break at base (kalcha).
9: Stairsteps (takilpus).
10: Wall niches (pukullutas).

▨ Wall
▩ Fill
▨ Platform
■ Water
◉ Stone

Fig. 9.5. Schematic view of bench terrace and associated features, Coporaque, Colca Valley, Peruvian Andes. From Treacy (1994*a*: 111)

out of the slope, without a fronting wall, as on loessic soils in central north-western Argentina (Field 1966: 481).

During construction of terrace walls, adequate drainage must be provided for in order to minimize wall collapse due to excessive saturation. Bonavia (1967–8) excavated pre-Inca sloping-field terraces in the Mantaro region of central Peru, uncovering evidence of carefully made drainage horizons. Builders placed a cobble fill layer below the topsoil, presumably to speed internal (through the wall) drainage when the topsoil filled to capacity and moisture seeped into the subsoil. Similar artificial cobble fills have been noted in Peru for bench terraces in the Urubamba Valley region near Cuzco (Keeley 1985), on the coast (Maldonado and Gamarra Dulanto 1978: 165), and in the Colca Valley (Treacy 1994*b*: 143, 147; Brooks 1998: 134). In Chile, terraces may have been drained internally by compacting the subsurface clayey soil into an inward and sideways-sloping platform so that water percolated out of the terrace solum (A. Wright 1963: 71).

The fill behind the fronting wall may come largely from the cut into the slope for the lower wall, as well as from the downward movement of eroding soil. The latter is referred to as 'self-filling' or accretional fill (*colmataje*). However, these cannot be easily irrigated, at least not until they are eventually leveled. On rocky slopes with thin soil, the fill and topsoil may have been transferred by hand from elsewhere, even from distant valley floors (Donkin 1979: 33), but this was probably not common. Stone may also have been carried a distance for terrace construction, as at Yucay (Niles 1999: 211). Planting surfaces range from about 5 to 15 m in width, being narrower on steeper slopes. Bench terraces are described and illustrated by Donkin (1979), and construction methods are examined by Treacy (1994*b*: 142–50) and Brooks (1998: 156–66).

Not all bench terraces are or were irrigated. Compared to sloping-field terraces, which receive runoff as well as direct rainfall, unirrigated bench terraces only receive direct rainfall and thus would seem to be at a disadvantage. On the other hand, only the lower portion of a sloping-field terrace may receive sufficient runoff for cropping, whereas the entire surface of an unirrigated bench terrace can be cultivated, assuming adequate rainfall. Thus an unirrigated bench terrace will utilize more available space than does an only partially cultivated sloping-field terrace (Sarah Brooks, pers. comm.)

Broad-field Terraces

Broad-field terraces are bench terraces broad cropping surfaces (up to 100 m or more wide) and low retaining walls. Valley-floor terraces are broad fields built in large river valleys, usually parallel to the rivers on alluvial fans (Fig. 9.6). They are irrigated by canal water drawn from the rivers higher up, and their main function is clearly to control irrigation water. The valley-floor river continues to flow without interruption by the terraces in contrast to cross-channel terraces. A variation is the silt-field, into which irrigation water is diverted and silt allowed to settle.

Broad-field terraces on mountain slopes are usually embanked by earth or stone walls. They may be irrigated bench or unirrigated sloping-field types. A variation is the *cuadro* in northern Chile (Field 1966: 119–23, 129–35, 148–52). *Cuadros* are rectangular or square fields, measuring about 5 by 8 m, although some are larger. As terraces they are distinctive in that they are embanked or semi-embanked by earth or stone ridges which are only about 12–28 cm high. The resulting basins are flooded to 12 cm or so depth by natural flooding or from irrigation canals. The *cuadros* are located on lower valley sides, and they consist of stepped groups dropping both toward the stream and downstream. They are fronted by low, massive stone walls. Some of the abandoned *cuadros* seem to be quite old, but pre-Columbian origin has not been confirmed.

Fig. 9.6. Valley-floor terraces on outwash fans, Pisac, Urubamba Valley, Peruvian Andes. From Donkin (1979: 109)

Terrace Origins

The origins of agricultural terraces, including specific forms and techniques, can be considered from the perspectives of earliest dates (above), causation, conceptualization, functions (below), processes, and diffusion. Immediate causal factors include demographic pressure, settlement relocation to mountain sides, risk minimalization, and climatic change lowering productivity. How farmers originally got the idea of terracing can only be speculated. Probably involved was the observation of soil accumulation and water concentration behind natural obstacles (rocks, logs) in dry season channels and on slopes already in cultivation, or behind artificial walls constructed for non-agricultural purposes (Brooks 1998: 375). The processes in initial terrace construction were trial and error based on observations of natural processes. For discussions of these various aspects of terrace origins, see Spencer and Hale (1961: 25–30), Patrick (1980), Doolittle (1990*b*), Williams (1990), Brooks (1998: 347–415), and Chapter 10 which follows.

Most farmers, of course, learned about terracing techniques from other farmers. However, diffusion is not necessarily a prerequisite 'for the spread of terrace culture . . . relatively little innovation is required for independent development' (Williams 1990: 82). Almost all terraces are pre-Columbian in origin, although many have been rebuilt or repaired over time. In only a few regions previously without terraces have new terraces been constructed. One instance is the Valles Altos Project by The United States Agency for International Development in the central Andes of Venezuela (Williams *et al.* 1986). AID also funded terrace construction projects in central and northern Peru based on earthen retaining walls similar to terraces cut into volcanic ash in Guatemala. The Peruvian government in the 1980s attempted to restore abandoned terraces in several regions of Peru, including at Coporaque in the Colca Valley (Treacy 1987*b*; 1989*b*). An independent restoration project was undertaken at San Pedro de Casta in the Río Rimac basin near Lima (Masson 1984; 1986; 1987). The Cusichaca Project has restored terraces between Cuzco and Machu Picchu (Kendall 1997).

Terrace Functions

A common assumption is that the function of terraces is the control of erosion. Actually, terraces have several functions, with erosion control usually being secondary (Donkin 1979: 34; Treacy and Denevan 1994: 93–5). For bench terraces, the main function is usually the control of irrigation water, which is brought by horizontal canals and dropped downslope terrace by terrace, or is fed laterally on to terrace cultivation surfaces by vertical canals. (The hydraulic functions of terraces are described and illustrated by Treacy 1994*a*.) However, a bench terrace is not required for irrigation. On a slope, a farmer can easily spread water from a canal above with simple hand tools (Williams 1990: 88). Furthermore, not all bench terraces are irrigated.[4] A flat surface also reduces erosion and facilitates cultivation activities. Terrace walls inhibit erosion and slow water movement thus assisting infiltration of moisture. A deep soil medium for planting is created behind a terrace wall, either through hand-filling of adjacent or distant soil, self-filling by natural downslope movement of soil, or a combination of both.

Terraces modify microclimate, including soil moisture, wind patterns, and temperature. Evans and Winterhalder (2000) report that: 'Other things equal, leveled (terraced) surfaces will have higher total solar insolation due to their having lower incidence angles during the midday period when direct beam transmissivity through the atmosphere is greatest.' Some additional solar radiation is reflected from adjacent terrace walls, but there is also some shading from the walls. The greatest growing season benefit in increased solar radiation to terraces is at low to mid latitudes on south-facing slopes of 30 degrees, with solar radiation being 15 per cent greater than on non-terraced slopes. The benefit is less on north-facing slopes and is negative on east- and west-facing slopes because of greater shadowing from upslope terraces. These differences probably help determine which slopes are terraced. While terraces themselves do not reduce night-time frost, mountain slopes have less frost than do adjacent valley floors which receive night-time cold-air drainage downslope.

A negative result of terracing is that the total area available for cultivation is reduced somewhat, compared to a non-terraced slope (Evans and Winterhalder 2000).

Terrace Abandonment

Numerous observers have noted that a high proportion of Andean terraces are abandoned; some are highly deteriorated, indicating long abandonment. Donkin (1979: 35) estimates one-third abandonment for the Americas. The portion is even greater for the central Andes. Masson (1986: 208) indicated 75 per cent abandonment, later changed (pers. comm. 1988) to 50 per cent in Peru. A. Wright (1963: 73) estimated 80 per cent for northern Chile. For the Colca Valley, measurements on air photos showed that 61 per cent of the bench terraces are abandoned (Denevan 1988: 22, 28).

The causes of terrace abandonment are varied and are debated (Donkin 1979: 35–8; Denevan 1986). Most abandonment is on upper slopes and in remote areas distant from present towns; however, abandoned terraces may occur in proximity with cultivated terraces. In other instances, extensive zones of terraces are completely abandoned.

A frequent explanation for terrace abandonment is climatic change, either colder temperatures causing abandonment at high elevations due to frost (Cardich 1985), or reduced rainfall causing abandonment of rainfed terraces (Donkin 1979: 36). For the Lake Titicaca Basin, Cardich (1985: 307–8) gives the present upper limit of cultivation as 4,050 to 4,200 m, with prehistoric field remnants up to 4,400 m, indicating a retreat downward of over 200 m. For the Mantáro Valley, Seltzer and Hastorf (1990: 402) for the cool period (Little Ice Age) from AD 1290 to 1850 give a crop depression of 150 m compared to the present. In the Río Japo Basin in the Colca Valley at this same time, Brooks (1998: 391) estimates that the maximum elevation of crops and terracing at 3,950 m was depressed downward 70 to 150 m.

Changes in settlement location could also lead to abandonment. The Spaniards relocated Indian populations from slopes to valley bottoms (Donkin 1979: 35–6). On the other hand, after AD 1300 in Peru many people were relocated from valley bottoms to higher slopes, which was during a cool period (Seltzer and Hastorf 1990: 408). Certainly, the massive population decline that came with the Spanish conquest resulted in terrace abandonment (Donkin 1979: 36–8). Social and economic factors help explain why abandoned terraces are not restored today, but inadequate technical knowledge is also a factor (Bonavia and Matos 1990; Masson 1987). Factors of terrace abandonment in the Colca Valley of Peru are considered in Chapter 10.

In the central Andes, agricultural terraces are the most visible manifestation of the pre-European cultural landscape. Most of the current terraces are of pre-colonial origin, but over half are now abandoned, for reasons that are not completely clear. A variety of complex and labor intensive forms of terraces reclaimed land that was otherwise excessively steep, rocky, dry, or cold, thus making possible permanent, highly productive, intensive cultivation.

Notes

[1] *Andén* means platform and includes non-agricultural terraces (Donkin 1979: 19).

[2] In the very arid Negev Desert in Israel, catchment areas for ancient, controlled, runoff agriculture ranged from 17 to 30 times the area cultivated (Evenari *et al.* 1971: 104). Catchments may have been smaller in the Andes.

[3] Donkin (1979: 111) mentions terrace walls at Yucay in Peru which are up to 9 m high.

[4] K. Wright *et al.* (1997: 46) maintain that bench terraces at Machu Picchu were never irrigated, although Donkin (1979: 116–17) saw irrigation canals there. Brooks (1998: 192–3) observed numerous unirrigated prehistoric bench terraces in the Colca Valley.

10

Terrace and Irrigation Origins and Abandonment in the Colca Valley

[T]he higher terraces that extend well up the slopes above the main valley floor are abandoned . . . [and] show only faintly in contrast to the fresh appearance of those on the valley floor and along the sides of the inner valley.

(Lt. George R. Johnson, USN, 1930: 9)

In the Department of Geography at the University of Wisconsin in the 1960s and 1970s the Latin Americanists were interested in the cultural ecology of shifting cultivation, agroforestry, traditional crops, and raised fields. We assumed that terraces and canals, so often photographed, were well studied; however Donkin's (1979) terrace survey showed otherwise. Furthermore, the magnitude of terrace abandonment in the Andes was astonishing given current demographic pressures on farmland. We were thus motivated to propose an interdisciplinary project which would examine abandoned terraces and associated irrigation and attempt to determine the cause or causes of abandonment.

John Treacy, who had worked with me on raised fields at Samborondón in Ecuador and on Bora agroforestry on the Río Ampiyacu in Peru, had lived for eight years in the southern Andes of Peru, and he wanted to return to carry out research for his doctoral dissertation. Chatting in my office in Science Hall (or was it over a beer on the lake terrace?), we agreed that a study of Peruvian terraces could be rewarding. Any one of many terraced valleys and basins could have served, but we were particularly impressed by the spectacular aerial photography of the Colca Valley appearing in the reports of the Shippee/Johnson aerial expedition of 1931 (Shippee 1932b; 1934; also G. Johnson 1930). Here was a long stretch of dense and varied terraces, in large part abandoned. The environment, archaeology, and history of the Colca Valley villages had been little studied, but there were detailed colonial *visitas* (censuses) available, recent air photography (1974), and a demographic history had just been published (N. Cook 1982). The dirt road from Arequipa to the Colca had recently been improved by the Majes Project which built a cement canal to transfer water from the Río Colca to an irrigation site on the coast.

In the summer of 1983, Treacy and I visited the Colca Valley, where we were assisted by anthropologist David Guillet, who already had initiated a study of water management in the village of Lari. We then assembled our own project (Denevan 1987*a*) with a team consisting of geographers Treacy, Hildegardo Córdova, and myself; archaeologists Máximo Neira, Dan Shea, Michael Malpass, and Pablo de la Vera Cruz; ethnohistorian Maria Benavides; and soil scientist Jon Sandor, in addition to North American and Peruvian students. Primary field research was conducted during the summer and fall of 1984, with individual subprojects continuing over the next decade. Our efforts were initially concentrated in the village of Coporaque, with additional archaeological excavations in Achoma and Cabanaconde and later in Chivay by Brooks (Fig. 10.1).

The main reports of the original Colca Terrace Project are found in Denevan (1986–8) and in Denevan, Mathewson, and Knapp (1987, Pt 1). Several theses and doctoral dissertations at Wisconsin and in Peru resulted directly or indirectly (Treacy 1989*a*; Vera Cruz 1988; 1989; Webber 1993; F. González 1995; Femenias 1997; Brooks 1998). During this period there were numerous ecological studies by other US scholars attracted to the Colca, including Guillet (1992), Gelles (1990), McCamant (1986), and Eash (1989). Many studies and reports were also produced by Peruvians and Peruvian institutions on history, agronomy, sociology, archaeology, and anthropology. The result has been that in only a few years the Colca Valley changed from a little-known backwater to one of the best-studied regions in Peru, accompanied by an upsurge in tourism and media attention, with resulting social change for the Colca villagers.

The Colca Valley

The valley of the Río Colca is located in the northern Department of Arequipa in southern Peru on the western slope of the Andean cordillera. It is but one of a series of deep valleys on the southwestern Andean flanks, most of which were terraced in pre-Hispanic times. The Colca is incised into a large, flat river terrace (*pampa*). The main cultivable portion of the valley is approximately 40 km long between the villages of Chivay and Cabanaconde at elevations between 3,200 and 3,600 m. Maize and quinoa were the primary prehistoric crops. Today, maize is planted mainly on terraces, while quinoa, broadbeans, alfalfa, and barley appear both on terraces and *pampa* fields. Potatoes are not as common as might be expected. Alfalfa has become a major crop as cattle feed. In addition to cattle, most farmers also have sheep, llamas, and alpacas which are grazed on the *puna* (plateau) above the valley, as well as on fallow fields.

The valley's arid climate has influenced the nature of terrace construction. The agricultural core has a semi-arid montane steppe climate, receiving an

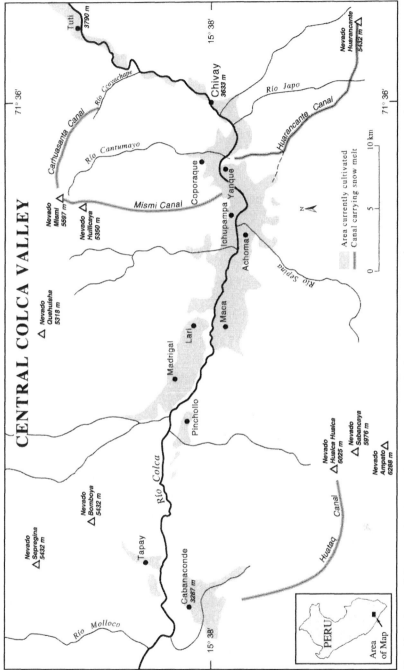

Fig. 10.1. Map of the central Colca Valley, Peruvian Andes, showing Coporaque, other villages and the Río Japo Basin. Pampa Finaya is located between Coporaque and Chivay. Cultivated land, mostly terraced, is shaded. From Brooks (1998: 6)

annual average of 385 mm of precipitation at the principal village of Chivay. Rains are highly seasonal, peaking in December and declining sharply during late May to early August. For this reason, almost all cultivable land is irrigated by canals bringing water either from natural springs or from streams flowing from mountains topped by permanent snow. The roles of irrigation are thus: (1) to extend the growing season by one to two months, allowing farmers to plant in August and harvest in May, and (2) to provide supplementary water in case rains slacken during the growing season. Soils are mainly of volcanic origin and are of good structure and fertility. Cultivated and abandoned terraces are located on the *pampas*, the slopes of the canyon below, and on the steep mountain slopes above to an elevation of nearly 4,000 m.

The Colca Valley has been occupied by terrace farmers since possibly 2400 BC (Brooks 1998: 270). The immediate pre-Inca culture was the Collaguas which dominated the valley. Little is known about pre-Collaguas people, although Brooks (1998: 313) has identified an earlier ceramic style she calls Japo. Lithic tools in caves indicate that pre-ceramic hunters were long present. The Inca interacted with the valley after 1471; however there is little evidence of a strong Inca presence in that no Inca administrative center has been located and Inca ceramics, while present, are not common. Also, there is little evidence (a few sherds) of the Wari (Huari) culture, which dominated much of southern Peru, *c.* AD 650–1000.

The Spaniards arrived in the 1540s and found numerous small villages and dispersed farmsteads which were then converted into *encomiendas* (land and labor grants). Later, as part of the survey of Peru under Viceroy Francisco de Toledo, 1571–4, the Colca Indians (Collaguas region) were consolidated into twenty-four Spanish style villages (*reducciones*) numbering an estimated 62,500 to 71,000 people projected back to 1530. European crops and livestock were introduced, transforming the economy; however population declined rapidly due to introduced diseases and removal of people to the mining centers. By 1721, the population had been reduced to about 8,600 (87 per cent decline). Population increased slowly to 27,534 in 1940 (Province of Caylloma), reflecting out migration to Arequipa and elsewhere (N. Cook 1982: 83–7). By 1993, the population had only increased to 37,600 (corrected from INEI 1993: 38–9).

In the main terrace sector of the valley from Tuti to Cabanaconde there are now twelve villages, each numbering between about 1,000 and 5,000 people, the largest being Chivay. Large units of once productive terraces are unused. This is in the Peruvian context of an increase of population from 6.2 million in 1940 to 23 million in 1993, inadequate food production, and apparent scarcity of arable land. Why has there been agricultural stagnation, or an earlier collapse without recovery, in the Colca Valley, as well as elsewhere in the Andes? Are the reasons environmental, such as climatic change or soil depletion and erosion, or are the reasons social? To consider this question we examined not only terrace abandonment but the history and prehistory of terracing in the Colca Valley, including the origins of terracing and irrigation.

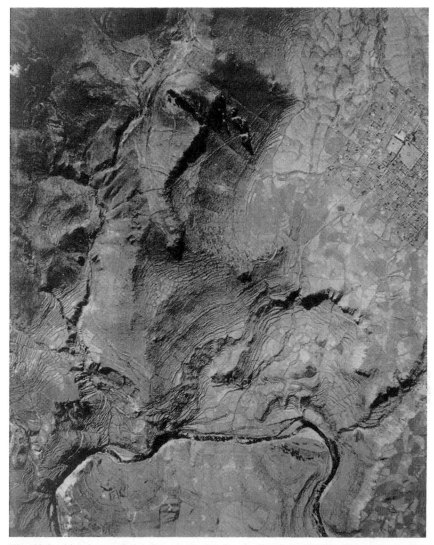

Fig. 10.2. Air photo of the Coporaque region in the Colca Valley, Peruvian Andes, showing cultivated and abandoned terraces. See Fig. 10.4. Photograph by Robert Shippee and George Johnson, 1931. Servicio Aerofotográfico Nacional, Lima

The Colca Terraces

The ecology of current Colca irrigated terraces has been best studied in the communities of Coporaque (Fig. 10.2) (especially Treacy 1989*a*; 1994*a*; 1994*b*; Treacy and Denevan 1994) and Lari (especially Guillet 1987*a*; 1987*b*; 1987*c*; 1992) on the north side of the valley. Most of the Colca terraces were originally constructed in prehistoric times; however with repairs subsequently, any given wall represents different construction dates and techniques.

The Colca Valley features a variety of terrace types that may be keyed into the terrace typology presented in Chapter 9, and described by Treacy and Denevan (1994) and by Brooks (1998: 126–36). The common and most visually impressive type is the contour or bench terrace on valley-side slopes. Bench terraces measure between 1 and 3 m in wall height, between 3 and 7 m in width depending upon degree of hillside slope, and from 40 to 60 m in length. Most Colca Valley bench terraces are irrigated by gravity-fed canals, and terraces often have water drops and channels built into walls and on the terrace platform. Most bench terraces are essentially irrigation platforms on steep slopes; their microclimatic virtues (frost reduction, moisture conservation) allow intensive maize cultivation. Brooks (1998: 192–5) reports finding numerous abandoned, unirrigated bench terraces in the Río Japo Basin between Chivay and Yanque, as well as elsewhere in the Colca Valley. Nearly half of all the terraces in the Japo Basin are of this type. Although not irrigated, the flat cultivation surface of such a bench terrace captures more rainwater than most of a sloping-field terrace, where rainfall runs off the upper surface to the lower surface, and hence, with adequate rainfall, bench terraces are more productive per unit of slope.

Builders used cut-and-bench methods to construct bench terraces. They excavated through A horizon topsoil into an indurated B horizon to emplace wall base stones ranging in size from 30 to 60 cm in diameter. Stability and interlocking fit were more important than the size of the wall stones; base stones may be smaller than stones placed on top of them. Next, soil was removed from the exposed vertical soil face in order to make the cobble and earthen drainage horizon which most Colca Valley bench terraces have. Builders then moved soil down the slope behind the wall,[1] while stonemasons built the wall higher to retain the soil. Mineralogical examination shows that terrace fill soil is composed of *in situ* slope soil; no exogenous soils were used (Sandor 1987*a*: 185). In extensively terraced regions such as the Colca Valley, with steep slopes, the costs in labor would preclude transport of soil fill from elsewhere, at least not in quantity.

Often, a bench terrace is damaged when waterlogged soil bursts the retaining wall and spills forth. Careless wall repair hastens further wall failure. Builders blame poor masonry techniques or dry-soil construction for broken walls. Builders prefer to terrace with wet soil since dry soil is porous and will quickly become saturated following heavy rains, thus provoking sudden, early bursting. Common mistakes cited include not fitting wall stones together tightly, or failure to press stones against the earthen fill behind. In the latter case, a gap forms between the wall and fill causing the wall to collapse when the gap fills with water.

Bench-terrace reconstruction duplicates the labor techniques of construction, except that soil spilled from ruptured walls has to be thrown back upslope and re-walled. Workers trench down to the original base to remove the fallen and buried stones from the original wall. The wall is rebuilt and the fill stones and soil are carefully repacked in behind the wall and tamped down. When reconstructing fields, masons replace up to 80 per cent of the ancient walls, and must laboriously throw spilled soils upslope, which demands more effort than moving soils downslope for new terraces.

An interesting aspect of Colca Valley terracing is the numerous *maquetas* (carved boulders) which depict patterns of irrigated terraces. Brooks (1998: 287–93) identified fourteen and describes several. They show primary canals, feeder canals, reservoirs, and end walls between terraces. Different functions have been suggested, but Brooks believes that the Colca *maquetas* served as general plans for terrace irrigation systems and also had a ceremonial role. Figure 1.1 shows the famous Sahuite *maqueta* in Cuzco, which depicts terraces and canals among other features. There is a full-scale model in the Museo de la Nación in Lima.

For Coporaque, Treacy (1994*b*: 81–4) described three types of sloping-field terraces: (1) linear, valley side, segmented, some with side walls (124 ha), (2) roughly rectangular, fully walled, near ridge tops (122 ha), and (3) 'stone-bordered', roughly rectangular, in nearly flat areas (460 ha). All were probably designed to trap runoff. Slope runoff was also directed by diversion walls into reservoirs and into canals which fed irrigated bench terraces. Treacy (1994*b*: 108) mapped five such sectors in the Coporaque area. Sloping-field terraces tend to measure between 1.3 and 2 m in wall height, 13 to 16 m in width, and between 30 and 50 m in wall length. The broad, sloping catchments of these terraces suggest that they were in part accretional or self-filling fields; they do feature artificial cobble drainage horizons. Most fields of this type are now abandoned in the Colca Valley.

Broad-field, irrigated bench terraces occur in places where steep upper slopes become more gentle, and they are often enclosed by stone walls (Fig. 9.2 and Fig. 10.3). Valley-floor broad-field terraces occur on the nearly flat, alluvial terraces (*pampas*) of the Río Colca. Walls are of stone or earth and are only 20–100 cm high, with very wide cultivation sectors between walls, compared to narrow bench terraces on slopes.

Simple cross-channel terraces and check dams, some no more than a rough line of stacked stones, are scattered thinly through parts of the valley. They generally cross gullies or intermittent stream channels on moderate slopes. Examination of aerial photographs shows that barrage-like terraces—stone walls flanked by long, perpendicular water-diversion walls—may have been constructed in some areas of the Colca Valley, but field ground checks are needed to confirm them. Cross-channel terraces, like most sloping-field terraces, are today unfarmed.

Non-terraced fields in the valley include large numbers of abandoned, stone-lined plots. There are 460 ha of these in eastern Coporaque in dissected lava-flow terrain between the Pampa Finaya plateau and the Río Colca. Smaller sectors occur in Tuti, Yanque, and Maca. The walls are of stacked stones 0.5 to 2 m high. They enclose mostly gently sloping fields, rectangular or irregular in

Fig. 10.3. Air photo of Chijra, Coporaque in the Colca Valley, Peruvian Andes, showing abandoned terraces, aqueduct (inverted Y), and prehistoric house walls. The terraces on a steep slope at the lower center and left are in the sector of Alto Cayra. Photograph by Robert Shippee and George Johnson, 1931. American Museum of Natural History

shape, ranging from 400 to 5,000 m² in size (Treacy 1994*b*: 86). The function of the stone walls is not known, possibly a means of moisture conservation.

There are now numerous unterraced, irrigated valley-floor fields of roughly rectangular shape on the *pampas*. They are surrounded by adobe or stone walls which serve to keep cattle out. These walls are rare on the Shippee/Johnson air photos of 1931, when sectorial, open-field grazing was still being practiced. Enclosure took place after 1940 as cattle greatly increased and seriously damaged crops (F. González 1995: 44–58). Other fields are house and garden plots, and recent (since the 1970s), unterraced but irrigated, valley-side sloping fields. The latter fields were made following new canal construction; such unterraced fields probably would have been unthinkable in pre-Columbian times.

The Soil Factor

We have considerable information about the soils of the Colca terraces and associated non-terraced terrain thanks in particular to the research by Jon

Sandor and Neal Eash (Sandor 1987*a*; 1987*b*; Eash 1989; Sandor and Eash 1995; Eash and Sandor 1995; also Dick *et al.* 1994). Most of the Colca terraces still utilized have been in near continuous cultivation since the early colonial period, given the descriptions of specific fields in the *visitas* (censuses)—fields we can still identify from their names. Cultivation of these same fields undoubtedly extended well into prehistory, given the dates we have from excavated terraces. Hence, a major question is what was the impact of many centuries of cultivation on the terrace soils? How was soil fertility maintained? Is soil decline related to terrace abandonment?

The region is underlain by volcanic andesitic and rhyolitic rocks, which generally weather into fertile soils. The soils consist of alluvium and colluvium derived from these materials. There are also lava flows and some volcanic ash present. The natural soils on slopes are mainly mollisols, which are grassland soils with abundant humus found in semi-arid regions. The degree of soil development in the Colca varies with geomorphic surface and age.

The terrace soils show significant differences from non-terraced soils as result of long-term cultivation. The thickness of the A horizon is greater, and phosphorus, nitrogen, and organic carbon levels are higher, as are earthworm activity, friability, and water capacity. These differences reflect centuries, possibly 1,000 years or more, of application of fertilizers (manure, hearth ash, garbage, crop residues); terracing which reduces erosion; and tillage, irrigation, crop rotation, fallowing, and use of legumes. These soil characteristics generally pertain to both still-cultivated terraces and long-abandoned terraces, as compared to terrain not terraced, thus indicating persistence of earlier, even ancient, soil management impacts. Phosphorus levels are less, however, in abandoned terraces, indicating partial depletion during the historical period with cessation of fertilization except from livestock manure. High levels of phosphorus appear in the B horizons of terrace soils, suggesting a very long period of downward movement of phosphorus from fertilizer (Sandor and Eash 1995: 177). Thus the 'data suggest that soil changes have taken the forms of increased fertility and tilth and that traditional agriculture management practices have conserved soils' (ibid. 178). This has made possible near-continuous cultivation of some sectors for as long as 1,500 years (ibid. 170).

Terrace Origins

The origins and evolution of the Colca terraces, particularly the irrigated bench terraces, may be related to circumstances that could help explain later abandonment of both irrigated and unirrigated (rainfed and runoff) terraces. Our archaeological excavations at Chijra (Chishra) and Chilacota (Ch'ilaqota)[2] near Coporaque, at Cabanaconde and Achoma, and later in the Río Japo Basin near Chivay (Fig. 10.1), are not conclusive but provide instructive insights. I will focus on Chijra and to some extent the Japo Basin, areas for which we have the best descriptions, ceramics, and Carbon-14 (C-14) dates.

The Chijra Area Terraces

The terraces of the Chijra area are located on a ridge just west of Coporaque along a slope extending from 3,752 m down to 3,272 m at the edge of the Río Colca (Figs. 10.2 and 10.4). The site of Chijra is an uplifted alluvial fan of moderate slope (Fig. 10.3). There are over 100 abandoned terraces, of which we measured and described thirty (Treacy and Denevan 1986). Of these all are valley-side bench terraces except for six broad-field bench terraces on the gentler lower slope. A large stone aqueduct crosses the terraces at a right angle, splitting into two arms before ending at the sharp break in slope at the lower edge of the broad-field terraces. The aqueduct is 5 m wide at the base and 1 m or more in height in places. On top are the remnants of a stone-lined canal, with a rock gate which could be closed by large stones to force water out of the canal laterally on to the terraces. Water was then dropped, terrace by terrace via stone water-drop grooves, down the faces of the back walls on to spreading stones at the base of the grooves (see Fig. 9.5). Small channels of stones across the bench surfaces led the water to the edge of the next wall below, or water was allowed to fall all along the edge to the next level, as is done today in the area. Farmers now use hoes to spread water evenly over a cultivation surface ('teaching water'—Treacy 1994*a*: 100) and then move it quickly to the next level to avoid over saturation. Stepping stones and benches in the walls are common, as are vertical interior drains. The aqueduct was fed by a canal leading from a reservoir above at Chilacota (Fig. 10.4). The bench terraces average 1.7 m in height, 7 m in width, and some were over 100 m in length. Terrace walls are single walls of uncut stone without mortar; they are underlain by cobble-fill to facilitate drainage. Vertical end walls may have served as divisions between household fields. (For details, see Treacy 1994*a*.) Also there are ruins of ten Collaguas houses (walls) at Chijra, most located on the broad-field terraces (Fig. 10.3).

Above Chijra at about 3,600–3,750 m on a gently sloping plateau is the site of Chilacota (D. Martin 1986). There is no evidence of any irrigation. There are three massive earthen walls across the plateau that may have served to trap runoff for crop growth. Between the walls there are fifteen roughly rectangular smaller walls (double stone walls vertically and single walls horizontally)—a form of sloping-field terracing. The abandoned Chilacotacocha reservoir (3,400 m^2) is just north of the plateau, separated by a *quebrada*. The south wall of the reservoir is 5 m high and has a water gate in it. From the gate a canal leads past Chilacota both to Chijra and to a *quebrada* which fed canals and terraces west of Chijra (Figs 10.2 and 10.4). The reservoir was fed by canals receiving meltwater from the snow line well above on the slopes of Cerro Mismi. Also, there is a long wall from a *quebrada* to the northwest which apparently diverted water to the reservoir.

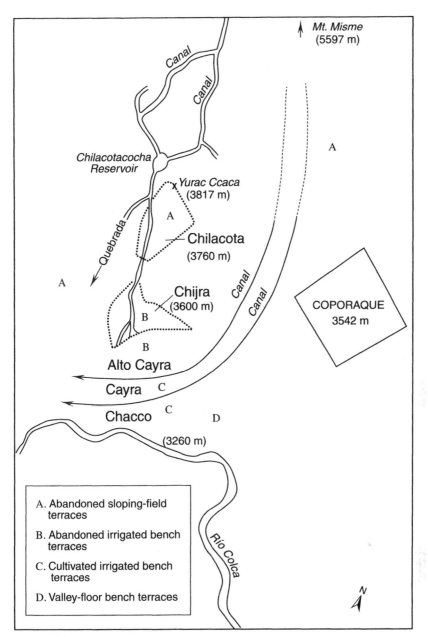

Fig. 10.4. Map of Chijra and environs, District of Coporaque, Colca Valley, Peruvian Andes. See Fig. 10.2. By W. M. Denevan

Fig. 10.5. Abandoned terraces (upper) at Alto Cayra and cultivated terraces (lower) at Cayra, Coporaque, in the Colca Valley, Peruvian Andes. They are separated by an active, horizontal irrigation canal, which does not show clearly. From Treacy (1994*b*: cover)

Below Chijra, the slope becomes very steep and is covered with abandoned, highly deteriorated, very narrow bench terraces (Alto Cayra). Some of these were irrigated by the aqueduct, while others were apparently unirrigated. Below this there are two active horizontal canals coming from the east side of the ridge. They irrigate active bench terraces between them at Cayra (Qayra) and bench and broad-field terraces between the lower canal and the Río Colca at Chacco (Chaqo) (Figs 10.4 and 10.5).

At Chijra, three terraces were trenched through sections of bench and wall, and two more terraces were excavated at Cayra and Chacco (Malpass 1987). Neira (1990: 155–66) excavated one of the house ruins. We obtained twenty C-14 dates from the terraces, the house floor, and an irrigation canal (Malpass 1987: 61). Ceramics, lithics, animal bones, and pollen were collected (Malpass and Vera Cruz 1988). At Chilacota, seven test pits and trenches were made in and near the sloping-field terraces. The accumulation over well-developed soil was only about 20 cm. No house walls were found, but a possible hearth was trenched near the terraces. Ceramic fragments were found, mostly undiagnostic, along with lithics, including very small (2–4 cm long) obsidian points

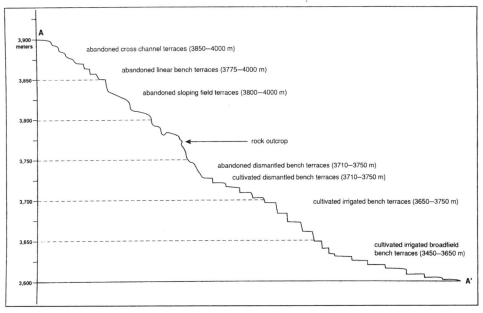

Fig. 10.6. Generalized slope profile showing approximate elevations of terrace types in the Japo Basin, Colca Valley, Peruvian Andes. The linear bench terraces were unirrigated. From Brooks (1998: 149)

used in hunting, and substantial quantities of deer and camelid bone. The only C-14 dates for the house, AD 300 ± 70 and 1570 ± 80 BC, were from soil below walls and are not good indicators of time of wall construction, but they do suggest a considerable antiquity.

The Japo Basin Terraces

In the Río Japo Basin between Chivay and Yanque, geographer/archaeologist Sarah Brooks (1998) carried out a detailed study of terrace history in the vicinity of the Collaguas ruins of Juscallacta. Here there are a variety of abandoned and still cultivated terraces. These include cross-channel, sloping-field, and bench terraces, all abandoned and never irrigated, between 3,775 and 4,000 m (Fig. 10.6). About 1,500 irrigated, mostly still-cultivated bench terraces occur between 3,450 and 3,750 m. Just above the irrigated terraces there is a lateral canal which feeds the terraces below. There is a total of about 394 ha of terraces, of which 200 ha are currently abandoned (Brooks 1998: 150); thus about 51 per cent of the terraces are abandoned, almost all of which are above the canal level. There are only ten or so sloping-field terraces surviving (compared to hundreds in the Coporaque area), and a few dozen cross-channel terraces, so most of the abandoned terraces are unirrigated bench terraces. Most of the lower, irrigated bench terraces are broken up into segments by vertical stone walls, whereas the upper, unirrigated abandoned terraces are not. There is much less abandonment of irrigated terraces at Japo than at Coporaque.

Between 3,710 and 3,750 m there is a band of dismantled (front walls removed) irrigated bench terraces, some cultivated and some abandoned. This destruction is of narrow terraces on steep slopes where the removal of terrace walls facilitates plowing with oxen. It may have begun in the colonial period and is continuing. The result is a probable increase in soil erosion and a reduced control of irrigation water (Brooks 1998: 170–6). Water can still be released from the horizontal feeder canal above on to the now unterraced sloping fields. Elsewhere, terraces are removed to plant forage for livestock (Benavides 1997).

Other terraces or locales in the Colca Valley were excavated at Achoma (Shea 1987; 1997), Cabanaconde (Vera Cruz 1987; 1988), and ceramics have been collected elsewhere (Vera Cruz 1989); however, no terraces or anything else were dated from these areas.

Terrace Chronology

We obtained several C-14 dates for terraces and canals from our excavations at Chijra near Coporaque (Malpass 1987: 61), as did Brooks (1998: 286) for the Japo Basin. These dates are from buried soil and charcoal from within terrace fills and beneath terrace walls. A word of caution is necessary before looking at these dates. Terrace walls are continually being rebuilt or repaired, and the soil fill behind the walls is disturbed by cultivation and also by both erosion and soil movement from above. Hence, anything dated in a terrace wall or fill, or an original soil surface beneath a fill or wall, may not be a reliable date for cultivation of that horizon or surface.

Brooks dated a sloping-field terrace at 3,790 m in the Japo Basin to AD 1510, and an unirrigated bench terrace at Pampa Funaya at 3,920 m between Coporaque and Chivay to AD 655. A date of 1570 BC for a sloping-field terrace at 3,760 m at Chilacota (Fig. 10.4) suggests that this type of terrace was also in use very early in the Colca Valley. There are no dates available for cross-channel terraces, but they are also undoubtedly very early. Probably both sloping-field and cross-channel terraces are older than unirrigated bench terraces, as the former are less complex and less labor-intensive but have a lower portion of cultivated space than do bench terraces. Probably all three types of unirrigated terraces continued in use until the development of irrigated bench terraces, and some continue in use today in the Colca Valley. Brooks dated an unirrigated bench terrace at 3,790 m in the Japo Basin to AD 1510 and one at Pampa Finaya at 3,880 m between Coporaque and Chivay at AD 655.

For irrigated bench terraces at Chijra (Fig. 10.4), the most reliable early dates are for the lowest levels of three excavated terraces: AD 510 and 660 (Trench 1), 550, and 570. For Trench 1, there are upper horizon level dates of AD 940, 1020, and 1370, and at Cayra there is a terrace date of AD 1340. In the

Japo Basin, Brooks has dates for irrigated bench terraces of AD 780, 1690, and 1847. There seems to have been terrace cultivation throughout the colonial period.

Thus the pattern of terrace types, elevations, and dates in the Japo Basin and in the Coporaque area indicates early unirrigated terraces (2400 BC or older) above 3,775 m (Japo) and 3,760 m (Chilacota) and a shift downslope to irrigated bench terracing by at least AD 500–600. Most of the irrigated bench terraces in the Japo Basin continued in use to the present. In the Coporaque area, the terraces at Chijra and Alto Cayra were abandoned at some time, whereas the lower terraces at Cayra and Chaaco have continued in use (Fig. 10.4). Settlement also moved downslope, from Juscallacta to the area of present Yanque and Chivay and from Chilacota and Chijra to the area of Coporaque.

Changing Terrace Forms

The shift of terrace cultivation downslope, with irrigation added, can be related both to changing climatic conditions and to increasing population pressure which necessitated dependable production in a region of undependable precipitation. Rainfall for Yanque at 3,417 m currently averages 419 mm (Brooks 1998: 65), which is inadequate for rainfed (unirrigated, non-runoff-watered) maize (*c.* 600 mm minimum) and potatoes (*c.* 500 mm minimum); however quinoa and cañihua, when well spaced, need as little as 300 mm, and were probably important crops on non-irrigated terraces (Treacy 1994*b*: 107). Rainfall in the Colca is somewhat higher at higher elevations, an estimated 518 mm at 4,000 m (Brooks 1998: 390).[3] Certainly there is sufficient moisture today for runoff fields (cross-channel terraces and sloping-field terraces), and even unirrigated bench terraces planted in quinoa. However, very few of these terraces are cultivated now, only 9 per cent in the Colca Valley, mostly in the community of Achoma and none in Coporaque or in Chivay which includes the Japo Basin (Denevan 1988: 22).

There were relatively dry periods in the central Andes of Peru during AD 540–560, 570–610, 650–730, and 1040–1490 with precipitation reduced *c.* 5–20 per cent, based on ice core measurements at Cerro Quelccaya east of the Colca Valley (Thompson *et al.* 1994; Brooks 1998: 74). Such a decline in rainfall could have contributed to the development of irrigated bench terraces at lower elevations where slopes are gentler and more easily irrigated. In addition, there is a frost factor. While there is greater precipitation above 3,750 m, there is also a greater frequency of frost and hence crop failure. The upper limit of agriculture, based on frost risk, was depressed as much as 150 m during the periods AD 650–850 and especially 1240 or 1255 to 1850 or 1880 (Little Ice Age) (Brooks 1998: 71, 384). The shift to irrigated terraces at lower elevations may relate to dry or cold periods or both. However, with sloping fields as high as 4,200 m above Chilacota and irrigated bench terraces being below 3,760 m at Japo and Cayra, a decline in the frost line would be of limited significance.

Thus, there were two climatic stress periods during the past 1,000 years: first, AD 540–730 (dry) and 650–850 (cold); and second, AD 1040–1490 (dry) and 1240–1880 (cold). The dates for irrigated bench terraces suggest that they were first being constructed during the earlier period. They may have been elaborated and expanded during the latter period. Malpass (1988) believes that the presently utilized irrigated terraces at Cayra and Chacco were not developed until the Inca period, but this seems much too late given such terracing in many parts of Peru well before then. The abandoned irrigated terraces at Chijra and Alto Cayra, fed by vertical feeder canals, probably date to the earlier dry period, whereas the still cultivated terraces lower down at Cayra (a single date of AD 1340) and Chacco, fed by horizontal canals, may date to the later dry period (Fig. 10.4). Also, fewer animal bones and lithics (obsidian arrow points) were found in the lower terraces compared to Chijra and Chilacota, indicating a greater emphasis on agriculture and less on hunting and grazing at the lower terraces.

In parts of the areas of irrigated bench terraces, there already may have been sloping-field and unirrigated bench terraces that were either destroyed by the new irrigated terraces or were converted into the new terraces. This is suggested by the sharp divisions between abandoned unirrigated terraces and irrigated terraces, the division marked by a lateral canal constructed to irrigate the lower terraces. Such a sharp lower boundary for the unirrigated terraces would not have existed prior to the canal, so unirrigated terraces may have extended below the level of the present canal. Our excavation of Trench 1 at Chijra did expose a buried stone rubble wall, dated AD 940, suggesting the construction of the present terrace over a previous one (Malpass 1987: 54–6). Both Treacy (1994*b*: 102–5) for Chijra and Brooks (1998: 392) for the Japo Basin argue for the conversion of unirrigated terraces into irrigated terraces below 3,750–3,760 m. The two terrace types could have coexisted for a time until the superiority of the irrigated bench terraces was clearly recognized. Some seemingly abandoned sloping-field terraces in Coporaque are cultivated today when adequate rainfall occurs or is anticipated.

This scenario, drawing on Brooks and Treacy, is an argument that the development of complex labor-intensive irrigated terraces resulted, in part at least, from climatic stress.[4] This is contrary to some popular and academic thinking that climatic deterioration (dryer, colder) was the cause of agricultural and social collapse at various places in the world. In the Colca Valley, climatic stress instead seems to have motivated the development of a more efficient, more productive, and less risky method of food production. There is another dimension to this story, however. Even without long-term climatic change, there was regular climatic variation, especially in precipitation, and thus great risk to crops during very dry years. For example, at Yanque with an average of 419 mm rainfall over thirty-one years, the wettest year, had 708 mm and the driest year had only 213 mm. Intermittent crop failure during dry years became progressively more of a problem to people as population grew. Irrigation is not fully dependent on immediate precipitation, but rather can also draw on reservoirs, springs, and meltwater from snowcaps. Irrigation can provide a more secure supply of water in terms of quantity and predictability, and it can extend the growing season. Thus it is a great advantage even when there are not decades- or centuries-long periods of reduced rainfall. Irrigation continues during wet periods, as we see in the twentieth century.

Finally, an expansion of terracing and a shift of cultivation downslope during late prehistory may have been associated with a greater interest in maize, which requires reliable moisture, a long growing season, and minimal frost. An important role of irrigation for Andean valleys, such as the Colca, is to lengthen the growing season for maize (Guillet and Mitchell 1994: 6–7).

Thus, shifts in terrace form and elevational level could have involved more than shifts in precipitation patterns and frost lines. This question is of more than local concern. The pattern of abandoned, unirrigated terraces at higher elevations and cultivated, irrigated terraces below them is widespread in the Andes (Malpass 1988). Firmer dating of climatic, agricultural, demographic, and other events in the Colca and elsewhere is essential if we are to better understand the history of Andean agricultural fields.

Terrace Abandonment

We have considered the abandonment of unirrigated terraces; however, numerous irrigated terraces have also been abandoned. Explanation is important given plans for restoring abandoned terraces.[5] There are actually two issues involved. One is the cause or causes of abandonment. The second is the reason or reasons why terraces are not restored to cultivation, given apparent land shortages in the Andes (Denevan 1986). Terrace restoration projects in Peru have paid Indian farmers to rebuild terraces and associated canal systems without knowing why the farmers themselves do not rebuild and use old terraces. It is quite possible that once terraces are rebuilt they will not be cultivated because of lack of sufficient irrigation water, conflicts over land rights, excessive labor costs, excessive distances from villages, or for other reasons. The situation is complex, involving environmental change, technology, social and economic organization, demography, and markets. Conditions change over time and from village to village, both for the Colca and for other terraced regions of the Andes (Treacy and Denevan 1994).

Mapping Abandonment

In order to determine the degree of terrace abandonment, cartographer Laura Hartwig and I mapped and measured both cultivated and abandoned fields for the Colca Valley (Denevan 1988). We used the 1974 overlapping vertical air photos which were taken for the Majes irrigation project, which diverts water from the Río Colca to the coast.[6] At a scale of 1 : 17,000 they are quite suitable for differentiating cultivated from abandoned terraces.

Our study included ten of the twelve Colca communities, Tuti and Tapay lacking complete photo coverage. Seven categories were mapped: Upland (sloping-field terraces) cultivated and abandoned; Terrace (irrigated bench terraces) cultivated and abandoned; Bottomland (valley-floor bench terraces and enclosed non-terraced fields) cultivated and abandoned; and Not Cultivated land.

For the mapped sectors of the valley (some of the upper zones are not on the air photos), we determined that the total area of fields was 14,356 ha, of which 6,071 ha, or 42 per cent, were abandoned. The total area of bench terraces was 8,962 ha, of which 5,426 ha, or 61 per cent, were abandoned. Of the sloping-field terraces, 91 per cent were abandoned, and of the valley-floor non-terraced fields, only 7 per cent were abandoned, possibly in fallow.

For the community of Coporaque, using the 1974 air photos at 1 : 17,000 scale, we measured 1,399 ha of fields, of which 726 ha were abandoned, or 52 per cent. Initially, Treacy (1987*a*: 153, 155), using the 1955 air photos at 1 : 55,000 scale, obtained a total of 1,445 ha, of which 970 ha, or 67 per cent, were abandoned in 1983–4 when 475 ha were reported in cultivation or in fallow. Using our 1974 air photo mapping units plus ground checking to revise our categories of field types and cultivated versus abandoned fields, we measured a new total of 1,456 ha, of which 784, or 54 per cent, were abandoned (Denevan 1988: 22). This was then adjusted by Treacy to give a total field area of 1,367 ha, of which 847 ha, or 62 per cent, were abandoned or semi-abandoned (Treacy 1989: 75–6; 1994*b*: 66). His 62 per cent abandonment for Coporaque, compared to our 52 or 54 per cent, suggests that for the entire valley our 42 per cent based on air photo analysis alone, without ground checking, may be too low.

In Coporaque, of the valley-side bench terraces (270 ha), 43 per cent are abandoned or semi-abandoned; 10 per cent of the valley-floor bench terraces (267 ha) are abandoned or semi-abandoned; 100 per cent of the sloping-field terraces (246 ha) and stone-bordered fields (460 ha) are abandoned; and none of the valley-floor walled, non-terraced fields are abandoned (Treacy 1994*b*: 66). Most of the bench terraces are or were irrigated; however at least 20–30 ha of the valley-side terraces are semi-abandoned and may never have been irrigated. The distribution of the various types of terraces and other fields in Coporaque is mapped by Treacy (1994*b*: 87). Figure 10.7 shows cultivated and abandoned terrace areas for most of Coporaque and Chivay and part of Yanque.

Our mapping and measurements from air photos demonstrate that: (1) The degree of abandonment is considerable. This supports undocumented speculation that over half the visible terraces in the Andes are abandoned. (2) The abandoned terraces tend to be on the high slopes and more distant from

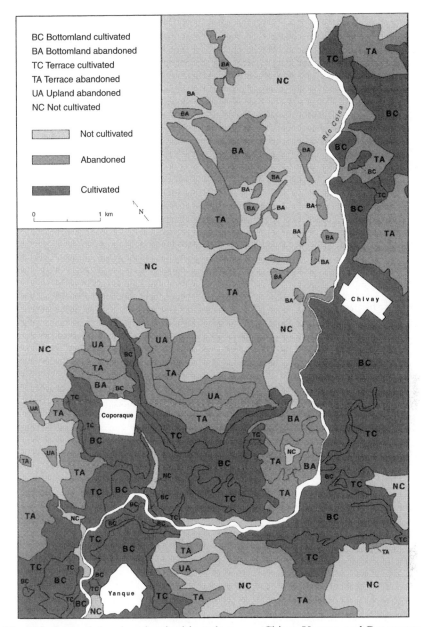

Fig. 10.7. Map of abandoned and cultivated terraces, Chivay, Yanque, and Coporaque, Colca Valley, Peruvian Andes. Based on 1974 air photos. By Laura Hartwig; corrected by John Treacy from ground observation. From Denevan (1988: 27)

current villages than the now cultivated bench terraces and bottomland fields. Accessibility is clearly a factor. (3) Most of the unirrigated, sloping-field terraces are abandoned—91 per cent overall and 100 per cent in some communities. Sloping-field terraces still cultivated may be irrigated from springs. (4) The least abandonment is on the *pampas* near the Río Colca—only 7 per cent overall and 0 per cent for some communities. Here, many of the fields are now enclosed by adobe walls; they are irrigated, but many are not terraces.[7] (5) Within the mapped sectors of the valley, which total 18,128 ha, 46 per cent is now cultivated, 33 per cent consists of abandoned terracing, and only 21 per cent is neither. Thus a very high portion (79 per cent) of the landscape is cultivated or was cultivated in the past.

Time of Abandonment

To understand why bench terraces were abandoned we need to know not only where abandonment occurred but when. For Coporaque, the sloping-field terraces at Chilacota must have been abandoned before the Inca period (AD 1470–1530), given the absence of Inca-influenced ceramics, probably during one of the preceeding prolonged dry and/or cold periods. The bench terraces at Chijra and Alto Cayra, now abandoned, were undoubtedly in use during the Inca period, given the presence of Inca ceramics. The *visitas* for 1591, 1604, and 1616 give the place names for terrace irrigation sectors, with data on crops grown in each. Many of these names (toponyms) persist today. From the three *visitas*, Treacy (1994*b*: 259–62) located 74 sectors for Coporaque of which twenty-three are not listed for 1616, suggesting their abandonment between 1591 (or 1604) and 1616.

The presently cultivated sectors of Cayra and Chacco were in cultivation in 1591, 1604, and 1616 (also in 1931). Chijra does not appear in the *visita* lists and apparently was abandoned by 1591. Other Coporaque terrace sites listed in the *visitas* were not identified and located by Treacy and probably represent now abandoned fields. While we cannot be more precise about the dating of terrace abandonment, much of it certainly occurred during the early colonial period. Unirrigated bench terraces were abandoned earlier, at least in the Japo Basin; Brooks did not find Collaguas sherds on most of the unirrigated bench terraces above 3,800 m. Undoubtedly, terraces went in and out of cultivation from place to place and from time to time depending on circumstances, in addition to short-term periods of fallow, and long-term abandonment.

Theories of Abandonment

The causes of irrigated terrace abandonment in the Andes have been the subject of considerable discussion and debate. Field (1966: 484–9) emphasized population decline and land-use change rather than climate. A. Wright (1963:

73–4) mentioned a 'drying landscape' and also out migration. Donkin (1979: 35–8) gave priority to colonial depopulation, but he also discusses climatic and other theories.

A major discussion of abandonment appears in the journal *Current Anthropology* (Guillet 1987*b*). Guillet's article, focusing on Lari in the Colca Valley, is followed by eleven commentaries, including by Andean cultural ecologists William Mitchell, Jeffrey Parsons, Jeanette Sherbondy, Gregory Knapp, and John Treacy; Guillet provides a final response. Guillet (1987*b*: 409) believes that terrace expansion and contraction are closely related to water availability, with abandonment in the short term due to 'constraints in the system of water distribution', and in the long term due to periodic droughts. 'During periods of relative water abundance, constraints are relaxed, allowing new terraces to be constructed and abandoned ones rebuilt.' The commentators elaborate on this thesis, raise objections, and suggest other significant factors. Treacy (p. 425) pointed out that 'one need not invoke climatic shifts alone to account for water scarcity'. J. R. Parsons (p. 423) says there needs to be 'a systematic consideration of the cultural forces that may be casual factors in terrace expansion and contraction'. Mitchell (p. 422) emphasizes that terrace farmers today are involved in a market economy, so that there is not a direct relation between population and amount of land in cultivation. Knapp (p. 421) reminds us that 'deintensification' is a more accurate term than 'abandonment', since no-longer irrigated terraces may still be grazed or dry farmed from time to time.

Factors of Abandonment

Prehistoric Abandonment We have mentioned that the unirrigated, sloping-field terraces and bench terraces were abandoned, either because of reduced rainfall or because of the greater reliability of irrigated terraces. There was a second period of prehistoric abandonment of some of the higher irrigated bench terraces such as at Chijra and Alto Cayra, for reasons unknown. There was a concomitant shift of settlement downslope, during one or both periods, and probably a shift from a staple of frost-tolerant and low moisture-tolerant quinoa to less tolerant maize. Thus there was considerable abandonment of terraces prior to the third period of abandonment during the early colonial period.

Colonial Depopulation We know from the ethnohistorical study by N. Cook (1982) that the Colca Valley suffered a major population decline in the sixteenth century due primarily to introduced epidemic diseases. For the Colca villages the estimated decrease from 1530 to 1721 was 87 per cent. For the community of Coporaque, the population changed from an estimated 5,957 in 1520 to 1,956 in 1604 (Treacy 1994*b*: 167), a 67 per cent decrease in less than a century. Commonly, under conditions of depopulation, the most intensive forms of cultivation and the least accessible agricultural lands are abandoned. In the Colca Valley, with Spanish-controlled Indian settlements being located on the *pampas* near the river, and the population reduced, the higher terraces were indeed abandoned. Population has partly recovered, but the 1981 total for the Collaguas villages was still only 68 per cent of the 1530 estimate. This compares with at least a 42 per cent reduction in land cultivated; however, much of the land now cultivated is in alfalfa for livestock for market; in addition, some food staples are brought in from outside the valley.

Climatic Change The relationship between terrace construction, use, and abandonment and climate is complex and inconsistent. Less precipitation (both rain and snow) would presumably mean a reduction of water available for irrigation and hence a reduction of terrace cultivation. On the other hand, decreased rainfall could have made rainfed agriculture impossible and labor-intensive irrigated terracing feasible. Farmers today say they would restore old, unused terraces to production if they had more water. However, abandoned canals reaching *quebradas* fed by snowmelt on the high slopes, such as Cerro Mismi, could be restored and irrigation water thus increased, but this has not happened. At Coporaque, there are three sectors of abandoned terraces, including Chijra, totaling 112 ha, that could be restored, each linked to unused canal systems (Treacy 1994*b*: 227).

I have already suggested that the shift to irrigated bench terracing first occurred during the dry period between AD 540 and 730, with later expansion during the later dry period AD 1040–1490. Large portions of both periods experienced colder temperatures which would have forced cultivation downslope where precipitation is even less, but only down *c*. 150 m or less. The subsequent period, AD 1500–1720, was wetter than average, but this was the time of probably the greatest abandonment of irrigated terraces.[8] The twentieth century has also been a wetter than average period, but without significant terrace restoration. (There has been some expansion of valley-floor *pampa* fields.) Thus climate change would not seem to be a primary factor in post-conquest terrace abandonment or in the lack of much restoration of old terraces today.

Springs Many terraces on slopes are watered from springs rather than from irrigation canals leading from *quebradas* fed by rainfall and snowmelt. These springs stop flowing or dry up due to unclear events, including depleted aquifers and tectonic activity; periodically they may revive and new springs may appear. Terraces are abandoned or restored accordingly.

Tectonic Activity Stanish (n.d.) has suggested that in the Moquegua Valley, tectonic uplift may have disrupted terrace/canal connections resulting in terrace abandonment due to the cutoff of irrigation canal intakes from stream channel sources which have downcut. David Keefer (pers. comm. 1997) reports that in the Moquegua area, landslides caused by earthquakes buried a section of a prehistoric canal resulting in terrace abandonment. Some terraces have been destroyed at Lari by landslides due to undercutting by the Río Colca.

Canal Water Loss Stanish (1987: 360) believes that in the Moquegua region of southern Peru prehistoric settlement may have been relocated up river in association with shortening of irrigation canals in order to reduce water loss from seepage and evaporation. This could have resulted in terrace abandonment in the original sectors irrigated downriver. This does not seem to be a factor in the Colca Valley where most canals have always been of relatively short length.

Canal Abandonment When canals are abandoned the associated irrigated terraces will, of course, also be abandoned. Canal abandonment may be due to a reduction of the amount of water available, but also for social and political reasons. Water may be diverted from one community or farming sector to another. In the Colca Valley, such conflicts are of long-standing and have led to bloodshed. Also, the labor required to maintain and repair a canal may become excessive in relation to productivity, especially on high, steep, and distant slopes. Figure 10.4 shows abandoned canals and a reservoir and aqueduct that once provided water to the terraces at Chijra and Alto Cayra. The abandoned 10 km long Carhuasanta Canal north and east of Cerro Mismi (Fig. 10.1) brought water from the Río Apurimac drainage (Amazon) to the Río Colca drainage (Pacific Ocean) on the western slopes of Pampa Finaya via another 30 km of *quebrada* and canal (Treacy 1994*b*: 116–20).

Water Availability and Management Efficiency How much water is actually available at the present time in canal systems in relation to crop needs? The study of the hydrology of the irrigation system in Coporaque by Waugh and Treacy (1986) suggests that there is nearly three times as much water available for crop production as is used by the present crops. This would seem to indicate that in terms of water availability the area cropped could be considerably expanded. We must keep in mind, however, that during some periods of the growing season there is more irrigation water available than can be used and the surplus is vented into the main *quebradas* and the Río Colca and thus lost to the crop system. Furthermore, no system of water management for crop production is 100 per cent efficient in capturing all the water brought into an irrigation system. Nevertheless, the loss of two-thirds of the water available seems excessive and raises the question of possible change in water-management efficiency since prehistoric times.

Variation over time in water-management efficiency would mean increased or decreased water supplies for irrigation. For example, if water is dropped too rapidly through a terrace system, more water runs off rather than being absorbed by the soil and being available for greater crop production. Indications are that water-management technology in pre-European times differed from current technology. For one thing, the Spaniards destroyed a social organization that probably contributed to efficient management of complex terrace/irrigation systems.

Maria Benavides (1997) points out that in the Colca Valley, there is insufficient irrigation water available to irrigate both the existing abandoned terraces and the currently cultivated terraces. Thus, she believes, when terraces were developed at lower elevations, where irrigation is more feasible, the terraces at higher elevations had to be abandoned. However, the higher terraces were not irrigated but were fed by rainfall and runoff. The moisture captured by the high terraces, which otherwise could have ended up in the canal system for the irrigated terraces below, would have only been a portion of the total water available for irrigation.

Vegetation and Soil Change Agriculture and overgrazing have drastically modified and reduced the natural vegetation cover, making slopes more susceptible to rapid runoff rather than water retention in the soil. Hence, the moisture available to crops from rainfall and runoff is reduced. Soil analyses by Sandor (1987*a*; 1987*b*) indicate that soil fertility was improved with terracing and thus was not likely a factor in terrace abandonment. Soils are mostly derived from volcanic and alluvial material and are relatively fertile. Furthermore, they are now maintained by the application of manure and compost and by periodic fallowing, and both were likely true prehistorically. Sandor surprisingly found fertility of the surface soils to be comparable in both cultivated and abandoned terraces. The latter are not actively fertilized, but do receive manure from grazing livestock.

Loss of Terrace Knowledge An anonymous reviewer has queried whether terrace abandonment could be the result of post-Columbian loss of knowledge of how to construct and maintain terraces effectively. Ancient terraces are still maintained and repaired and some new bench terraces are being constructed. The quality of terraces has indeed deteriorated in some places, resulting in premature wall collapses and washouts (Treacy 1987*b*: 53–4). A higher frequency of breakdowns occurs where previous breeches were not adequately repaired. However, terrace farmers are well aware of good versus poor practices, so deficiencies are not necessarily a matter of loss of knowledge. And regardless of their knowledge about maintenance and rehabilitation, farmers today know little about how their terraces were initially developed, most present terraces having been first constructed hundreds of years ago (L. Williams 1990: 91). Involved is not only information about construction, but also decision-making as to location, dimensions, soil and water management, etc. Knowledge which may have been lost could have been replaced by adequate similar or different techniques. So we do not know if loss of traditional knowledge has contributed to terrace abandonment, but this seems unlikely.

Labor Inputs today are clearly larger for terraces on higher slopes compared to those for lower slopes. This is due to: (1) greater distances from settlements to fields and hence travel time, (2) the difficulty of working on high, steep slopes, and (3) the higher level of maintenance necessary for terraces on steep slopes. Also soils may be poorer on higher, steeper slopes, resulting in lower crop yields per investment of labor. Nevertheless, farmers in Coporaque say that high labor input is not the reason why they do not restore and farm abandoned terraces.

Changes in Land Use The introduction of European crops and cattle caused the breakdown of a fragile agroecology. Cattle rapidly damage terraces, increasing maintenance, and at times leading to terrace abandonment. The walls of bench terraces, both abandoned and used, have been ripped out in some areas in order to facilitate both plowing and grazing. The Spaniards brought about a greater emphasis on pastoralism, which has low labor requirements, at the expense of agriculture. The recent shift towards planting alfalfa for cattle feed has reduced the land in subsistence crops. Thus, the relation between land terraced, food production level, and number of people supported is less direct than in the past. Finally, segments of terraces, abandoned and cultivated, have been destroyed by the road and canal-building activities of the Majes water diversion project.

Social Access to Land and Irrigation Water Communities as well as individual households do not have equal access to irrigation water. (This can result in violent conflict or in efforts to renegotiate access to water.) One community or some farmers, as a result, can have sufficient water to cultivate all the land available, whereas adjacent communities or neighbors do not have enough water to irrigate all their terraces. Furthermore, if more water were available it would be very difficult for people in a community to agree on who would have access to it as well as to the new land which would be irrigated. Additional land for additional irrigation would not likely come from private holdings but rather from unused land that is almost all communal land used for pasture. J. Treacy (pers. comm. 1988) indicated that in Coporaque in the 1980s there were prospects of obtaining additional water and opening up an abandoned canal (the 'Inca' Canal east of the village), but agreement could not be reached on who would get the abandoned terrace land for restoration and the new water to irrigate it, and hence the plan was dropped.

As Treacy (1989a: 342) pointed out, farmers are most concerned with the timing of water availability, 'making sure that supplies are distributed in an orderly fashion among many users during the time scarce supplies are available to seed crops. The total amount of water [available] is an abstraction of little interest to many Coporaqueños'. Timing has to do with crop needs in relation to climate, water storage capacity, the irrigation system, and the social allotment of water. 'Few would be willing to give up their water rights and privileges to risk watering new lands under unfavorable conditions.' Water security is more important than maximizing water supply in order to increase land area cultivated. For a discussion of the problems with attempts to restore abandoned terraces at Coporaque, see Treacy (1987*b*).

Alternative Livelihoods Under conditions of population growth, and even with population stability, there may be attractive alternatives to farming terraces under difficult conditions as well as to subsistence farming in general. These alternatives include seeking wage labor locally and migration in search of jobs in Arequipa or on the coast. The attractions may be food, or cash, or urban excitement and opportunities, or hopes of a better life for the next generation. Low market prices for farm staples is another factor. Rural to urban migration is occurring throughout the South American Andes for these reasons.[9] For the twelve central Colca villages the population increased from 15,201 in 1940 to only 21,517 in 1981[10] (Denevan 1987*a*: 17), a period during which the population of Peru nearly tripled. The young people of the Colca Valley have chosen to go elsewhere rather than restore and farm abandoned terraces.

By the completion of our Colca terrace project, the focus of research had shifted from explaining bench terrace abandonment, which was primarily related to sixteenth-century population decline, to explaining why abandoned terraces had not been restored and brought back into production. We realized that in our emphasis on ecological research we had neglected socio-economic matters which are critical to understanding why abandoned terraces stay abandoned—resource management policy (especially water rights), land tenure, alternative livelihoods. These conditions, of course, are themselves affected by the expansion and contraction of land under cultivation. Subsequent studies of the Colca Valley gave greater attention to water management as the key to land use, particularly terrace use: Treacy (1989*a*; 1994*b*) on Coporaque, Guillet (1987*b*; 1992) on Lari, and Gelles (1990; 1994) on Cabanaconde, as well as elsewhere in the Andes (Mitchell and Guillet 1994).

The short- and long-term variability in the environment, especially climate, creates conditions which the Colca people have more or less adapted to by changing the form and location of terracing. However, demographic growth, or lack of growth, or decline seem to be the most important reasons for terrace expansion or contraction. Today, there is demographic stagnation without terrace restoration. There is demographic stagnation because people perceive better opportunities elsewhere. As the population continues to grow and external opportunities decline there may be an indigenous movement back to the semi-arid Andean mountain valleys such as the Colca. In the Colca, at least, environmental conditions do not seem to preclude the restoration of now-abandoned irrigated terraces. What was done in prehistoric times should be able to be done today, and without resorting to energy- and capital-expensive modern technology.

Notes

[1] Treacy (1987*a*: 53; 1994*b*: 147) believed that the Coporaque bench-terrace fills were done by hand, whereas sloping-field terraces were self-filling via downslope movement of soil. Sandor (1987*a*: 185) believes the Coporaque bench terraces were filled by natural slope processes.

[2] These local names are not villages but rather locales, specific habitats with clusters of fields, whose present toponyms can usually be found in the early colonial *visitas*.

[3] Based on Winterhalder (1994: 47, 60). Winterhalder demonstrates why irrigation is required in the Colca Valley on the western side of the southern Peruvian Andes, whereas in the Sandia Valley on the eastern side rainfed cultivation of terraces is possible due to greater total rainfall and other factors.

[4] Cardich (1985) argues that shifts of cultivation to lower slopes in the Andes was due to climatic changes toward colder temperatures. He believes it was warmer from 300 BC to AD 500; colder from AD 500–1000, warmer from AD 1000–1350; and colder from AD 1320 through the Inca period. This might explain an abandonment of the Chilacota fields after 1320; however temperature would not explain the often sharp division between irrigated and unirrigated terraces elsewhere.

[5] Proponents of the restoration and expansion of traditional terracing in Peru include Peruvian ecologist Marc Dourojeanni (1983: 66, 70) and several contributors to the volume on *Andenes y camellones* (Torre and Burga 1986). The United States Agency for International Development had a major project to improve traditional terrace agriculture and to construct new terraces in Venezuela, Guatemala, and Peru (L. Williams 1986).

[6] The 1931 vertical photos at a scale of 1 : 13,000, taken by the Shippee/Johnson Expedition (Denevan 1993), are of excellent quality and we used them to determine the change in extent of terracing for the community of Coporaque between 1931 and 1974. The difference was very small— an increase of overall cultivated area of 30 ha or *c*. 4 per cent. This increase was on sectors of *pampa* (river terrace) land which had been uncultivated in 1931.

[7] Most of the enclosure process occurred between 1930 and 1980 in order to prevent animal infringement into fields, thus eliminating open-field, communal grazing (González 1995).

[8] This is true for most of the Colca Valley, but not for the Japo Basin where very few irrigated terraces have been abandoned.

[9] Depopulation or stagnation has occurred in many Andean villages in recent decades, despite higher fertility and survival rates (Preston 1996).

[10] The 1993 Census gives a lower total of 18,344 (Brooks 1998: 104).

Part IV

Raised and Drained Fields

Clearly, other men at other times have found other ways of living in this tropical floodplain environment with tools inferior to ours.

(James Parsons and William Bowen 1966: 343)

11

Lost Systems of Cultivation

The Indians who lived and still live in the grassy savannas, unhindered by forests, obtain their crops with less work . . . , because with *macanas* they can throw up earth in wetlands along both sides of a ditch, covering the grasses with the excavated material, and then planting maize, manioc, and other roots, and considerable pepper.

(Padre José Gumilla [1745] 1963: 429)

Discovery

Prior to 1963, the term 'raised field' was not to be found in either the agricultural or archaeological literature. And the term 'drained field' was mostly used for modern wetland fields. In 1961, geologist George Plafker and I became aware of what seemed to be pre-Hispanic agricultural fields in the form of ridges, mounds, and ditches in the seasonal swamps of the Llanos de Mojos, an enormous but little known savanna in Bolivian Amazonia (Plafker 1963; Denevan 1962; 1963*a*). Aerial reconnaissance, ground survey, and examination of air photos convinced us that these features were indeed man-made and ancient. We initially could find no published mention of such fields in Bolivia or anywhere else in South America. Thousands of these fields could be seen in Mojos, many in very good condition, some over 350 m in length and over 20 m in width. So their unexpected discovery was dramatic. They were indicative of intensive agriculture and dense populations sometime in the past in a region where, in 1961, savanna cultivation was non-existent and populations sparse. This led to my doctoral dissertation (Denevan 1963*b*; 1966*a*) on these fields and associated prehistoric causeways, habitation and ceremonial mounds, and artificial canals.

We considered the Mojos raised fields to be unique in South America, a lost system of wetland cultivation that had been abandoned with the Spanish conquest in the late sixteenth century. However, similar relic fields began to be reported and described in other tropical savannas and also in the Andes, especially around Lake Titicaca (Figs. 8.2, 11.1). Reports in the *New York Times* (27 July 1966) and in *Scientific American* (Parsons and Denevan 1967) generated international attention and further sightings of raised fields elsewhere in the Americas, as well as in New Guinea and in tropical Africa. An

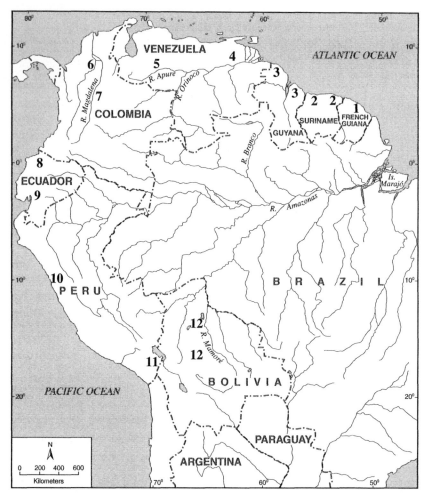

Fig. 11.1. Map of principal raised- and drained-field locations in South America. 1, French Guiana; 2, Suriname; 3, Guyana; 4, Karinya; 5, Barinas; 6, San Jorge; 7, Bogotá; 8, Highland Ecuador; 9, Guayas; 10, Casma; 11, Titicaca; 12, Mojos. By W. M. Denevan

archaeological feature that had been mostly unknown in 1961 had become commonplace less than a decade later. General descriptions of New World raised and drained fields subsequently include Denevan (1970*a*; 1982), Darch (1983), and Turner and Denevan (1985).

Early Reports

Most of the South American raised fields are pre-European, now abandoned, and only recently discovered. However, there are some minor exceptions to these generalizations, and these merit examination to provide perspective on the explosion of attention since 1963 and the now-recognized importance of these fields.

There are actually several colonial reports of raised fields which were observed under cultivation, so at least in some areas they had continued in use after the conquest. In 1745, the Jesuit missionary José Gumilla (1963: 429) said that in the Orinoco savannas past and current Indians piled up earth along ditches for cultivation, using wooden spades (*macanas*). Earlier, in 1589, in the same region, the Spanish explorer Juan de Castellanos (1955: 1: 539) reported causeways in association with abandoned raised beds (*labranzas viejos camellones*). In both instances no indication is given as to specifically where in the Llanos these fields were located. Gumilla (1963: 430–1) also reported that the Otomac Indians in the savannas between the Orinoco and the Río Apure planted crops on the grassy margins of lagoons as the lagoons dried up ('recessional' cultivation). There are several reports of *camellones* in highland Ecuador between 1573 and 1668 (Caillavet 1983: 12–14), and in the Sabana de Bogotá in the sixteenth century (Knapp 1991: 147); descriptions are poor, however.

Another early report of active raised fields was in Mojos. When I was doing research there in 1961–2 I diligently searched the historical literature for early mentions of raised fields without success. Later, however, the Jesuit historians Tormo and Tercero (1966: 97) found a brief mention of apparent raised fields in a Jesuit document dated 1754: 'they plant their crops in the *pampa*, making ditches and heaping up earth.'

The prehistoric raised fields around Lake Titicaca should have been obvious to any farmer or traveler in the vicinities of Puno, Juliaca, Desaguadero, Pomata, and other towns. However, there are only a few scattered mentions of them in the early twentieth century. Actual floating gardens consisting of potato crops grown on lake-bottom muck placed on totora reed mats by Uru Indians were described by Balthasar Ramírez in 1597 (Maurtua 1906: 1: 295), and the practice has continued to the present. However, these are not really raised fields.

Prior to 1961, there were only a few mentions by modern scholars of raised fields in South America, and recognition of them as being agricultural. The great Swedish anthropologist Erland Nordenskiöld, who explored Mojos in the early twentieth century, wrote that: 'In some parts of Mojos people have attempted to make seasonally flooded soils useful by drainage. In particular, it was brought to my attention that between San Borja and San Ignacio there are drainage works consisting of trenches which are often parallel'; also, 'I have seen remains of drainage works in the region of San Ramón which are often parallel and which were probably excavated' (Nordenskiöld 1916: 149–50). These are the only mentions of drainage features I know of in Nordenskiöld's various writings on Mojos, and apparently he had not seen many raised fields and was not impressed with them.

The French anthropologist Alfred Métraux (1942: 19) in his classic monograph on eastern Bolivia said that: 'In the region inhabited today by the Chiman Indians, especially between San Borja and San Ignacio, there are remains of large canals, dikes, and raised earth platforms built to drain the vast marshes and to convert them into fields.' Métraux did not give a source for this information (probably Nordenskiöld), and he apparently did not do any field work in Mojos himself. In 1962, I traveled from Bolivia to an international anthropology congress in Rio de Janeiro specifically to meet with Métraux and discuss the Mojos raised fields with him. This was shortly before his death. I showed him some spectacular aerial photos, but to my disappointment he showed little interest.

Max Schmidt (1974: 64–5), in 1914, briefly mentioned 'mound cultivation' (*aterrados*) in the upper Paraguay region, Mojos, Marajó Island, the delta of the Río de la Plata, the Antilles, and Lake Titicaca.

In Suriname, Dutch agronomist J. de Kraker reported apparent raised fields in the coastal savannas in 1939. Several clusters were described by Dutch geographers and archaeologists in the 1950s, but these studies received little attention (Boomert 1976). In most of these instances the anthropologists involved were unaware of the extent of the fields they had reported or their potential significance as evidence of intensive pre-European agriculture in marginal habitats.

The geographer Carl Sauer mentioned wetland mounds and ridges in his classic *Agricultural Origins and Dispersals*, but no locations were given: 'Here [New World tropics] . . . planting is done by setting out cuttings, usually in mounds or ridges of top soil thrown up by spade-like tools. . . . In contrast to the Old World there are no wetland domesticated plants. All thrive best with good drainage. Where drainage was poor the cultivators built the mounds high to provide aeration' (C. Sauer 1952: 45). Clearly, Sauer was aware of this well before the rest of us.

It was not until 1963–7 that raised fields began to receive serious attention, as well as media interest. Why so late, given the extent of the fields and the numerous regions where they occur? First of all, raised fields are not conspicuous at ground level, especially when covered by tall grass and when relief has been considerably reduced over many years by sediment accumulation. Many are completely buried and are observable only at river or road cuts. Various observers must have seen them—locals, travelers, scholars—both on the ground and from above. However, they apparently assumed that they were natural features or some sort of modern earthworks, and the seemingly obvious questions of age, origin, and function were not asked. It is noteworthy that geographers, trained to interpret landscapes, have been more instrumental in the discovery and initial reporting of raised fields than archaeologists.

Another factor in the lateness of discovery of raised fields is that almost all of them ceased being cultivated in pre-European times or early in the colonial period as a consequence of population shifts or reductions or possibly environmental change. Finally, many of the areas with raised fields, such as in Mojos and the Orinoco Llanos, are remote and had not been carefully examined by scholars until recently.

It was observation from the air, subsequent examination of aerial photography, and the publication of low-level photos that brought attention to the raised fields in South America. The availability of all of these has accelerated since the mid twentieth century.[1] From the air, raised fields stand out distinctly in times of flooding, and even in dry periods the ditches between fields retain more moisture than adjacent field tops and thus are greener, outlining the drier fields. Surveys of the use of air photography in archaeology have given considerable attention to the discovery of raised fields in South America and Mexico (Deuel 1969: 256–62; Reeves 1975: 2: 2003–8).

There are a few examples of raised fields still being cultivated in South America. In the 1960s a small sector of raised fields was being farmed near the town of Batallas on the Bolivian side of Lake Titicaca. These were about 2 m wide, based on a photo given to me by geographer David Preston. Clark Erickson (n.d.) briefly mentions some others. In the Sabana de Bogotá in highland Colombia, mounds (30 cm high, 1 m in diameter) and ridges (*camellones*) (up to 2 m wide and 30 m long) are cultivated (Eidt 1959: 386–7; 1981: 37).

A raised feature which is very common in the Andes today is a low, narrow crop ridge or garden bed (under 1 m wide), similar to the European 'lazy bed'. They are called *eras* in Colombia (R. West 1959; Bruhns 1981) and *huachos* (*wachus*) in Peru (Denevan 1970*a*: 651; Zimmerer 1996*a*: 130–1). They appear in long, parallel groups both on slopes (often aligned vertically) and on flats. In both instances their primary function seems to be to improve drainage, but they also improve the cultivability of heavy grassland soils. They are typically turned over with the *taclla* (footplow) in Peru and Bolivia and with the *chuzo* (a type of hoe) in Colombia. They are usually planted in potatoes and other Andean tubers. Most of these active raised fields are small and ephemeral in that they do not survive for centuries as do the large pre-European raised fields. They are rebuilt every few years as new fields.

Characteristics of Raised Fields

We struggled at first to decide on appropriate terminology for artificially elevated earthen fields, most of which improved drainage in wetlands (lake edges, seasonally inundated savannas, floodplains, swamps). Plafker (1963: 372) referred to 'furrow-ridge' fields where ditches and ridges alternated; 'mound fields' which consisted of circular mounds; and 'furrows' for ditching without raised surfaces. 'Furrow', however, is an inappropriate term as it implies plowing, and these fields were not plowed. Parsons and I (1967) used 'ridged fields', which wasn't satisfactory because both ridges and flat platforms occur. In 1963 and 1966 publications on Mojos, I used 'raised fields' just for platform fields. Then I used 'drained fields' in an article in *Science* (Denevan 1970a). One problem is that not all drained fields are raised and not all raised fields are drained.

In 1974 in an article on Old World elevated fields, Bill Turner and I defined 'raised fields' as 'any prepared land involving the transfer and elevation of soil above the natural surface of the earth in order to improve cultivating conditions' (Denevan and Turner 1974: 24). Since then, 'raised field' has been a general term for any kind of artificially elevated field—ridges, platforms, mounds—regardless of size, shape, and function. Siemens (1998: 43) prefers 'platforms and canals' for prehistoric wetland fields in Veracruz, Mexico. Some prefer 'ditched field', but not all ditched fields are raised fields. In Spanish, the common terms used are *camellón*[2] (planting bed), *campo elevado* (raised field), and sometimes *raya* (stripe).

The building of earth ridges and platforms serves numerous cultivation functions, of which several are usually operative for a specific field system. It may not be easy to determine the primary function or functions of abandoned fields. These functions include drainage by means of field-raising and field-ditching, soil aeration, reduction of root rot, increased nitrification, pest reduction, pH increase, moisture retention (in ditches), fertility enhancement (by application of humic soil, organic residues as compost, and muck from ditches), easier weeding, easier harvest, and increase in soil and water temperatures (Denevan and Turner 1974; Sillitoe 1996: 378–92; Erickson 1998; Kolata 1993: 183–90). All the large, prehistoric raised fields in South America are on sites with seasonal flooding or permanent standing water or severe waterlogging, which suggests that drainage was the primary function. However, soil improvement functions may have been equally as important.

Examples of shapes, profiles, and patterns of raised fields are shown in Figs 11.2 and 11.3. Some fields are over 1,000 m long (Orinoco), as wide as 25 m (Mojos, Bolivia), and as high as 2 m (San Jorge, Colombia). There is a very wide variety of sizes, forms, and arrangements or patterns of raised fields, and we can only speculate on the reasons for these differences. Factors to be considered include length and depth of flooding, slope and soil, land tenure, and simply cultural or individual differences. Some patterns of fields and ditches suggest an intentional effort to either run water out of the fields or to retain water within the fields as a means of irrigation when rainfall and water-table levels are insufficient. In both Mojos and around Lake Titicaca, we observed embanked clusters of raised fields, with outlets, that may have served for water retention (Plafker 1963: 375; C. Smith *et al.* 1968: 357; Erickson 1993: 395).

The areas covered by raised fields and associated ditches range from over 120,000 ha at Lake Titicaca to only 1,550 ha at Caño Ventosidad in Venezuela. Productivity can only be determined from contemporary or experimental raised fields. Comparisons can be made with yields from non-raised fields in the same locales. The best data comes from northern Lake Titicaca where potato production on raised fields (including ditches) ranged from 8 to 16

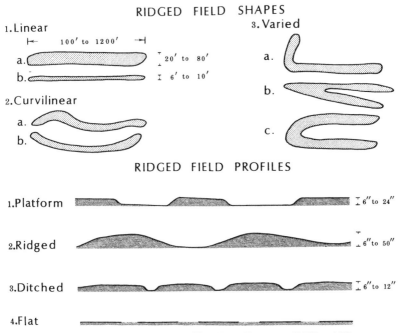

Fig. 11.2. Raised-field shapes and profiles. From Denevan (1970*a*: 649)

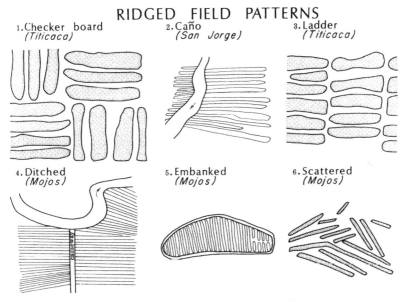

Fig. 11.3. Raised-field patterns. From Denevan (1970*a*: 650)

metric tonnes/ha, double to triple the harvest for non-raised fields (Erickson 1993: 407). For experimental fields in Mojos, agronomist Julio Arce obtained 25 metric tonnes/ha of manioc, well above local yields (Erickson 1994*a*). Given measured field surfaces and feasible crop productivity, it is possible to estimate the pre-European populations supported per hectare of raised fields. For example, for the Lake Titicaca raised fields Erickson (1992: 297) estimates 37.5 people per ha of cultivated surface, and Kolata (1993: 201) obtains 39 per ha.

The Raised and Drained Fields of South America

Chapters 12 and 13 describe prehistoric raised fields in Mojos in Bolivia and at Lake Titicaca in Peru and Bolivia. Chapter 14 discusses prehistoric and contemporary ditched fields. Following are brief descriptions of other raised fields in South America and elsewhere.

San Jorge, Colombia

The next large group of raised fields to be identified after Mojos is located in the seasonally flooded Mompos Depression of northern Colombia between the Cauca and San Jorge rivers (Fig. 11.4), and also in the basin of the lower Río Sinú (Plazas *et al.* 1993: map 2). They were discovered, initially on air photos, by James Parsons in 1965 (Parsons and Bowen 1966). There are at least 90,000 ha of surviving field and ditch surfaces (J. J. Parsons, pers. comm.).[3] Many of the ridges occur on natural levees at right angles to stream channels (Fig. 11.3), whereas others are in checkerboard patterns. They average 6–7 m in width and a meter in height. Some have become grown over in forest since they were abandoned. Others are now buried under sediment, with a few exposed along river banks where channels have cut into buried ridges and swales. The Mompos Depression was already known for its prehistoric mounds and elaborate gold work created by the Sinú (Zenú) chiefdom. The ancient fields have since received major attention from Colombian archaeologists Clemencia Plazas and Ana María Falchetti (1981; 1987; 1990; Plazas *et al.* 1993). In a 1,440 ha sample area of raised fields, with 400 habitation platforms, they estimate that a population of 2,400 people and a density of *c*. 160 per km^2 had been supported (Plazas and Falchetti 1987: 498). They obtained possible dates for canals as early as 810 BC and 330 BC (Plazas *et al.* 1993: 61). After AD 1000, there was a gradual abandonment of the wetland area, and colonial sources do not mention drainage systems.

Venezuelan Llanos

The traditional view of indigenous subsistence in the Venezuelan Llanos is one of shifting cultivation confined to the gallery forests along the major rivers and

Fig. 11.4. Raised fields in the Río San Jorge floodplain of northern Colombia. Photograph by William Bowen, 1965

to the larger islands of forest within the savannas, with the great grasslands serving only for hunting. This is the basic pattern today, except that the grasslands are used for cattle grazing; only a few attempts have been made to cultivate the savannas with modern technology.

With the reporting of ancient raised fields in several savanna regions of South America in the 1960s, a number of people began looking for similar evidence in the Orinoco Llanos. Several reports were made of apparent ridged-field remains in the Colombian Llanos but none have been studied. The Reichel-Dolmatoffs (1974) discovered mounds in the savannas of the Upper Río Meta, which they believed were agricultural.[4]

The discovery of the Caño Ventosidad raised (ridged) fields in the Venezuelan Llanos was made in 1968 by James H. Terry, who was then working as a geodesist for the Inter-American Geodetic Survey. On an aerial photograph, he noted a pattern of parallel, linear features which resembled photos of raised fields shown in our *Scientific American* article (Parsons and Denevan 1967). He wrote of his discovery in May 1968 to Parsons and later corresponded with me. In October 1968, I examined the air photos of the Ventosidad area in the office of the Dirección de Cartografía Nacional in Caracas and confirmed the probable artificial nature of the ridges. Communication with Alberta Zucchi, the leading authority on the archaeology of the western Llanos, led to a joint

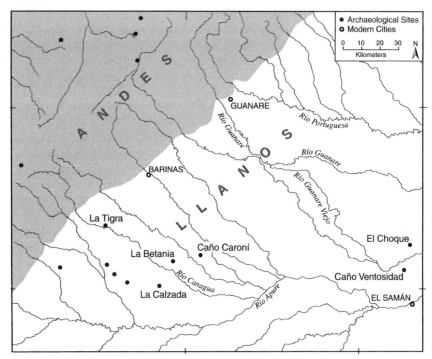

Fig. 11.5. Map of the western Llanos of Venezuela, showing archaeological sites. Adapted from C. Spencer *et al.* (1994: 120)

project[5] to study the prehistory and ecology of the Ventosidad raised fields in 1972–3. Zucchi had excavated non-agricultural mounds in Barinas previously at La Betania and later at La Calzada (Fig. 11.5). Our research is reported in Zucchi and Denevan (1979) and Denevan and Zucchi (1978).

Most of the fields extend from the edge of the Caño Ventosidad gallery forest and natural levee down the gentle backslope to the upper edge of Estero (swamp) Guajaral (Fig. 11.6). Most are arranged in pairs with a ditch between the ridges and with open savanna between pairs. The length of the ridges is considerable, with many over 1,000 m and up to 2,000 m. Such lengths are extreme and suggest to some observers that the ridges may not have been agricultural. However, we know of no other explanation. The average width of the ridges is 15.5 m, and the ditches average 4.4 m, while the average space between ridge pairs is 48.8 m. Present relief from ridge top to ditch surface is 0.25–0.50 m, but ditch depths have been considerably reduced by sediment accumulation (Fig. 11.7).

We excavated three sets of fields and ditches plus several trenches in adjacent savanna in order to obtain information on structure, function, antiquity, and cultural associations (Zucchi and Denevan 1979). No adjacent settlement sites were discovered, and only one ceramic sherd was found in the ridge

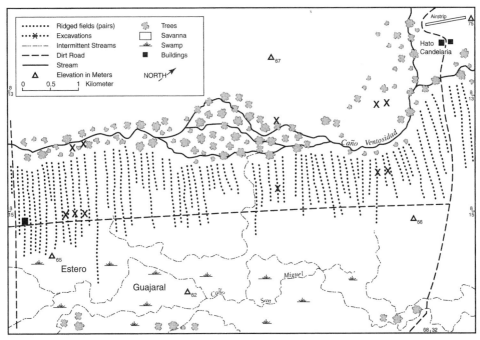

Fig. 11.6. Map of Caño Ventosidad raised fields, Orinoco Llanos, Venezuela. Adapted from Denevan and Zucchi (1978: 238)

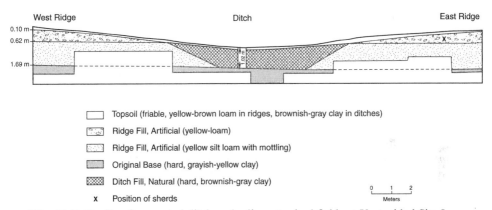

Fig. 11.7. Profile of excavated ditch and adjacent raised fields at Ventosidad Site I, Orinoco Llanos, Venezuela, showing ditch fill and original surface (1.69 m). From Denevan and Zucchi (1978: 241)

excavations, at Ventosidad Site I (Fig. 11.7). This sherd plus a settlement site in a small forest island 10 km distant suggest that the field construction was by the El Choque culture, *c.* AD 1200–1400. (I well remember our excavation of this forest island because after a half day of digging by the archaeologists, I looked up into the tree overhead and saw a large jaguar lying on a branch quietly observing us.) Zucchi relates the El Choque ceramics to expansion of Arauquinoid people in the middle Orinoco region. Good ethnohistorical evidence indicates large Indian populations in the central Llanos in the sixteenth century (Morey 1975: 306–23).

The Ventosidad raised fields are relatively few in number (500 to 525) in comparison with the tens of thousands elsewhere in South America. They only cover about 1,550 ha of ridge surface. My original examination of air photos of other parts of the Llanos, as well as the Maracaibo Basin, produced no additional fields. Air photo survey and aerial reconnaissance still have been far from complete. It would be unusual if raised-field cultivation was practiced in only a few small areas of the Llanos. The rate of sedimentation is relatively rapid in the Orinoco Llanos, compared with other savannas, and it may be that most other former ridges have now been buried.

Subsequent to our excavations at Caño Ventosidad in 1972–3, two other small drained-field sites were discovered in the 1980s in the Department of Barinas. These provide some different perspectives. The first find was by Adam Garson (1980: 129–30, 327; also see Denevan 1991: 237–9) at Hato La Calzada west of Ventosidad, a well-known site with numerous causeways and mounds (Fig. 11.5). The fields are ridges covering an area of 10.3 ha. They were only seen on air photos and were not examined on the ground. Ceramics in associated causeways and mounds date to after AD 500. The other site (ditched fields) is LaTigra (see Chapter 14).

The intriguing question remains; if raised and ditched fields were a known technology and viable ecologically and in terms of labor efficiency and productivity, why was field drainage restricted to a few small sites in the Venezuelan Llanos? Or have other such fields all been destroyed by people or by nature?

Guianas Coast

As mentioned earlier, raised fields were described by Dutch scholars in the 1950s in the poorly drained coastal savannas of Suriname. Geographer J. I. S. Zonneveld (1952: 44) wrote that: 'There are patterns of strips and dots visible on the [air] photo, which make one think that they are remnants of human earthworks . . . It is not yet known what the origin is of these remnants and how they may be related to the possible early presence of Indians and/or maroons [fugitive slaves] in these areas.'

The Suriname fields were brought to the attention of a broader audience by a paper in English by Aad Boomert at the Sixth International Congress for the Study of Pre-Columbian Culture of the Lesser Antilles, held in Guadeloupe in 1975. He described some fields as being almost square, about 4 by 4 m, whereas

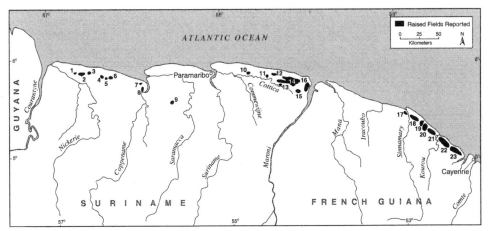

Fig. 11.8. Map of of raised-field clusters (numbered) in Suriname and French Guiana. Adapted from Boomert (1976: 142) and from Rostain (1994: fig. 75)

more rectangular fields were several tens of meters long; the ditches were 0.5 to 1.0 m wide and the field heights were about the same (Boomert 1976: 136). Boomert mapped the locations of the Suriname raised fields known as of 1975. One cluster is around a large artificial mound named Hertenrits near Caroni, dated at about AD 700 (No. 2 on Fig. 11.8).

There are several sectors of mound and ridge raised fields in French Guiana (Guyane). These have been mapped and studied in detail by Stéphen Rostain (1991; 1994: 126–44; 1995). They were photographed in 1957 by pilot R. Kappel (Boomert 1976: 138), but were initially thought to be the work of the French penal colony. Most are in poorly drained coastal savannas between Cayenne and the Sinnamary River (Fig. 11.8). The mounds are 0.2–0.8 m high, 0.3–5.0 m in diameter, and the ridges are 0.3–0.8 m high, 1.0–3.0 m wide, and up to 30 m long (Fig. 11.9). Associated canals, probably for drainage, are up to 150 m long. The earliest fields date to about AD 200. Palikur Indians just south of French Guiana in Amapá, Brazil were using raised fields (mounds and short ridges) in the eighteenth century (Grenand 1981: 25), so such fields continued in use after the conquest in the Guianas.

In 1992, anthropologist Neil Whitehead (pers. comm.) reported seeing similar features in the Fort Nassau savanna near the Berbice River in Guyana (also see Rostain 1995: 127). Most of the fields in the Guianas seem to be short and oblong, or lines of mounds, in contrast to the long, continuous linear fields found elsewhere. As the only known examples of prehistoric raised fields in eastern South America, the Guianas fields need more study.

Guayas Basin, Coastal Ecuador

From the base of the Cerro [de Samborondón] all the way to the Los Tintos River and well beyond, the landscape has been sculpted into

Fig. 11.9. Linear raised fields, coastal French Guiana. From Rostain (1991: 19)

> intricate patterns of raised fields with occasional mounds a few meters
> higher occupied with several houses on stilts.
>
> (Mathewson 1987*a*: 77)

One of the largest concentrations of prehistoric raised fields in South America is in the Middle Guayas Basin east of the city of Guayaquil in coastal Ecuador. Here there are large savannas which are periodically flooded by the overflowing of the Guayas, Babahoyo, Daule, and other rivers rising in the Andes. The area has been undergoing mechanized wet-rice development, and the old fields and associated earthworks are rapidly being destroyed and with them the opportunity to study the evolution and ecology of intensive agriculture in a lowland floodplain. This was the objective of the University of Wisconsin Samborondón Project in 1979–80 (Denevan and Mathewson 1983).[6] Figure 11.10 shows the distribution of the relic raised fields and of the morphologically differentiated field complexes in the Guayas Basin.

Modern archaeological study of the Guayas prehistoric earthworks began

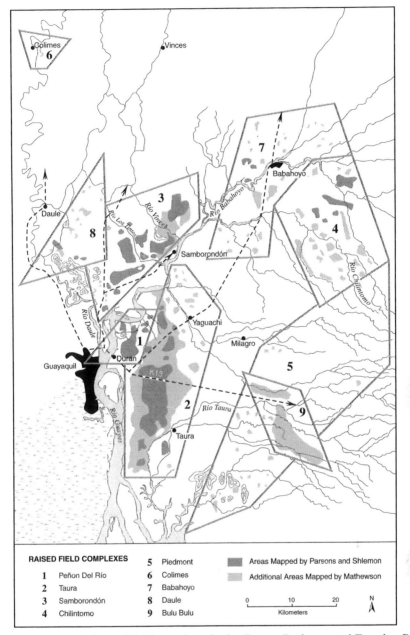

Fig. 11.10. Map of raised-field complexes in the Guayas Basin, coastal Ecuador. By Kent Mathewson; from Denevan and Mathewson (1983: 170)

in the 1950s with the Guayaquileño archaeologist Emilo Estrada along with Betty Meggers and Clifford Evans (Estrada 1957). In summarizing their work, Meggers (1966: 132) suggested that the artificial mounds that 'occur by the hundreds . . . in the southern Guayas basin . . . where the land is inundated every year for several months during the rainy season', were built for habitation sites or for ceremonial and/or funerary purposes, but an agronomic function was not considered.

In 1965, James Parsons (1969) recognized raised fields while flying out of the city of Guayaquil on a commercial plane on his way to visit me in Lima. He immediately suspected the agroeconomic nature of the thousands of mound, ridge, and platform structures that occur throughout the lower Guayas Basin based on the similar complexes we had just recently discovered in Bolivia and Colombia. He returned to Guayas in 1976 with geomorphologist Roy Shlemon to further map the fields and to attempt dating (Parsons and Shlemon 1982; 1987).

We identified and named nine distinct complexes of raised fields in the Guayas Basin, each with markedly different patterns. These are located on Fig. 11.10 and are illustrated in Figs 11.11 and 11.12. For details, see Mathewson (1987a: 230–42).

Three of these complexes were identified in 1965 and partially mapped by Parsons (1969) (Peñon del Río, Taura, and Samborondón), and in 1975 he and Shlemon mapped the location of additional fields, including part of the Chilintomo complex, for a total of 24,000 ha of field and ditch surface (Parsons and Shlemon 1982; 1987). Based on a more extensive examination of air photos, Mathewson added five more complexes and expanded the total area of field and ditch surface to a minimum of 50,000 ha (Denevan and Mathewson 1983). Many fields were buried long ago by natural siltation, and numerous surviving fields are obscured by forest or water.

For our graduate assistants, the Samborondón Project had several spin-offs. Kent Mathewson returned to Guayas to undertake a broader dissertation study of 'Landscape Change and Cultural Persistence in the Guayas Wetlands, Ecuador', in which the raised fields received further attention (Mathewson 1987a). He also did a study on 'Estimating Labor Inputs for the Guayas Raised Fields' (Mathewson 1987b). Another important member of our team, Richard Whitten, was unable to complete analysis of the large collection of ceramics he had excavated for us; he later drifted out of archaeology. David Stemper, on the other hand, was motivated by his experiences with us to carry out doctoral research on the archaeology of the Río Daule chiefdoms. Excavations of the raised fields at Yumes on the Daule and examination of air photos of the Colimes-Cerritos raised fields (Fig. 11.10) are a major component of his dissertation, which was later published as a monograph (Stemper 1993). The final member of our field team, John Treacy, was later to be a primary participant on the University of Wisconsin projects in Peru on Bora agroforestry in the Amazon and Colca Valley terracing in the Andes.

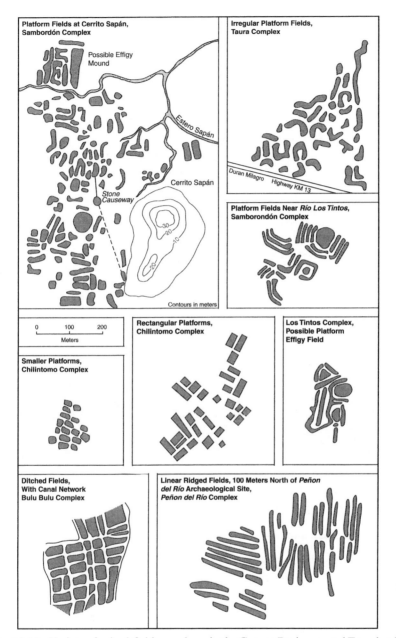

Fig. 11.11. Variety of raised-field complexes in the Guayas Basin, coastal Ecuador. By Kent Mathewson, from Denevan and Mathewson (1983: 174)

Fig. 11.12. Air photograph showing raised fields at Samborondón, along the Río Los Tinto in the Guayas Basin, coastal Ecuador, August 1966. From Denevan and Mathewson (1983: 175)

Another follow-up of our research was the Piñon del Río Ancient Agricultural Technology Project directed by Jorge Marcos and funded by the United States Agency for International Development in the early 1980s (Marcos 1987). Four of the resulting studies were published (Buys and Muse 1987; Muse and Quintero 1987; V. Martínez 1987; and Pearsall 1987).

One of the sub-projects of the Peñon del Río research was a set of experimental raised fields constructed in 1983–4, which provide us with some of the little data available on crop yields from raised fields in the tropical lowlands in South America. Manioc yields were 12,900 and 11,250 kg/ha/yr. For maize, the production was 3,750 kg/ha/yr for the wet season and 1,979 for the dry season, for a total double crop of 5,729 kg/ha/yr (Muse and Quintero 1987: 259, 264). The single-crop wet season maize yield exceeds the 3,000 kg/ha/yr (single crop) that Sanders (1979: 373) obtained for *chinampas* in Mexico and the 2,000 kg/ha/yr (single crop) obtained from experimental fields in Mojos (Erickson 1994*a*). The alluvial soils of Guayas are clearly superior to those of Mojos, so this lower figure for Mojos is not unexpected.

Unfortunately, most of the Peñon del Río raised fields have now been destroyed by urban growth of the town of Duran and by modern agriculture. The expansion of mechanized padi rice production, with bunds, dikes, reservoirs, and water pumps, is rapidly destroying raised fields throughout the Guayas Basin. A new system of wetland reclamation is replacing an ancient one, the agricultural engineers apparently being oblivious to the existence and long-term sustainability of the prehistoric raised fields.

Sabana de Bogotá, Colombia

Another region of Colombia with raised fields is the vicinity of the Andean city of Bogotá, at 2,600 m elevation, the location of the great Chibcha chiefdom. The Sabana is poorly drained, but is rich farmland (wheat, barley, maize) today, made possible by large, elaborate drainage canal systems. Archaeologist Sylvia Broadbent (1968; 1987) discovered raised fields near Suba 20 km north of Bogotá. I visited these fields with her in 1968. They had been leveled, but we could make them out from a small airplane because of different heights and densities of wheat crops overlying the former fields and ditches. Small patches of remnant prehistoric fields occur elsewhere. It is possible that much of the large Sabana de Bogotá was covered with raised fields prior to being ploughed up and cultivated by Spanish settlers.

Northern Highland Ecuador

Max Uhle (1954: 86) in a lecture in Quito in 1923 briefly mentioned raised fields near Nabón in northern Ecuador. Raised-field remnants have since been found by several geographers in wetlands in highland basins, initially on air photos by Roy Ryder (1970) at Cayambe and nearby Paquiestancia in 1969. Additional fields were found by Ryder and Alfred Siemens at San Pablo, at the south edge of Quito (fields now destroyed) by Greg Knapp, and at Otavalo and farther west at and near Sigsicunga and at San José de Minas by Knapp and by Gondard and López. These fields are described by Batchelor (1980), Knapp and Ryder (1983), Gondard and López (1983: 145–61), Knapp and Denevan (1985), and Knapp (1991: 147–62).

Some of the raised fields are buried under many meters of volcanic ash deposits, resulting in excellent preservation and dating (Knapp and Mothes 1998). Vertically overlapping ridges and swales have been exposed by excavations for large buildings in downtown Quito, such as the Banco Central at the intersections of Amazonas and Villalengua streets. At Cayambe and in central Quito, large permanent raised fields are as old as possibly AD 600. Smaller temporary raised fields date to at least 300 BC in central Quito. At Chillogallo at the southern edge of Quito, the eruption of the Quilotoa volcano *c.* AD 1280 buried raised fields which were never again used, so we know the probable date and cause of abandonment there. Other small raised fields continued to be cultivated north of Quito.

The known raised fields and associated ditches total 1,934 ha, by far the largest group being at Cayambe with 1,237 ha. Site elevations range from 2,400 to 3,100 m. The fields are associated with large artificial mounds (*tolas*) with ramps, which have been identified as late pre-Inca. The 'wave length' or distance from ridge center to adjacent ridge center varies from 3 to 9 m, with ridge and ditch width being about the same. The original height of the ridges was 80–100 cm or more, but heights of surface fields are now under 50 cm. At Cayambe, large semicircular embankments oriented *en echelon* perpendicular to the gentle slope may have served to impound water for dry period irrigation (Batchelor 1980: 678–82). Experiments by Knapp (1991: 159) indicate that raised-field surfaces can be 1.3° C warmer than adjacent flats on nights with frost, the apparent result of the development of tubes of warm air in the ditches between raised fields. Thus frost reduction may have been a major function of these fields.

None of the raised fields in these basins are known to have been in use in historical times. However, modern farmers have dug drainage ditches to reduce flooding in crop fields and in pastures. Current small potato and maize ridges (*wachunkuna*) have a wave length of only 0.75–1.0 m and do not last very long.

Casma Valley, Coastal Peru

Since they usually occur in regions of poor drainage, raised fields were not expected to be found in the Peruvian coastal desert. In 1970, following one of the most destructive earthquakes in modern Peru, the US Geological Survey did an air photo survey of the quake area along the central coast and inland. Some time later I received a set of infrared air photos of the lower Casma Valley from George Plafker of the US Geological Survey, the geologist who had discovered the Mojos fields in 1961. As he pointed out, the photos clearly showed patterns of abandoned raised fields. They are located near the mouth of the Casma River in a zone where the river periodically overflows, creating a large seasonal swamp. These fields were later examined by archaeologists Tom and Shelia Pozorski and colleagues (1983) and Jerry Moore (1988). The surviving fields cover 240 ha, and measure 2–3 m in width, 40–130 cm in height, and up to 80 m in length. Apparently they were constructed by the Chimú culture, *c.* AD 1300–1470.

Also in South America

Pre-European raised fields have been reported in several highland basins in the Andes, but most appear to be narrow (under 1 m) beds such as the contemporary *eras* and *huachos* mentioned earlier. Somewhat larger (1–4 m width) early raised fields have been described by Warwick Bray and colleagues (1987) in the mountain basins of Colima in Colombia and by Hastorf (1993: 136–7) near Jauja in the Mantaro Valley in the central Andes of Peru.

It is likely that further discoveries of pre-European drained fields will be made. People who have flown to Argentina report seeing linear features in the swamps of the central Río Paraná, but these have not been confirmed. Marajó Island at the mouth of the Amazon would seem to be an obvious place to encounter drained fields. The eastern half of the island is seasonally flooded savanna. Large settlement and ceremonial mounds occur in open savanna far from forested high ground. Was the savanna cultivated? The mounds were constructed by the spectacular Marajoara chiefdom (*c*. AD 400–1300). Some mounds are as high as 20 m, with an area up to 90 ha, large enough to have had villages numbering over one thousand people on each (Roosevelt 1991*a*: 31, 38, 65). Roosevelt (1991*a*: 26) suggests that there may have been raised fields which could now be buried under sediment. Another potential location is the coastal savanna of the state of Amapá, which lies between Marajó Island and the Guianas. Also, raised fields have been reported on the north coast of Ecuador near Esmeraldas, but these have not yet been described.

A very different type of raised field is the platform garden (*azotea* or *horta*) (R. West 1957: 143, 145–6, 241). These are soil-covered platforms of palm wood slats or old canoes, elevated 2–3 m above the ground and planted in vegetables and medicinals. They serve to avoid moist surface soil and leaf cutter ants. They are common today in eastern Brazil and in the Pacific lowlands of Colombia where they were reported as early as 1593. West believes they are an African introduction.

Raised Fields Elsewhere

The only New World raised fields that were well known and studied prior to the Mojos discoveries were the *chinampas* of the Valley of Mexico. These are of prehistoric origin, established by AD 800 (J. R. Parsons 1991: 37), some of which are still in cultivation in the shallow waters of Lakes Chalco and Xochimilco, others in relic form only on the drained bed of Lake Texcoco on the outskirts of Mexico City. Smaller clusters have been found recently in several other highland Mexican basins. In the tropical lowlands of Mexico, raised fields and drainage ditches were found initially in Campeche by geographer Alfred Siemens, and then in Quintana Roo, Vera Cruz, and Belize, all since 1968 (Siemens 1998). These fields have received systematic, interdisciplinary attention by means of extended archaeological field projects. They are believed to have been important means of intensive, sustained food production for the Aztec and Mayan civilizations (Sluyter 1994).

In North America large raised fields were not constructed. However, in the upper midwest, mainly Wisconsin and Michigan, narrow cultivation ridges were common (Gallagher 1992). They have been known since the eighteenth century, and have usually been called Indian garden beds. Very few sites have survived to the present, however, most having been destroyed by plowing. These features are about 0.5 m high, 0.5 to 1.50 m in width, and of variable length. They occur in linear and curvilinear clusters on a variety of surfaces, mostly well drained, apparently serving a variety of functions. None were ever observed in use by early travelers and settlers and so most may have been abandoned prior to European presence. However, another minor form of raised field, the corn hill, was in common use by Indians in the east and midwest until the early twentieth century. A few relics have survived.

As more and more prehistoric drained-field sites were reported in the Americas, we marveled at how they had been ignored for so long. When we turned to the Old World tropics to see what parallels might occur, we found little evidence for ancient raised and ditched fields, but we did find many instances of such fields currently being constructed and cultivated in Asia, the Pacific, and Africa (Denevan and Turner 1974). However, while such fields have been reported, descriptions are brief, and very little agro-ecological research has been done on them. Today, for the most part, they are less studied than the abandoned New World raised fields.

Some of the largest and most spectacular of the Old World drained fields in use occur in highland New Guinea basins such as the Baliem Valley in Irian Jaya and in Kuk Swamp in Papua New Guinea. Buried drains also occur in Kuk Swamp dating to possibly 7,000 BC (Golson 1989: 679). There are abandoned raised fields on Frederick Hendrick Island off the south coast, and also on islands in the Torres Strait (Barham and Harris 1985). The present Kuk fields, which have been studied by Pawel Gorecki (1982), and the Baliem fields provide insights into the construction and management of the abandoned New World fields, such as the practice of mucking to maintain soil fertility.

In Europe, drainage by means of ditching and raised platforms is a very old practice dating to at least medieval times. Well-studied examples include the polders of the Netherlands, the fens of England, and the Pontine marshes of Italy.

Raised-field Experiments, Restoration, and Destruction

An obvious question, given the extent of abandoned drained fields in the Americas and their survival in a few places, is why are traditional systems of wetland agriculture not being revived given food demands in Latin America? I suggested this possibility long ago (Denevan 1966*a*: 139; 1970*a*: 653), along with others, but there was little interest by the agricultural establishment until recently. There were abortive attempts at experimental fields in the 1970s in the Guayas Basin, Belize, and Tabasco, Mexico for the purpose of better understanding ancient fields as well as for considering the possibilities for re-establishment (Mathewson 1987*a*: 223, 260–1; Muse and Quintero 1987; Puleston 1977: 455–6; Gómez-Pompa *et al.* 1982). Now, successful programs of revival of raised field cultivation are under way around Lake Titicaca in both Peru and Bolivia, in both cases initiated by archaeologists studying pre-European fields (Chapter 13). One of these archaeologists, Clark Erickson, has also established the first modern raised fields in the Llanos de Mojos of Bolivia, and there is considerable local interest in expanding them (Erickson 1995: 92–3; 1998).

On the other hand, the ancient raised fields that have survived to the present, and which provide models and inspiration for revival are threatened by modern human activity. The Guayas raised fields are being rapidly destroyed by wet-rice irrigation systems, and many of the Titicaca fields are being leveled by ploughing. A large wheat irrigation project in the Department of Puno in Peru wiped out an entire sector of ancient fields. In 1986, the wheat was destroyed by the greatest lake-plain flood of the century, while some of the restored raised fields nearby remained productive.

Certainly a reason for the lack of interest in cultivating wetlands in South America is ignorance of indigenous antecedents and their sustainability and productivity. Also, these were very labor-intensive systems, both to construct and to maintain. However, high labor inputs could have been offset by very high production returns (Erickson 1995: 246). Today, of course, hand labor for construction could be replaced by mechanization, if done in an ecologically sound way.

Wetlands have had a negative image for travelers, settlers, and farmers (Siemens 1998: 26–7). Poorly drained swamps, grasslands, and lake margins have standing water, muck, and hordes of insects. Such places are uncomfortable and even dangerous. They have been avoided more than utilized. The modern approach to utilization is complete and permanent drainage, conversion from wetland to dryland. The evidence of pre-European raised fields, however, indicates that earlier people had a very different, more positive, perspective about wetlands. They lived in and around wetlands, hunted and fished in them, and cultivated them. Archaeologist Charles Stanish (1994: 312), who has studied the Lake Titicaca raised fields, believes: 'that [wetland] raised-field agriculture is among the most important intensification strategies utilized by Prehistoric farmers in the Americas.'[7]

Notes

[1] Curiously, the use of air photography to study prehistoric agricultural fields developed much earlier in Europe, particularly in England (Deuel 1969: 40–58).

[2] Caillavet (1983: 13), in a discussion of raised fields in highland Ecuador, indicates that *camel-*

lón does not appear in Spanish lexicons until after 1492, that the term probably was created to refer to the raised-field systems in the Americas, and that it is derived from *camello*, or camel's humps. However, Burgoa (1934: 295) in 1674 used '*camellones*' for agricultural terraces in Mexico.

[3] Parsons and Bowen (1966: 320) earlier estimated 65,000 ha (160,600 acres) of fields and ditches. Plazas and Falchetti (1987: 485) estimated 500,000 ha of raised fields in the region, but this apparently includes large interspersed sectors without fields and ditches.

[4] These mounds, located in the seasonally flooded region of the Río Manacacías, have an average diameter of 3 m and an average height of 60 cm. They cover over 100 ha, with a sample count showing about 1,000 mounds per ha.

[5] University of Wisconsin–Madison and Instituto Venezolano de Investigaciones Científicas (IVIC), Caracas. Field assistants were Marshall Chrostowski (physical geography), María Lea Salgado de Labouriau (pollen analysis), and Alejandro Barazarte (archaeology).

[6] Field research was primarily by Kent Mathewson, with assistance from University of Wisconsin graduate students John Treacy (geography) and David Stemper and Richard Whitten (archaeology).

[7] Two additional studies of raised-field functions should be noted: Vasey *et al.* (1984) on waterlogged soils, and Thurston and Parker (1995) on plant disease management.

12

The Mojos Raised Fields

> There are remnants of cultivated fields, roadways, and canals . . . inter-
> preted as indicating the former existence of a highly organized, populous,
> agrarian society that antedates the culture of the Indian tribes now in the
> area.
>
> (George Plafker 1963: 372)

The Llanos de Mojos (Moxos) is an immense savanna occupying most of the
Department of the Beni in the Amazon Basin of northern Bolivia (Fig. 12.1).
Most of this savanna is subject to flooding for four to six months each year.
Pure grasslands (*pampas*) are interrupted in places by palms and scrub, as well
as by forested patches or islands (*islas*) and gallery forests along the rivers—the
Río Guaporé, Río Mamoré, Río Beni, and their tributaries. The various forms
of savanna and swamp cover about 90,000 km², with an additional 20,000 km²
of intermingled forest.

When the Spaniards entered Mojos in the late sixteenth century from Santa
Cruz de la Sierra to the south, in search of the El Dorado of Paititi, they found
complex chiefdom societies with large populations. The first settlements, Jesuit
missions, were not established until 1682, by which time the principal Indian
cultures (Mojo, Baure, Cayuvava, Movima) and numerous others had been
reduced in size by slave raiding and epidemics (Denevan 1966a: 112–20; Block
1994: 78–89). The Jesuit accounts provide the first good descriptions of the
Mojos people, including various forms of earthworks (mounds, causeways,
canals); however, the only known report of raised fields is in the previously
mentioned Jesuit document from 1754 (see Chapter 11). When the Jesuits
departed in 1767, ranchers took over, and the region remained an isolated
backwater of Bolivia until the mid twentieth century. The first road to reach
Trinidad, the capital of the Beni, was completed in the 1980s.

In 1961 when I went to the Beni as a graduate student, bush planes and DC-
3 passenger/cargo planes provided the best means of transport, and it was from
such planes that the great quantity of indigenous earthworks became apparent
and raised fields were discovered. The aerial photography of the Bolivia
California Petroleum Company made it possible for George Plafker (1963) and

Extracted, with revisions and additions, from: William M. Denevan, *The Aboriginal Cultural
Geography of the Llanos de Mojos of Bolivia*, Ibero-Americana 48, Berkeley: University of California Press, 1966a: 84–96.

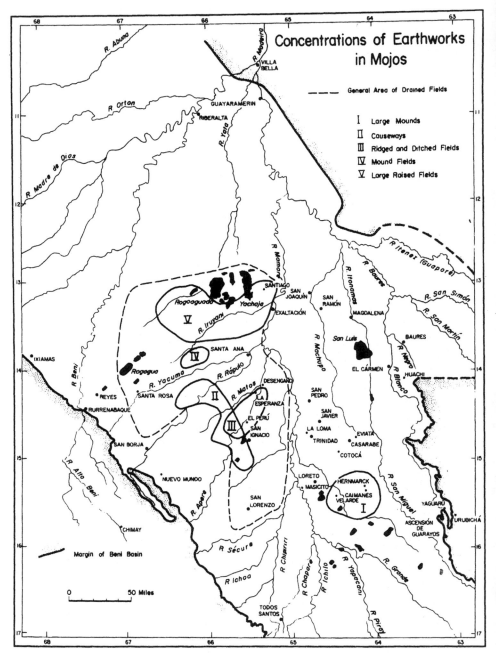

Fig. 12.1. Map of Llanos de Mojos, Bolivian Amazonia, showing concentrations of Pre-European raised fields and other earthworks. Adapted from Denevan (1966*a*: 56)

myself (1962; 1963*a*; 1963*b*) to document these features and to confirm the extent of the remarkable raised fields.

Today, agriculture in the Llanos de Mojos consists of shifting cultivation using slash-and-burn techniques and is mostly confined to forest areas. There is little evidence of savanna cultivation since the conquest. Beniano hoe farmers and foreign agronomists give the following reasons for not farming the Mojos savannas: flooding or poor drainage, low fertility, heavy clay soils, and grass competition. Nevertheless, the savannas were once farmed, as is indicated by the still visible remnants of tens of thousands of drained fields.

Types of Fields

In Mojos there are four main types of savanna fields that were raised[1] and drained for cultivation: (1) platform fields where earth was piled up to form low, rectangular flat surfaces, (2) narrow, ridged fields, (3) fields consisting of regularly spaced small mounds, and (4) fields in which ditches were dug to provide drainage (see Chapter 14). Less common are fields ditched in rectangular or gridiron patterns. The different types of fields generally do not occur in the same area. Apparently most of these features are located between the Río Beni and the Río Mamoré; however, I had a few brief glimpses from airplanes of what may have been old savanna fields east of the Río Mamoré between Magdalena and Trinidad. The main areas containing drained fields are shown in Fig. 12.1.

Despite their numbers and significance, the drained fields on the Mojos savannas received almost no attention from travelers and scholars until the early 1960s. The fields are usually indistinct and unimpressive on the ground, and for this reason and because they are away from commercial flight lines they remained little noticed (see Chapter 11) until the recent availability of aerial photographs and the ground and aerial operations associated with oil exploration.

Between 1956 and 1962, the Mojos drainage features came to the attention of several of us: geologist George Plafker (1963) of the Bolivia California Oil Company, Professor Mary Key (1961), then with the Summer Institute of Linguistics in Riberalta, engineer Kenneth Lee (pers. comm. 1993), and myself (1962). Plafker's study was based on air photo interpretation and sketchy reports from geophysical crews operating out of San Lorenzo, San Ignacio, El Perú, and La Esperanza. My own work was based on aerial photos, ground studies at Santiago, La Esperanza, and the San Ignacio area, and reconnaissance flights in light planes in the areas of San Lorenzo, San Ignacio, La Esperanza, Santiago, Exaltación, the Río Iruyani, Lago Rogoaguado, and Lago Yachaja. Mary Key flew many times over the fields between the Río Yacuma and Lago Rogoaguado. Lee made many flights starting in 1957.

Fig. 12.2. Platform raised fields north of Santa Ana, Mojos, Bolivian Amazonia. Photograph by Clark Erickson, May 1994

Platform Fields

The most spectacular of the drained fields are the large platforms found from just south of the Río Yacuma to just north of Lago Rogoaguado and from the Río Mamoré at Exaltación west to about 67° W (Fig. 12.1). These fields average 5 to 20 m in width and some are as much as 335 m long (Fig. 12.2). A few cover as much as 1 ha. They are spaced anywhere from 3 to 6 to 100 m or more apart and occur in clusters of up to several hundred; some are in parallel alignments and others are at irregular angles to one another. The height of those I saw on the ground at Santiago northwest of Exaltación ranged from 0.15 to 0.60 m, which is sufficient to place them above normal floodwaters for most of the year; however, some years there may be a meter or so of standing water on parts of the savannas. Although the platform fields, and other earthworks, are constructed of clay soils that resist erosion, heights have been reduced by the actions of precipitation and flooding and by the accumulation of sediment between fields at a rate of possibly 0.1–0.2 m a century.

On US Army trimetrogon photos I counted about 2,600 large platform fields. Bolivia Shell Oil Company photos west of Lago Rogoaguado plus my reconnaissance flights beyond the areas of photo coverage indicated a total of at least 5,000 large platform fields in the northwestern Beni.

Mary Key (1961) described the platform fields as follows (minor editing):

They were oblong in shape with somewhat of an appearance of cultivated plots lying parallel to each other . . . as few as four and as many as twenty . . . Then another series would angle off obliquely or at right angles, and another series off from that, and so on.

They seemed to completely cover the area, though some patches were obliterated or covered with water so deep that the pattern was lost . . . These formations are of uniform width and the usual length is from three to five times as long as the width. We noted that the channels or narrow strips between them were usually green . . . definitely deeper, and at this time of the year were filled with water. The reflection in the water gave back the sun's rays. The oblong shapes were higher and drier . . . We estimated that each oblong was about 10 meters wide.

The raised fields stand out clearly from the air because of their light color, a result of exposed earth under a sparse grass cover, in contrast to the darkness of the dense and usually more verdant grass cover of the surrounding lower ground. At Santiago in December 1961, I found a few centimeters of water in the swales while the platforms were dry. The break between the fields and the swales was quite abrupt. The grass on the platforms was paja cerda (*Sporobolus indicus*) and that in the swales was gañotillo (probably a *Panicum*). On the fields the surface soil was a light-gray clay loam (acidic, pH 5.5), but below 0.75 m the subsoil was very hard, yellowish mottled, white clay (neutral, pH 6.1). The soil in the swales was a deep, gray clay loam with no mottling (acidic, pH 4.9). There was a fairly high organic content in the topsoils of the swales but very little in that of the fields. The difference in the two soils probably reflects the organic preservation associated with poor drainage in the swales in contrast to the well-drained raised-field soils which are old enough to have experienced some leaching, mottling, and hardpan formation.[2] The poor quality of the grass cover on the fields and the presence of trees mainly on the edges of the fields are probably related to these soil differences. While the raised-field soils are poorer than the swale soils, this was not necessarily true when the raised fields were first constructed.

Most of the platform fields are in *pampa*, but those at Santiago are in open *arboleda* (scrub) with scattered chaaco (*Curatella americana*) trees and some tajibo (*Tabebuia* sp.). Most of the trees are on the better-drained ground of the fields, but some grow in the swales. Much of this scrub had invaded the area following a decline in the number of cattle beginning about 1950.

Many platform fields are lined with small, round termite or ant mounds. The termites obviously prefer the better-drained ground of the fields.

The *vaqueros* at Santiago were oblivious to the existence of artificially raised platforms, although they were aware of slight differences of vegetation, relief, and drainage. Two men who were given an opportunity to see the fields from the air were quite amazed by the rectangles that covered the area where they had lived most of their lives. Thus it is not surprising that these large raised fields were not mentioned by the Jesuits or later travelers. Cayuvava Indian families living at Santiago and nearby Exaltación also said they knew nothing of the fields and had no legends of their ancestors farming the savannas.

Ridged Fields

South of the Río Yacuma narrow ridges are separated by ditches. These fields are found in association with causeways, artificial mounds, ditched fields, and mound fields, but not with the large platform fields that only occur further north.

The variety of the shapes and sizes of the ridged fields can be seen in Fig. 12.3. The fields range from 6 to 300 m in length and average 1.5 to 6 m in width. The only ridged fields that I saw on the ground in 1961–2 were a group of ten about 5 km south of San Ignacio; they were 0.3 to 0.9 m high and 1.2 to 1.5 m wide, while the ditches were about 0.3 m deep and 0.9 m wide. (I was to return thirty years later in 1993 to the San Ignacio area to observe archaeologist Clark Erickson excavating raised fields, canals, and causeways.) Oil company personnel who drove swamp buggies over similar fields said that the ditches averaged between 0.3 and 0.6 m in depth (Plafker 1963: 376). The fields are not always straight, and there is no apparent orientation with respect to slope or natural drainage features. The object seems to have been to keep the water off the fields, but not to lead the surplus water out of the area of cultivation. The fields occur in groups which are sometimes surrounded by sinuous ridges that may have functioned as dikes (see Fig. 11.3) (Plafker 1963: fig. 12.4).

The ridged fields occur in scattered but dense concentrations. For example, in an area 15 km west of El Perú there are several thousand ridged fields covering about 40 ha including an abandoned stream meander (Plafker 1963: fig. 12.4).

Mound Fields

Regularly spaced, small mounds, sometimes referred to as *montones*, were observed by Plafker (1963: 376) and by myself from the air and on photos but not on the ground. Plafker (1963: fig. 8) shows a group of oblong mounds that appear to be about 1.2 to 1.5 m long, 0.6 to 0.9 m wide, and spaced about a meter apart. Most of the mounds, however, are circular, 3 to 6 m apart, and about 1.5 m in diameter. The heights are unknown but are probably no more than 0.3 to 1 m. The mounds are evenly spaced in straight or nearly straight lines that form orchard-like blocks. From high altitudes it is easy to confuse fields of artificial mounds with termite mounds built on parallel raised fields; however, termite mounds are not evenly spaced.

Mound fields that are definitely artificial are not found in the areas of other types of drained fields, but are found in open *pampa* far from large *islas* and gallery forests. Plafker (1963: 376 and figs 7, 8) saw mound fields 'in a roughly elliptical area from about 65 to 125 miles [100–200 km] northeast of Rurrenabaque'. Figure 12.1 shows the main location of groups of mound fields between Santa Ana and Lago Rogagua. Presumably these 'mound orchards' were built to provide drainage for crops. Their location far from rivers and large *islas* and their proximity to small, possibly artificial *islas* suggests that they may have been constructed by the people of small villages who were forced by population pressures into open savanna subject to deep flooding. Small mounds may have been built for crops instead of large raised fields because of

Fig. 12.3. Ridge and mound raised fields north of San Ignacio, Mojos, Bolivian Amazonia. From Denevan (1966*a*: 178). Photograph by Bristow Helicopters, Inc., *c.* 1960

the lack of manpower to build raised fields, while ditched fields were not practical because of deep flooding.

Gridiron Fields

Another, but apparently rare, type of old field pattern in Mojos consists of what might be called gridiron fields. These consist of rectangles enclosed on three or four sides by ditches. About 85 km west of Santa Ana aerial photographs show ditches which are 60 to 120 m long. In the same area there are mound fields and small round *islas* that may be artificial. From the air I also saw what seemed to be gridiron patterns near the northeastern side of Lago Rogoaguado, about 10 km east of Magdalena, and just north of San

Fig. 12.4. Experimental raised fields, Beni Biological Station east of San Borja, Mojos, Bolivian Amazonia. Student Juan Carlos Román (left) and agronomist Julio Arce, June 1992. Photograph by Clark Erickson

Joaquín. Bill Key, a Summer Institute of Linguistics pilot, saw similar patterns between San Joaquín and Magdalena and near Benecito on the Río Benecito west of the Río Yata. It is difficult to say how much raising of earth there is between the ditches. I tried to locate the gridiron fields near Magdalena on the ground, but was hopelessly confused by a maze of deep cattle paths through the *pampa* soil.

Numbers of Fields

I saw from the air or on aerial photographs an estimated 5,000 large platform fields, 6,000 ridged fields, and 24,000 ditched fields for a total of 35,000 individual drained fields, not to mention a dozen mound fields each containing hundreds of mounds. The total area covered by these fields, not including the ditches, is about 2,600 ha. The numbers of fields are based both on counts and group estimates, and the area is based on the average size of each type of field. I only flew over a portion of the region having fields; and identifying fields on aerial photographs is difficult, even when the exact location of the fields is known. Consequently, there are many more drained fields than represented by the above figures. A total of 100,000 linear drained fields occupying at least 6,000 ha of field surface spread unevenly over the western Beni is a minimum estimate.[3] There could be several hundred thousand fields.

Cultivation of Raised Fields

There are no eyewitness accounts of the abandoned fields of Mojos being built or cultivated. Nevertheless, the only logical explanation for them is that they are artificial features constructed to drain the savannas and to elevate platforms, mounds, and ridges above floodwaters for the purpose of cultivation. Remarkably similar features that do have an agricultural function are currently found in use in Latin America and elsewhere in the world (see Chapter 11). There is little likelihood that the ditches were cultivated rather than the fields, since this would have required dry season irrigation or the use of aquatic crops during the wet season, and there is little evidence for either. The fields do not seem to be oriented to the slope in such a way as to facilitate irrigation.

All the drained fields that I saw were on terrain subject to flooding. However, at Nuevo Mundo, south-southeast of San Borja in dense forest a seismic crew reported seeing 'furrow-ridge' fields (Plafker 1963: 375). Raised fields have been reported in the forests between the Río Chapare and the Río Sécure (Ray Henkel, pers. comm. 1970). Those forest fields may represent the persistence in a dry habitat of a cultural trait that originated in a wet savanna habitat; or the forest site may once have been in savanna; or the forest site may be poorly drained itself. Indians of the Chimane group are found at Nuevo Mundo as well as in the southwestern area of the savanna fields, but the location of the Chimane in the sixteenth century is uncertain. Erickson (1994*a*) observed raised fields in forest near San Ignacio and in the region of San Borja.

Little is known about the techniques, tools, or rotations used in savanna cultivation in Mojos. Savanna farmers in West Africa use hoes, and those in central New Guinea use wooden digging sticks with flattened ends for piling up mounds and ridges, for tillage, and for weeding, but there is no evidence of any Mojos Indians using any tool for cultivation other than a digging stick.[4] However, the early sixteenth-century Taino of Hispaniola were able to heap up their *montones* using only a digging stick; weeding was done without hoes, and the collection of manioc and sweet potato tubers growing in the mounds was done with digging sticks (Sturtevant 1961: 73).

Fiber mats and baskets were probably used by the Mojos people to move earth. A sketch of a pot from Mojos of unknown age and authenticity shows Indians moving earth piled on a hyde (Pinto Parada 1987: cover). Some idea of the manpower and time required to build a large raised field is provided by the construction of an airfield on low ground at Baures. A strip 365 m long, 23 m wide, and about 1.0 m high was built by six men, led by missionary pilot Walter Herron of Magdalena, working steadily for six weeks. The work was done mainly without machinery but with shovels, picks, and wheelbarrows. The use of only baskets and digging sticks to build a similar sized raised field would certainly have required more workers and/or a longer period of time.

The negative conditions of poor drainage, clay soils, weedy grasses, and low fertility all can be alleviated somewhat by mounding or ridging, as is presently true in savannas in New Guinea and West Africa. What is especially critical is the maintenance of fertility. While the mineral content of the savanna and forest soils of Mojos is similar, both being low in phosphorus and calcium and moderate in potash, the organic level of the topsoil under savanna is considerably lower than under forest.

Thomas Cochrane (pers. comm. *c.* 1975) of the British Tropical Agriculture Mission to Bolivia indicated that in southwestern Mojos the fertility of the *pampa* soils is actually greater in and below the impervious claypan layer due to reduced leaching. Consequently, the construction of agricultural ridges probably had the important function of exposing superior subsoils and burying the poorer topsoil.

The fertility of the soils of mounds and ridges can be stretched, as well as improved initially, by various means of fertilization. For example, in central New Guinea a grass mulch is composted into mounded garden beds, a fallow tree crop of nitrogen-fixing casuarina is often planted, and in one instance peat was burned to produce potash to fertilize mounds (Brookfield 1962: 247–8). Organic-rich mud is thrown up from adjacent canals onto the *chinampas* in Mexico, thereby improving fertility and also increasing the height (M. Coe 1964: 93). On Hispaniola, Taino women fertilized *montones* with urine and with ash from burnt trees (Rouse 1948: 522). The soil of the cultivated *era* ridges or *camellones* of the upper Cauca Valley area of Colombia has vegetable matter mixed with it to improve fertility (R. West 1959: 281). The Mojos Indians, of course, did not have significant quantities of animal manure available.

Cultivating the savannas and savanna settlement would have been encouraged by the opportunities for exploiting a rich terrestrial and aquatic wildlife. During the long period of flooding fish swarm on to the grasslands, and enormous numbers of migratory birds appear. Other game includes turtle, mollusks, swamp fox, tapir, caiman, deer, rhea, anteater, jaguar, peccary, and capybara.

The relation between numbers of drained fields and population size in Mojos cannot be determined when little is known about the length of time the different types of fields were used or the time period during which they were built. The examples of raised fields cultivated today, as in New Guinea and central Mexico, suggest that near-permanent cultivation is possible by means of fertility improvement, such as mucking organic canal mud. High labor investments are offset by high productivity. Population densities averaging 22 per ha of raised field surface are possible with a maize staple (single crop) based on 3,000 (*chinampas*), 3,750 (Guayas experiments, Chapter 11), or 4,000 (Mojos experiments) kg/ha/yr (Denevan 1982: 190–1; Muse and Quintero 1987: 259; Erickson 1994*a*). Thus, 2,600 ha of permanently cultivated raised field surface in Mojos could have supported 57,200 people, or proportionally less if not all the fields were in cultivation at one time. For a manioc staple supporting 16 persons per km^2, the population would have been 41,600 (Denevan 1982: 190–1).

The known raised-field area, however, is small and may have been much larger in prehistory.

Invasion of fields by savanna grasses may be just as much a reason for field abandonment or fallowing as loss of fertility. In San Ignacio, I talked with a man who once planted rice, corn, melons, and sweet potatoes early in the rainy season on *pampa*. Most of his crops matured before flooding, but he said he had to weed frequently, and the grass was so bad the second year that he gave up. The native grasses of Mojos are not as aggressive and troublesome as the *Imperata* grasses of the African and South-East Asian savannas, but weeding was undoubtedly necessary. Padre Castillo (1906: 309) in 1676 noted that the Mojo Indians carefully weeded their manioc patches.

Cultivation of grassland unquestionably involves more work than forest cultivation, but it can be done using only simple tools. In New Caledonia dense grassland turf is turned over with long digging sticks by several men walking side by side, and then the clods of turf are broken up with clubs (Carneiro 1961: 59).

The main crop grown on the drained fields of Mojos may have been manioc, which was the staple in Mojos in the seventeenth century; maize and sweet potatoes are other possibilities. Large numbers of maize cribs (*percheles*) (see Chapter 5) were reported in 1617: 400, 500, and 700 in individual fields, not necessarily raised fields (Denevan 1966a: 98). The Taino planted manioc and sweet potatoes in their *montones* but not maize. The main crop on the New Guinea mound fields is the sweet potato, and in Africa manioc is grown on ridges and mounds. The reasons why maize is seldom planted on ridges and mounds in savannas are unclear; however, maize does require more nitrogenous soils than does manioc. Maize did grow well on experimental raised fields in Mojos (see below) and in the Guayas Basin of Ecuador (see Chapter 11).

There are several possible explanations why the Mojos savannas were farmed when apparently there was adequate forest, with superior soil, available within the savanna region for cultivation. Stone tools were rare in Mojos, since stone had to be brought from long distances. Until metal tools were introduced by the Spaniards, it may have been easier to build up earth platforms in the savanna than it was to clear forest (see Chapter 7). Of relevance is the fact that in eastern Mojos, where there are a number of rock outcrops of the Brazilian Shield, there are very few drained fields, although there are large numbers of mounds, causeways, and canals. On the other hand, population may have become so dense that there was not enough forest land available, although this seems unlikely. Also, tribal territorial control may have prevented migration and thereby necessitated intensive cultivation of whatever land was available locally. All of these possibilities were eliminated by the Spanish conquest which introduced metal tools for clearing forests,[5] drastically reduced the population, and established new groupings of the Indians. These events began taking place well before the first missions were established and probably even before the first explorers reached Mojos.

The Antiquity of Savanna Farming

The abandoned drained fields of the Llanos de Mojos are all prehistoric, not Spanish. There is no conclusive evidence concerning when they were first built. Since the people who built large mounds and causeways did not always drain fields for savanna cultivation, the fields are not necessarily contemporary with them. That the fields are not extremely old is indicated by the fact that they have not been completely destroyed by erosion and silt deposition. Erickson (1995: 90) suggests the fields evolved over at least 1,000 years.

More interesting than the date of origin is the date that the construction and use of drained fields was abandoned. Was this abandonment a result of natural or social factors long preceding the Spaniards or was it an indirect result of the Spanish conquest? This question is of some significance in view of claims that certain seasonally flooded savannas in South America could not continuously support advanced native cultures (Meggers and Evans 1957: 30–2; Leeds 1961: 26).

The early explorers and the Jesuits did not mention drained fields or savanna farming,[6] except for the one brief report in 1754 (see Chapter 11). Raised fields were probably all abandoned by the nineteenth century. That the Jesuits did not even notice abandoned fields is not surprising since earlier and subsequent travelers did not notice them either. The first explorers and Jesuits in Mojos possibly got into the main areas of drained fields, but these areas were not described until much later when the missions of San Ignacio (1689), San José (1691), San Borja (1693), and Exaltación (1704) were founded. In the period intervening between the first Spanish exploration and the founding of the missions there were only a few expeditions and no colonization in Mojos. Nevertheless, there was sufficient direct contact between Spaniards and savanna Indians, both in Mojos and in Santa Cruz, and indirect contact through other tribes to disrupt the savanna tribes numerically, socially, and technically by: (1) introducing disease and thereby considerably reducing a population that may have once numbered several hundred thousand, (2) slave raiding activities which brought about tribal displacements and breakdowns throughout the interior of South America, and (3) the introduction of metal tools and other implements, initially through traders. As already pointed out, these factors eliminated the most likely reasons for savanna cultivation.

John Walker (1998) has carried out archaeological excavations in the area of the Río Iruyañez (Iruyani) north of Santa Ana de Yacuma in northern Mojos. He has radiocarbon dates of AD 410–620 and AD 1275–1645 for raised-field associated settlement sites. He believes that final abandonment of the fields occurred during the early colonial period as result of depopulation, new tools facilitating forest clearance for swidden cultivation, and the Jesuit mission introduction of cattle and emphasis on new economic crops (sugarcane, cacao, rice).

Recent Research

Most of the above was written in 1963 as part of my doctoral dissertation, which was published in 1966. I was very excited about what I had seen in Mojos, believing that the raised fields that Plafker and I had discovered contradicted much of what I had learned about Indian agriculture in Amazonia. I tried to communicate this insight through lectures, interviews, and publications, including articles in *Scientific American* and *Science*. The responses tended to be 'how interesting', but with little follow-up research, although our reports did lead to discoveries of raised fields elsewhere (see Chapter 11). In Bolivia, there was little interest. For Bolivians, the great prehistoric achievements were in the Altiplano, not in the tropical lowlands. My 1962 article published in Bolivia describing the Mojos fields was ignored; likewise my suggestion that raised fields once were and again could be a means of intensive savanna food production.

In 1976, however, the Mojos earthworks were rediscovered following a report in the *Times*, UK (3 August) that made headlines not only in Bolivia but throughout the world. The source was the freelance reporter Ross Salmon, who claimed evidence of 'a vast prehistoric civilization that cultivated the great savannas east of the Andes'. Salmon's main source was the American engineer and long-time Beni resident Kenneth Lee (1977), who had been investigating the local earthworks on his own for about 20 years. Influenced by the resulting publicity, several archaeologists began research in Mojos, in particular the Chilean Victor Bustos (1976) and the Argentines Bernardo Dougherty and Horacio Calandra (1981; 1984). However, their excavations concentrated on the large prehistoric mounds rather than on the raised fields and causeways. Interest in prehistoric Mojos again faded, except for a few general essays. A curious by-product was a novel by Beniano historian Rodolfo Pinto Parada, titled *Pueblo de Leyenda* (1987), about ancient Indians in Mojos who cultivated raised fields, with modern insertions in which real archaeologists, geographers, and especially Kenneth Lee appear.

In 1990 Clark Erickson, with a team of Bolivian and American archaeologists, initiated a long-term project on the Mojos raised fields and associated earthworks and settlement, the first sustained research on Mojos prehistory. Erickson had previously studied the Titicaca raised fields in Puno, Peru (see Chapter 13) and had done an earlier survey in the Beni (Erickson 1980). Finally, twenty-seven years after our reports in *American Antiquity* (Plafker 1963; Denevan 1963a) the Mojos raised fields were being given serious attention by archaeologists. Erickson's project was concentrated near San Ignacio, but surveys were conducted elsewhere in the Mojos savannas.

Erickson has shown that the Mojos raised fields were much more extensive and important than I had ever realized with my limited resources in 1961–2 (Erickson 1994a; 1995). His examination of air photos combined with ground surveys indicates that large numbers of raised fields cannot be seen from the air or on air photos, or even on the ground when grass cover is high. Many fields have been destroyed by erosion or buried under sediment; others are under forest.

Erickson's team excavated a series of trenches in raised fields, causeways, and canals. Radiocarbon dates from one field site indicate raised-field use between AD 1 and AD 1200 (Erickson 1994*a*; 1995: 90–1). Pollen has been identified for cultivated *Xanthosoma*, annatto, and guayusa. Erickson (1995: 73) argues that at least some causeways and canals functioned not only for transportation but also for water management. Causeways may have served as dams to impound water within field blocks, to control water levels for the raised fields, and possibly to concentrate fish.[7]

Experimental Raised Fields

Erickson's project has included the construction of experimental raised fields (Fig. 12.4). They provide 'important insights into how the system functioned, the kinds of crops grown, labor inputs in construction and maintenance, nutrient production and cycling, dynamics of field hydrology, crop productivity, potential carrying capacity, sustainability of the system over time, and other important issues' (Erickson 1995: 92). The first small block of fields was constructed by local farmers at the Biological Station of the Beni near San Borja and was later expanded as agronomy student thesis projects (Erickson 1994*a*; 1995: 92–3). Excellent yields were obtained—25 metric tonnes/ha (field and ditch combined) of manioc and 2 metric tons/ha of maize. Sweet potatoes and plantains also did well. 'Several older farmers remarked that it was the first time in their lives that they had seen the *pampas* produce agricultural crops' (Erickson 1995: 92–3).

Subsequently, additional experimental raised field plots were established in two small communities near San Ignacio on both forest and *pampa* sites. The success of these fields is still being monitored, and it remains to be seen whether this 'new' form of agriculture will become widely established in the Beni. Currently, most of the savannas are in the control of cattle ranchers who may not be interested in savanna cultivation. The necessary labor inputs seem high. 'The social, political, economic, demographic, and environmental co text of present farming in the Beni is very different from that when the fields were originally constructed and used by prehistoric farmers' (Erickson 1994*a*).

Having been involved in the discovery of the Mojos raised fields many years ago, it is gratifying to see the interest in them today by scholars, agronomists, and farmers.[8]

Notes

[1] Terms that sometimes are used in the Beni for raised fields are *camellones, aterrados, tablones,* and *huertas antiguas* (old gardens).

[2] While there is some deposition of alluvial silt from overflowing rivers, most of the flooding of the savannas is from standing rainwater which is poor in nutrients. The Mojos savannas are not a great alluvial basin.

[3] The estimate of 15,000 ha given in Denevan (1982: 190) should have been 15,000 acres, which converts to about 6,000 ha.

[4] Nordenskiöld (1924: 37, 65) said that the Moseten in the foothills west of San Borja used 'digging spades' or 'simple digging clubs', but provided no details.

[5] According to Padre Marbán (1898: 148), who was there, by 1676 Mojo Indians were traveling to Santa Cruz to trade cotton goods for 'machetes to cut and clear their chacras'.

[6] Padre Eder (1985: 75) wrote in 1791 that the ashes of burnt grasses provided good fertility, but that, nevertheless, the *campos* could not be cultivated because of the waters that flood them.

[7] Nordenskiöld (1916: 150) mentioned seeing a dam near San Borja which he thought had been built for irrigation purposes, but he gave no details. This may have been a causeway. Kenneth Lee (pers. comm. 1976) strongly believed that the 'causeways' served primarily to control water for cultivation, and that the many rectangular lakes in Mojos were artificial reservoirs (also in Pinto Parada 1987). More likely, the lakes are geological.

[8] John Walker (1999) recently completed a dissertation on the archaeology of large raised fields in northern Mojos. He discusses causes of raised-field abandonment, field characteristics and dates, associated settlement sites and forest islands, and social organization. This material was received too late for consideration here.

13

The Titicaca Raised Fields

Traces of ancient cultivated fields of a special type, in the form of slightly elevated beds more than a meter in width, are preserved in many places from ancient times. There are great extensions of them, for example, on the Bolivian shore of Lake Titicaca ... The beds follow different plans; straight, curved, circular, spiral, and so forth.

(Max Uhle [1923] 1954: 86)

The prehistoric reclamation of wetlands for agriculture in the New World was practiced not only in the tropical lowlands but also by the large states in the mountain basins of Mexico and the Andes. The largest concentration of highland raised fields is in the region of Lake Titicaca in Peru and Bolivia at 3,800 to 3,900 m above sea level. These fields are of pre-Inca origin and were a major source of food for the Tiwanaku civilization. Their existence helps to substantiate other indications of a dense pre-Colombian Indian population in the region. They are now used mainly for pasture, but efforts are being made to revive them. Today, only the higher and drier parts of the lake plain are cultivated.

The first mention of the Titicaca raised or ridged fields that I know of was in 1901 by the German anthropologist Karl Kaerger (1901: 346), who briefly described fields 4–6 m wide that he had seen from a train near Juliaca in Peru. He said that there was no human memory of anyone farming these lands, only grazing. Local tradition was that the old fields were Inca, and the great German/Peruvian archaeologist Max Uhle agreed, according to Kaerger. Uhle, himself, later described Titicaca raised fields in a lecture in 1923 in Quito. At the American Museum of Natural History, on an anthropology-office wall, I saw an air photo taken in the 1930s by archaeologist Wendell Bennett of the area around the Tiwanaku ruins which clearly shows some raised fields. Bennett, an early authority on Tiwanaku, missed these fields and thus their significance for the Tiwanaku civilization.

The first modern recognition of the extent and significance of the Titicaca raised fields was by geographer Clifford Smith in 1966, initially on air photos and then in the field in the Puno region of Peru with Patrick Hamilton. I

Portions extracted from C. T. Smith, W. M. Denevan, and P. Hamilton, 'Ancient Ridged Fields in the Region of Lake Titicaca', *The Geographical Journal*, 1968: 134: 353–67. Revisions are by W. M. Denevan.

initially told Smith, then a neighbor in Lima, that he must be wrong, that such fields only occurred in the tropical lowlands, such as in Mojos; however, his photos convinced me. Later that year, I visited the Puno fields with geographer Pierre Stouse, and I also went to La Paz to examine air photos there and found that additional raised fields are found on the Bolivian side of the lake. The largest cluster is near Aygachi in Pampa Koani. I examined these fields briefly with Bolivian archaeologists, including Carlos Ponce Sanjinés. My efforts to convince both scholars and developers of the need for major research on the Titicaca fields were ignored for the next fifteen years, except for dissertation study by Thomas Lennon in 1976–7 (1982; 1983) on water management of the Peruvian fields. The first systematic archaeological excavations were finally undertaken in the 1980s by Clark Erickson and Alan Kolata.

This chapter is a revision of our original regional survey of the Titicaca raised fields (C. Smith *et al.* 1968). For the recent work by Erickson and Kolata on the specific areas of Huatta (Huata) and Tiwanaku, the reader is referred to their publications. Some of their conclusions are presented here.

The Distribution of Raised Fields near Lake Titicaca

The distribution shown in Fig. 13.1 is based on field study[1] and aerial photographs taken 1995 at scales of 1:65,000 and 1:15,000. At the smaller scale, raised fields of a few meters in width are only perceptible if contrasting vegetation patterns on ridges and troughs help to distinguish microfeatures of relief, or if the troughs are partially flooded. Field observation showed that some patterns could not be discerned on the aerial photographs available, and some fields which were easily identified on photographs at the scale of 1:15,000 were not always found on 1:65,000 photographs of the same area. It is certain that Fig. 13.1 and Table 13.1 underestimate the true extent of surviving ridge patterns. It is also impossible to estimate what has been destroyed as a result of subsequent cultivation and other disturbance and by burial under lake sediment.

The distribution of the areas of raised fields around the margins of Lake Titicaca is markedly uneven. They do not occur on the rugged hilly terrain which fringes the lake itself for much of its northeast and some of its southwest shorelines, and they are clearly restricted to areas of poor drainage on level or nearly level terrain developed on lacustrine or alluvial sediments. The largest area of this kind is the *pampa* of Paucarcolla-Juliaca northwest of the lake, and it is there that the largest single cluster of raised fields is to be found; the ground is covered with large sectors of raised fields of various kinds scattered over more than 500 km². At the southern end of the lake four important groups of fields are located on marshy ground at Pomata, Desaguadero, Pampa Koani in the Río Catari Valley south of Aygachi, and on the north side of the Río Tiwanaku Valley. (For a map locating the Bolivian fields, see Kolata and Ortloff 1996: 116.) But there are also many small and scattered pockets, none

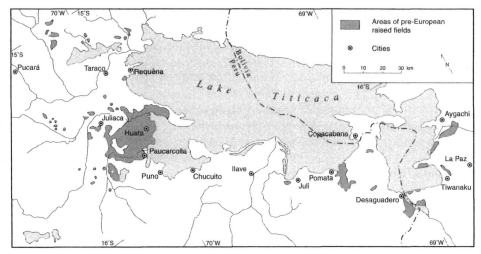

Fig. 13.1. Map of the distribution of pre-European raised fields in the Titicaca Basin. Based on air photography at scales of 1:65,000 and 1:15,000. Mapped primarily by Patrick Hamilton. Adapted from C. Smith *et al.* (1968: 354)

Table 13.1. Location and extent of the Titicaca raised-field patterns[a]

Location	Hectares		
Peru: (C. Smith *et al.* 1968: 355)			
Main block: Paucarcolla-Juliaca plain	56,533		
Scattered: northern area	7,770		
Pomata	5,108		
Scattered: southern area	2,192		
Desaguadero: includes fields in Bolivia	6,501		
Total	78,104		
Peru: added by Erickson (n d.)	31,196		
Total Peru		109,300	
Bolivia (Kolata 1991: 124)			
Río Catari Valley (Aygachi/Koani)	7,000		
Río Tiwanaku Valley	6,000		
Río Desaguadero Valley (included under Peru)	–		
Total Bolivia		13,000	
Total Peru and Bolivia (Erickson n d.)		122,300[b]	(1,223 km^2)

[a] This is a revision (upward) of the C. Smith *et al.* (1968: 355) total of 82,056 ha.
[b] Erickson (2000) also gives a 'conservative estimate' of 120,000 ha.

larger than 1,000 ha, occurring mainly in wet depressions or on valley floors, often set back from the river courses adjacent to the sharp break of slope at the valleyside.

It is clear that raised-field patterns are to be found only on level terrain which is or was marshy or subject to periodic inundation. It is also evident that most of them occur within a short distance from the lake. Over 92 per cent of the fields are within 30 km of the lake. Although the lake itself has a mean level of

3,803 m, 98 per cent of the fields are below 3,850 m and the highest fields are at 3,890 m above sea level, northwest of Pucará.

These elevations are well within the limits of the indigenous food crops of the region: tubers such as potatoes, oca, ulluco, and grains, such as quinoa and cañihua. All of these may be cultivated up to 4,000 m under favorable conditions. However, there are very few areas of raised fields in wide marshy valleys at any distance from the lake. It may be that differences of local climate had a significant effect on crop possibilities, and therefore on the practicality of ridging. The moderating influence of Lake Titicaca on the climate of the area is certainly of considerable importance in present agriculture. Liability to damage by frosts during the growing season is now, and must always have been, a major hazard. Near the lake this hazard exists but is at a minimum for the region. At Puno (3,850 m) there were 123 days during which frost occurred, but at Chuquibambilla (3,910 m), northwest of Pucará, frost occurred on 197 days during the year.[2] During the rainy season (also the season of plant growth) there were only four occasions on which frost occurred at Puno (October to March), but there were fifty-two days with frost at Chuquibambilla. At Copacabana, a favorable lakeside peninsula location in Bolivia, there were no frosts from 18 August to 1 May. From November to March, mean monthly minima averaged −2.7° C at Puno, −0.1° C at Capachica (1957–64), and +0.5° C at Juliaca. All of these stations are on or fairly near the lake, but at Chuquibambilla the corresponding average was −2.5° C. The Puno weather station, it should be noted, is (or was) in a frost hollow although quite near the lake itself. Even close to the lake there are considerable frost risks resulting from cold-air drainage down to low-lying land, sufficiently so to make the terraced slopes valuable for crops susceptible to frost. Away from the lake this effect must be accentuated, and in remoter regions it is precisely the valley bottoms that carry maximum frost risks.

It seems anomalous that there should be few traces of raised-field patterns in two quite large areas of low-lying, badly drained land near the lake itself, the *pampas* of Taraco and Ilave, in contrast to their abundance in the Paucarcolla-Juliaca plain. The distribution of raised fields may be understood to some extent in terms of relief, drainage, and local climates, but the *pampa* of Taraco and the delta of the Ilave River are environments very similar to that of the Paucarcolla-Juliaca plain, yet are almost devoid of raised fields. This plain is, however, distinct in terms of modern land tenure, land use, and population. Most of this area is in large haciendas devoted to extensive pastures and carrying a sparse population. Some of these haciendas are of long-standing origin (e.g., Buena Vista Hacienda south of Juliaca), though it is impossible to say with any certainty which of them were established in early colonial times. Clearly, much of this land has been under permanent pasture for a considerable period of time. The *pampas* of Taraco and Ilave, on the other hand, are occupied by indigenous Indian communities and carry some of the highest rural population densities of the Peruvian sierra. Almost their entire surface is now under cultivation, with an average of 0.25 to 4 ha available to each family. It seems likely that any traces of former field systems would have been destroyed. Present farmers use the Spanish wooden plow and modern plows, as well as the traditional *taclla*, or Andean foot plow. The use of the modern plow has certainly destroyed some of the raised fields in better-drained areas near Taraco since the air photographs we used were taken in 1955. In the Pomata area plows are still in the process of erasing ancient ridge patterns, and this may have happened in the past on a much larger scale.

The Classification and Description of Field Patterns

The patterns of raised fields vary greatly over the Titicaca region and within specific areas. They are difficult to classify, even on morphological grounds, though it seems worth attempting if classification might help to throw light on the chronology of development, types of site association, and agroecological functions. Platform and ridged fields alternate with flat-floored troughs with a range from 15 cm to 1.25 m in height. Ridges are generally arranged in bundles of strips, and they are usually parallel, though not always. There are many variations in width, length, shape, and arrangement, and it is on these variations that the following classification is based.

Open Checkerboard (see Fig. 11.3)

This is a term used by Parsons and Bowen (1966: 329) to describe a similar pattern in Colombia, and it is worth adopting here. The most common type of pattern is the grouping of approximately five to twenty ridges into bundles, each bundle more or less at right angles to the next. The term 'open' is used simply to indicate that troughs between ridges are open-ended, so that water may circulate freely from one bundle of strips to the others. The ridges are not continuous and there are no surrounding embankments. The average width ('wave length') of ridge and trough combined (from the center of one ridge top to the center of the next ridge top) varies from 5 to 20 m, and the length is from as little as 2 m to 40 m or more. Between Vilque and Atuncolla (Hatuncolla) northwest of Puno the average width of ridges is 5 m, the troughs 1.5 m, and length from 2 to 14 m, but only a few kilometers further north, the wave length increases to about 15 m. Near Hacienda Machacmarca south of Atuncolla, the norm is 10 to 12 m and on the southern shore of Lake Titicaca 10.5 to 14 m. The open checkerboard pattern is found mainly on poorly drained or seasonally flooded land at the margin of existing lakes and in plains well away from the foothills. It is the most important single type, and includes most of the raised-field patterns on the Paucarcolla-Juliaca plain, as well as at Desaguadero and at Requeña (Fig. 13.2).

The construction of each block of field strips may have been done by indi-

Fig. 13.2. Open 'checkerboard' raised fields near Requeña at the northwest end of Lake Titicaca, Peru. Taken in July 1955 when the lake level was unusually high, submerging or partially covering many of the fields. Servicio Aerofotográfico Nacional, Lima

vidual households cooperating to reclaim new pieces of land in marshy areas to supplement land on slopes or better-drained terrain elsewhere. The irregularity of the dimensions of ridges and of the number of strips suggest that reclamation could not have been highly organized or stringently planned. Similarly, the absence of well-defined major canals, with some exceptions, suggested to us that the primary concern was to build up ridges of cultivable land and that an effort to effect any kind of integrated drainage or irrigation system was secondary. Exceptions to this are discussed later.

Irregular Embanked Pattern

Quite distinct from the open checkerboard pattern described above is one which occurs in much more limited areas near Pomata, near Huatta close to Lake Titicaca on the Paucarcolla-Juliaca plain, and occasionally between Vilque and Atuncolla (Fig. 13.3). In these areas, groups of ridges are sometimes enclosed or partially enclosed by low embankments which in some cases are circular or near-circular and in others highly irregular.[3] Irregularity is further accentuated by the continued existence of ponds of standing water and small patches of marsh interspersed among enclosures. In the Pomata area, discontinuous embankments enclose irregular groups of ridges on either side of the Pomata River. Their irregularity contrasts markedly with the relatively orderly patchwork quilt of the open checkerboard pattern nearby. Here, as elsewhere in the region, the embanked types occur in particularly wet locations near lakes or streams, and in the Pomata case, at least, it is hard to resist the conclusion that they were created at a later stage than the open checkerboard, which is here to be found on slightly better-drained land. In the Juliaca area, embankments also occur in the wettest locations, and the open checkerboard extends much further inland from the lake margins. But this is not always the case, and the open checkerboard group near Requeña lies at the very margins of the lake and even today is partially flooded (Fig. 13.2).

The existence of embankments around groups of ridges does indicate an attempt at water control. However, the absence of irrigation ditches or drainage canals in the Paucarcolla-Juliaca plain is surprising if embanking represented an attempt at coordinated water control. It may therefore be that embanking was sometimes a simple means of resisting occasional or seasonal or excessive inundation (Lennon 1982: 189), and an effort made by the individual farmer against personal disaster, for there is no trace of long, continuous earthworks to provide protection for a large area, and bundles of unbanked ridges often lie adjacent to embanked ridges.

Riverine Pattern

Near Atuncolla northwest of Puno there are narrow ridges more or less at right angles to modern or ancient and abandoned watercourses. As in the San Jorge

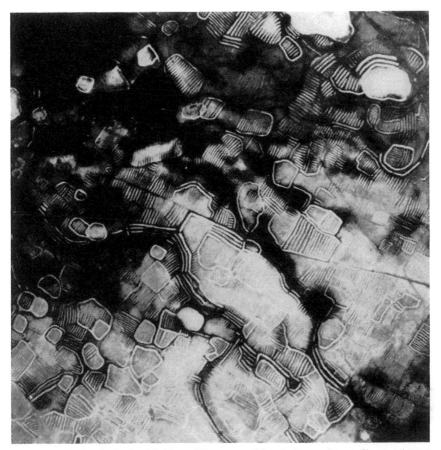

Fig. 13.3. Embanked raised fields and linear canal (center) near Atuncolla, northwest of Puno, Lake Titicaca, Peru. Each field is about 15 m wide and 65 m long; the dark edges are water. Servicio Aerofotográfico Nacional, Lima, July 1955

floodplain of northern Colombia, the ridges run down slight natural levees so that effective drainage is secured. But this type of pattern is much less common in the Titicaca region than it is in Colombia, where it has been labeled as the *caño* pattern (Parsons and Bowen 1966: 329).

Linear Pattern

Another oriented pattern consists of long, narrow ridges that are generally parallel and which may be up to 400 or 500 m long. The long axes are usually at right angles to the break of slope between hill and plain. On the Capachica Peninsula east of Huatta, near Desaguadero, and northwest of Puno these linear fields are on the lower skirtlands below steeply rising hills on which staircases of narrow agricultural terraces indicate a history of intensive settlement.

Houses and farms are commonly located at the break of slope, and it may be that these long, narrow ridges, often partly located on gently sloping land at the upper ends, represent a plainward extension of linear holdings that extended from cultivated hill terraces through the skirtland to the marshy ground beyond. Measurement of 206 examples in the Pomata area gives a mean of 8.7 m for the width of field strips.

Ladder Pattern

Along the southern shore of Lake Titicaca, especially near Pomata, a fifth type of raised pattern may be differentiated which does not appear further north (Figs 11.3 and 13.4). Fields are frequently wider than elsewhere, averaging 15 to 25 m but occasionally reaching 35 m. The lengths are variable, though most are between 30 and 70 m, and the mean for a sample of 48 measurements is 56 m. These raised fields are actually platforms rather than ridges. They are sometimes so wide in relation to their length that they approach squarish or rectangular forms. One set of such groups has a mean width of 45 m. These platforms are arranged into long, ladder-like systems. Parallel groups of 'ladders' are separated by troughs and occasionally by stone or adobe walls, though many of the walls cut across the platforms in a way which suggests that they post-date the field patterns. Similar ladder-like arrangements of ancient fields occur with staircase terraces in the Pomata area. Thus, the organization of platform cultivation in a ladder pattern on the periodically flooded plains may simply represent the adaptation to the lake plain of practices of land division on the adjacent uplands. If this is so, the construction of the raised-field platforms may have been roughly contemporary with the creation of the hillside and hilltop fields.

Combed Fields of Pampa Koani

Finally, there is one other type of pattern which does not easily fit into the classifications adopted above. At Pampa Koani near Aygachi in Bolivia, long, roughly parallel and curvilinear ridges occur which are frequently quite narrow (2 to 6 m in width) and grouped in bundles of five to as many as thirty-five strips. They vary greatly in length from 20 to 150 m. Some of the ridges converge to pinch out the intervening troughs completely, giving the impression of a 'combed' pattern.

The Cultivation of Raised Fields

The only large area of raised fields for which soils had been fully mapped and described at the time of our study was that north of Puno. All of the fields (ONERN 1965, vol. 3) of the Paucarcolla-Juliaca plain are classified as gley

Fig. 13.4. 'Ladder' pattern of raised fields near Pomata, Lake Titicaca, Peru. Taken in October 1955. At the left are some rectangular platform fields. At the top center are some modern cultivated fields, undrained. Servicio Aerofotográfico Nacional, Lima

humic Andean planosols. These are lacustrine *pampa* soils consisting of fine sediments and characterized by poor drainage and neutral to strong alkaline reaction (pH 6.8 to 8.5). This soil is of moderate fertility where not too alkaline. It is significant that the high pH decreases with depth. It is likely that the raising of fields served not only to reclaim poorly drained land and to ameliorate unfavorable climatic conditions, but also to bring less alkaline soils to the surface. Closer to the lake the soils are even more poorly drained, often with a pH of over 8.0, but even here remnants of raised fields exist as at Requeña. The ONERN resource survey states that agriculture is limited by poor drainage, and the recommended land use is grazing. On one air photograph the Requeña fields (Fig. 13.2) are included in a land-use category mapped as good only for fishing and wildlife (ONERN 1965, vol. 3, map 39). Perhaps it is fortunate that the pre-Hispanic raised-field cultivating farmers of this lakeshore area lacked the benefits of modern land-use surveys!

No early descriptions have been found of the methods of cultivating the raised fields or the specific crops raised, but some insight can be gained from current agricultural practices on similar terrain and even on the old fields themselves, although most of the surviving ridge and trough areas are now used for pasture. The field tops commonly support *Stipa* spp. (ichu) and various other grasses, with species of *Halofitetum* on the more alkaline sites. Even the roots of some of the alkaline grasses provide feed for pigs, and saline soils in the raised fields of Requeña have been badly ripped up by rooting pigs.

Where soils are not excessively alkaline or flooding too serious, as in the Paucarcolla-Juliaca plain, some of the raised fields are still cultivated today. Where peasant communities are still using the footplow, as around Huatta, *huachos* or 'lazy beds' (narrow potato ridges) are laid out on top of the old ridges both at right angles to field length and longitudinally (Fig. 13.5). Thus the ancient raised fields are themselves now ridged by modern *huachos*. The lazy-bed system of cultivation is widespread through the Andean highlands. *Huachos* can readily be distinguished from the ancient raised fields by their much smaller size, since they are seldom over a meter in width and average about 0.25 m in height.

Fig. 13.5. Two broad pre-European platform raised fields separated by a narrow trough at Huatta, Lake Titicaca, Peru. Narrow potato ridges (*huachos*) of recent origin are superimposed on the two platforms separated by a ditch. From C. Smith *et al.* (1968: pl. 4)

At Requeña in 1966, Indian farmers had the tradition that the old ridges were planted with potatoes, quinoa, and cañihua, and they said that their great-grandparents occasionally grew the same crops on the ridges in the nineteenth century before soil salinity became too bad. Throughout the areas of raised fields where there is cultivation today, these are the traditional crops, with the important addition of barley. Invariably the 'bitter' or *chuño* potato is grown, as it is much more tolerant of both alkalinity and frost than the larger, non-bitter potato. Closer to Lake Titicaca where temperatures are warmer, other tubers are grown such as oca, as well as wheat, and some dwarf maize (*confite puneño*) is cultivated on the lower lakeside slopes as high as 3,900 m.

The usual rotation pattern today on lazy beds built on top of raised fields near Huatta (3,850 m) is one crop of potatoes on the freshly turned soil, quinoa or cañihua the second year, and barley the third year, followed by at least three years of fallow. Probably only about 5 per cent of the densely ridged *pampa* around Huatta was cultivated in 1966, the main reason given being the likelihood of a considerable crop loss from frost. Huatta itself is on a small hill about 30 m high, the warmer slopes of which are intensively cultivated without regular fallowing. Near Juliaca, where the *pampa* soils are less fertile and often very alkaline as a result of evaporation of shallow pools of water, the rotation is the same as at Huatta but the fallow period lasts for six or seven years. However, it is probably land tenure and the extensive pastoralism of the large haciendas rather than drainage, poor soils, and frost hazards, which are responsible for the absence of cultivated crops over much of the lacustrine plains near Titicaca. Among peasant communities land is also set aside for communal pastures, although these lands are cultivated from time to time.

The Functions of Raised Fields

The distribution of raised fields in the Titicaca region suggests that they were constructed in order to reclaim wetlands subject to seasonal inundation by rainfall or from overflowing rivers. What is much less certain is whether there was also any conscious attempt to drain via canals, to conserve water, or to irrigate.

The arrangement of raised fields and intervening troughs in relation to the lie of the land or in relation to existing watercourses does not, usually, suggest that there was an attempt to carry away excess water. Most types of ridging would definitely have hindered systematic drainage or would have done nothing to assist it. The open checkerboard and the ladder types still hinder effective drainage; the embanked fields show no signs of canal drainage systems; the riverine and linear types, and the 'combed' patterns of Pampa Koani, would all assist drainage to a certain extent wherever the linear axes coincided with the direction of slope, as in fact they usually do.

In the areas of surviving raised fields, there are, on the other hand, so many instances in which patterns can be picked out by the existence of standing water or wet, marshy ground that one is tempted to see a conscious aim towards water conservation. In the Pomata area the organization of ridges into a 'ladder' pattern hinders drainage but could have been expressly designed to retain water on the fields. The design of the embanked fields in the same area, but not, apparently, in the Paucarcolla-Juliaca plain, was often intended to guide water into partial enclosures rather than to exclude flood waters. And the 'combed' ridge pattern of Aygachi still has the effect of distributing water extensively over the area of raised fields.

In the region as a whole there are positive advantages to a practice of water conservation. Precipitation is highly seasonal and relatively unreliable. At Puno the mean annual precipitation is 623 mm; at Capachica (east of Huatta) 828 mm; and at Juliaca 650 mm. Precipitation is strongly concentrated in the six months from October to March, with a marked maximum in January and February (at Puno, 87 per cent of the annual total falls in October to March; at Capachica, 89 per cent; and at Juliaca, 88 per cent). At all of these stations, however, there is a 50 per cent probability of deficient rainfall in any month. Drought is one of the two major agricultural hazards of the area, and there is a close relationship between low potato yields and years of deficient precipitation. Modern agronomists strongly recommend the use of irrigation on a much larger scale than is now practiced (ONERN 1965/1: 186). The conservation of water is clearly of great importance, especially in years of deficient rainfall, but it may also assist in reducing hazards from frost. In March and April, towards the end of the rainy season, night frosts are frequently the result of excessive radiation into a clear atmosphere, and constitute a serious hazard for the potato crop. For this reason hill slopes with good air drainage are often preferred for cultivation rather than the valley floors and the flat *pampa* surface. The high specific heat of water is one of the few defenses available against night frosts in these areas, and it may be that crops grown in the raised-field system in areas of waterlogged or flooded troughs could better resist frost hazards (see below).

Few areas of raised fields show any sign of systematic irrigation, but the scale at which the air photos are available for much of the area (1 : 65,000) makes it difficult to discern the patterns of water distribution in any detail, and it is fairly certain that most fields in the immediate neighborhood of the lakes were never part of an organized irrigation system (see below). However, in the Pomata area, for which photographs of 1 : 15,000 are available, there are several abandoned irrigation channels which appear to be related to the systems of embanked fields. The 'combed' fields of Aygachi (Pampa Koani) would have been excellent distributary systems for irrigation water taken from the available natural streams of the area. Garcilaso de la Vega (1966: 113), not always the best of authorities, in the early 1600s made reference to the establishment of irrigation in his account of the Inca conquered province of Chucuito: 'As long as the conquest [of Chucuito] lasted, he [the Inca] had been engaged in visiting his kingdom, seeking to improve it by extending the cultivated lands, for which purpose he ordered new irrigation channels to be dug, and such necessary works as barns, bridges, and roads for communication to be executed.' However, it would be unwise to assume that this passage necessarily refers to the Pomata area. Also, irrigation systems established long before the Inca conquest may well have been wrongly accredited to the Incas, as they often have been in other parts of Peru. Furthermore, Cieza de León (1959: 271) in 1553 denied the existence of any irrigation around Lake Titicaca, although it did and still does occur.

Conclusions about the purposes of raised-field systems must necessarily be tentative. It is certain that they served to extend the area of cultivable land; it is likely that in some areas it was water conservation rather than effective and rapid drainage that was sought; and in a few areas it is clear that raised fields were associated with irrigation systems.

Finally, the distribution of raised fields and their functions are also relevant in a consideration of present and past levels of lakes and rivers in the region as a whole. Lake levels could not have been very different from their present position at the time the ridges were being made and used, for most raised field areas are still more or less at water level. Records of the level of Lake Titicaca from 1914 to 1989 show significant seasonal and overall fluctuations. Seasonal fluctuation is rarely less than 50 cm or greater than 100 cm, and averages 80 cm. The extreme range so far recorded is 6.28 m, from a minimum of −3.72 m in December 1943 to a maximum of +2.56 m in April 1986 (R. Hill 1959; Roche *et al.* 1992: 79).[4] When the lake is high, as it was in July 1955 when the photos shown in Figs 13.2 and 13.3 were taken, troughs near the lake are filled with water, whereas the same troughs were completely dry when I visited them in July 1966 when the lake was at a low level. The air photographs show a few offshore linear features which may be submerged ridges and troughs. Some ridges at the edge of Lake Umayo northwest of Puno and neighboring lakes were partially submerged in 1966, so the level of this lake may have been a little higher than at the time the fields were constructed.

The Antiquity of the Titicaca Raised Fields

The raised fields around Lake Titicaca originated and were used in prehistoric times (see below). Did their construction and/or use continue into colonial and later periods? There are few mentions of them in the historical and travel literature. Few instances have been found in which raised fields played an integral part in recent farming,[5] prior to field construction and restoration beginning in the 1980s. Where prehistoric raised fields are cultivated today by means of superimposed lazy beds, the underlying ridges are regarded as irrelevant and a hindrance. This is the same kind of discontinuity as there is in the modern use for arable farming of land that lies in pre-modern ridge-and-furrow in the English Midlands.

In the field, we repeatedly asked local farmers and authorities if they knew anything about the origins, antiquity, and functions of the raised fields. The usual replies were that nothing was known about them, that they were very old, or that they must be natural since they have always been there. The most positive information was obtained at the community of Requeña. Requeña has communal pasture, much of which consists of large raised fields subject to deep flooding (see Fig. 13.2). The local Quechua people have the tradition that these fields were made by the Incas, although one man said that they were made by the pre-Inca Aymara.[6] It was agreed that they were no longer used after the Spaniards came. Large fields were said to have been made by and belonged to the men and smaller ones by and for the women. The crops were said to have been potatoes, quinoa, and cañihua in rotations similar to those that can be found today in the area.

There are other indications, though none of them is as concrete as one would wish, that the creation and use of raised fields was primarily pre-Columbian, though it is difficult to determine exactly when most of them were abandoned. First, the silence of Spanish colonial travelers argues against their continued use into colonial times, though this is a negative point and inconclusive. Inattention to agricultural features is commonplace. Second, many of the best examples of raised-field patterns occur on the rough grazing land of large ranching haciendas established in the early colonial period. However, there is some evidence that the growth of large haciendas and the displacement of Indian populations was still going on in this area at the end of the nineteenth century and in the early twentieth century (Chevalier 1966; H. Martínez 1962). Third, the shape and size of the ridges is such that they must have been produced, not by the plow of Spanish introduction but by hand methods of Indian cultivation. Fourth, it may be argued from the form and distribution of the raised fields that they must be of considerable antiquity. Many are seriously eroded and have rounded or irregular sides (Fig. 13.5); troughs have been enlarged at the expense of the ridges and some of the troughs at the edge of Lake Titicaca are now covered by several inches of salt formed by evaporation. Raised-field patterns are notably absent (buried?) from areas where there has been recent alluviation by a prograding river, as on the lakeward side of Paucarcolla north of Puno where the Río Ylpa enters the lake.

When we conducted our survey in 1966 there had been no archaeological dating of raised fields. We noted the association of raised fields with *chullpas* (circular stone burial towers) at Lake Umayo, which mainly date to AD 1100–1450. The Aygachi (Pampa Koani) fields on the southern shores of Lake Titicaca in Bolivia are only 10–15 km from the ruins of Tiwanaku, and we suggested that raised fields may have helped sustain that large site and associated pre-Inca state (AD 300–1100). Ceramic styles of surface sherds on raised fields at Pampa Koani ranged from Chiripa (pre-Tiwanaku) to Inca. Given recent dating of Titicaca raised fields by Erickson (1987) and Kolata (1991) (see below), our original discussion of the dates of field origins is no longer pertinent.

Many of the raised fields were probably abandoned in pre-Inca times or as result of Inca period disruption, but the final decline likely came with the massive population reduction after 1492, the associated abandonment of crop land, and the establishment of Spanish livestock haciendas on the lake plain (Erickson n.d.). Some of the raised fields, however, may have continued in cultivation for an unknown time into the colonial period.

Recent Archaeological Research

Since our discovery and survey of the Titicaca raised fields in 1966, there have been two major archaeological projects examining the fields. Clark Erickson worked at Huatta north of Puno in Peru from 1981 to 1986 (Erickson 1987; 1988*a*; 1992; 1993; 1994*b*; 1996; 1998; 1999; 2000; n.d.). Alan Kolata and colleagues have carried out archaeological excavations and ecological field work on the raised fields near Tiwanaku in Bolivia starting in 1979 (Kolata 1986; 1991; 1993; 1996*a*; 1996*b*; Kolata and Ortloff 1989; 1996; Ortloff and Kolata 1989; 1993; Binford *et al.* 1997; Graffam 1989; 1990; 1992; Carney *et al.* 1993). Both Erickson and Kolata established experimental fields and promoted the restoration or construction of new raised fields by local farmers. Additional studies were undertaken by Thomas Lennon (1982; 1983) in the Paucarcolla-Juliaca area and by Stanish (1994) at Juli and Pomata. Briefly, some of the major findings and conclusions of these archaeologists are as follows.

Functions (Fig. 13.6)

In the 1960s, we originally believed that there had not been intentional construction of canals to either bring in irrigation water during dry periods or to carry away excess water. Lennon (1982; 1983), however, argues that canals were intended for both functions in the areas of Paucarcolla and Atuncolla north of Puno. Ortloff and Kolata (1989) interpret Tiwanaku aqueducts and canalized *quebradas* to have been drainage canals for reducing seasonal inundation of both agricultural fields and urban areas, including diverting water from raised fields toward Lake Titicaca. In contrast, at Huatta Erickson (1992: 293–4) found that: 'large canals were constructed to bring additional water to low lying areas, to expand the natural wetlands rather than drain them.' Also, the hydraulic system of 'canals and embankments, probably functioned to remove, dilute, or separate water with high levels of salts and alkalinity from fresh water' (Erickson 1992: 294).

The study of raised fields north of Puno by Lennon (1982; 1983) involved intensive pedestrian survey, detailed examination of aerial photography,

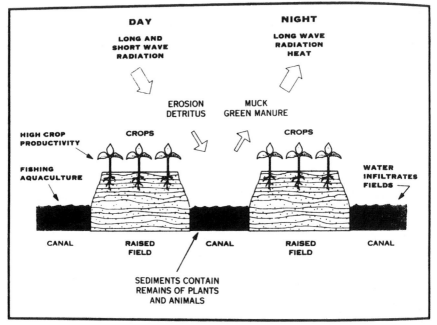

Fig. 13.6. Diagram of Titicaca raised fields. From Fagan (1995: 224); based on Erickson (1992: 292–6)

and statistical analysis of raised-field patterns and characteristics in relation to water regimes. He concluded that the different patterns of raised fields functioned to most effectively control water for drainage, moisture conservation, and irrigation. Associated canals served to both remove or divert excess water and to move water into water-deficient areas. The result was a complex and systematic water-management system designed to maximize the amount of land that could be cultivated under highly variable drainage conditions.

The Huatta and Tiwanaku projects have confirmed our brief speculation in 1968 regarding the microclimatic effects of raised fields. Night frosts are reduced by the released heat from daytime-captured solar radiation in the water and vegetation in ditches between fields. At Huatta, temperatures were 1–2° C higher on raised fields than on the surrounding *pampa*, and night frost was reduced from 6 to 4 hours (Erickson 1992: 296). In 1988 on Pampa Koani, there was a severe late frost; crop losses were 70–90 per cent on non-raised field plots, but were minimal on nearby restored raised fields (Kolata and Ortloff 1989: 259).

Excavations of raised fields and experimental fields also confirm that productivity is increased and maintained by the addition of nitrogen-rich sediments and organic material from the ditches between fields. The ditches concentrate aquatic macrophytes, other vegetation, and water-borne silt and other material. These were transferred to the planting beds either directly from the water or were dredged from the ditch bottoms (Kolata 1991: 102; Erickson 1992: 292–3). Of particular importance are nitrogen-fixing *azolla filiculoides* (water fern), as well as blue-green algae, which are especially effective when the ditches retain water all year (Biesboer *et al.* 1999). The natural soils on the lake plain have low nitrogen levels, except in marshes.

Productivity

The present average potato productivity, without commercial fertilizer, in the Titicaca region is between about 1,500 and 6,000 kg/ha/yr (Kolata 1991: 107); the average for the Department of Puno, 1955–64, was 3,050 (Christiansen 1967: 82). In contrast, experimental raised-field plots at Huatta obtained yields between 8,000 and 16,000 kg per hectare of field and ditch per year (Erickson 1993: 407). In Bolivia, for rehabilitated raised fields, Kolata (1996*b*: 213, 227, 230) obtained average yields of 21,000 kg/ha/yr for 1987–8 and 14,850 kg/ha/yr in 1990–1. Ditches are not included, so a reduction would have to be made in order to compare Kolata's yields with those of Erickson. Approximately 50 per cent of the total ditch and field area is actual planting surface.[7] In both places chemical fertilizers were not used. Potato yields on raised fields without chemical fertilizers were still double the yields on traditional non-raised fields supplemented with chemical fertilizers (Biesboer *et al.* 1999: 264). These averages are only for a few years, and a longer time span of testing is needed. Still, the yield differences between non-raised fields and raised fields is impressive.

Labor Inputs

We originally believed that raised-field construction and maintenance would have been very labor-intensive compared to non-raised fields. However, we were not aware of the tremendous increases of productivity with raised fields. Erickson (1988*a*: 237, 249) believes that the labor expended in raised-field construction was modest when prorated over the life of the fields and that the labor required for maintenance was low. Graffam (1992: 891–3) demonstrates that production per unit of labor is greater for raised fields than for non-raised fields—somewhat greater even in the first year (counting construction costs) and twice as great in the following years (counting maintenance costs).

The Extent of Raised Fields

On air photos, we originally measured 82,056 ha of visible raised fields and associated ditches (C. Smith *et al.* 1968: 355). Certainly, additional fields have been buried under sediment or destroyed by erosion or by recent farming and other activities. Based on ground survey and additional discoveries on air photos, Kolata (1991: 124) increased our field area for the Bolivian side of the lake from 3,952 to 13,000 ha. This total includes *c.* 6,000 ha for the vicinity of the Tiwanaku ruins, which we completely missed, and *c.* 7,000 ha for Pampa Koani (Catari or Aygachi) compared to our 3,000 ha. Erickson (n.d.) has added 31,196 ha on the Peruvian side of the lake. The revised total area of existing fields for the Titicaca Basin comes to 122,300 ha (Table 13.1), or 1,223 km^2 (Erickson n.d.). This includes both field surface and ditches. For field surface alone, the total would be reduced by about 50 per cent.

Population Estimates

Depending on calculations of yield, total field area, and population density, several different estimates have been made of the population that could have been supported by the Lake Titicaca raised fields (Table 13.2). The totals range from 215,000 to 2,293,000.

Using the measured 122,300 ha of known raised fields and ditches, reducing this by 50 per cent to 61,150 ha for field surface only; using a population density of 37.5 per hectare based on potato yields from experimental raised fields in Puno, Peru (Erickson 1992: 297; 1999); assuming that no more than 75 per cent of field surface was productive given periodic fallowing and crop attrition to disease, pests, spoilage, frost, and flooding (Kolata 1993: 200); and assuming one crop per year; then the resulting potential population supported by raised fields on the Titicaca lake plain would be 1,720,000. However, the population density of 37.5 per hectare may be too high. On the other hand: (1) if a second or partial second crop was obtained, which Kolata (1993: 201) finds feasible, then the total estimated population would be higher, (2) an unknown quantity of raised fields has been destroyed, and (3) other forms of cultivation were present, such as terracing on adjacent slopes and non-raised fields on the higher lake plain, so that the total regional population would be higher

Table 13.2. Estimates of Titicaca populations supported by raised fields

Density (per ha of cultivated field surface)	Total cultivated field surface (ha)	Total estimated population	Source
5.7[a]	37,746[c]	215,000	Denevan (1982: 190)
14.1–23.7[b]	41,000[d]	578,000–970,000	Graffam (1992: 891)
39.0[b]	30,771[e]	1,200,000	Projected from Kolata (1993: 203)
37.5[b]	41,028[f]	1,539,000	Erickson (1992: 297)
37.5[b]	61,150[g]	2,293,000	Erickson (n.d.)
37.5[b]	45,863[h]	1,720,000	Projected from Erickson (n.d.)

[a] Based on recent non-raised field yields of potatoes. [b] Based on experimental raised field yields of potatoes. [c] 46% of 82,056 ha of fields and ditches, 100% productive. [d] 50% of 82,000 ha of fields and ditches, 100% productive. [e] 50% of 82,056 ha of fields and ditches, 75% productive (25% fallow or attrition). [f] 50% of 82,056 ha of fields and ditches, 100% productive. [g] 50% of 122,300 ha (Table 13.1) of fields and ditches, 100% productive. [h] 50% of 122,300 ha of fields and ditches, 75% productive (25% fallow or attrition).

than that supported only by raised fields. A peak prehistoric west-side Titicaca Basin population of between 1.5 and 2 million people at times is reasonable in my opinion.

For just the Bolivian side of Lake Titicaca (Pampa Koani, Río Titicaca Valley, plus all of Desaguadero, a total of 19,000 ha), Kolata (1993: 203) obtained a population of 555,750, using a density of 39 per km; however he did not reduce the total field area to cultivated surface only, which at 50 per cent would reduce the population by half. Likewise, Kolata's (1993: 203) calculation of 204,750 for Pampa Koani should also be reduced by half. Regardless, it is significant to note that the present population of Pampa Koani, now lacking raised-field technology, is 'only about 2,000 people at a level slightly beyond bare subsistence' (Kolata 1993: 204).

Raised Fields and the Tiwanaku State

An important conclusion reached by Kolata and colleagues is that the Tiwanaku state to a large extent was supported by a circum-Titicaca reclamation of wetlands along the edge of the lake by means of raised fields which fed large and dense concentrations of people on a continuous basis (Kolata 1986: 760). Furthermore, Kolata (1986: 760) believes that: 'the construction, maintenance, and production of these fields were managed by a centralized political authority.'[8] However, Erickson (1993) argues that the raised-field system functioned at the community level independent of the state. 'It is important to remember that raised fields were used in the Titicaca Basin before, during, and after the Tiwanaku phenomena. There is no necessary relationship between the Tiwanaku state and the construction and management of raised field systems' (Erickson 1999; n.d.). These differences remain to be resolved, but the productivity of raised fields was undoubtedly a significant asset to the state.

Chronology of Changing Raised-field Importance

Archaeological surveys in the 1980s at Huatta in Peru and Pampa Koani in Bolivia have provided accurate dating. In the northern Titicaca Basin, raised-field farming began about 1000 BC, was largely abandoned from AD 300 to AD 1000, resumed between AD 1000 and AD 1450, with final abandonment during the Inca and early colonial periods (Erickson 1987; 1992: 290–1). Erickson (n.d.) believes that the first raised-field farmers were a group 'practicing a wetland-oriented economy such as that of the ethnohistoric and modern Uru'. In the southern Titicaca Basin raised fields probably date to 800–200 BC (Chiripa), continuing and expanding until about AD 1000 (Tiwanaku I–V), with an apparent decline thereafter attributed to regional drought (Kolata 1991: 113, 115; Ortloff and Kolata 1993: 212–14).

Kolata and colleagues argue that not only was there abandonment of raised fields because of a prolonged dry period, evidenced by the Quelccaya ice cores, but that this abandonment caused the collapse of Tiwanaku civilization

(Kolata 1993: 282–302; Binford *et al.* 1997; Ortloff and Kolata 1993). However, these publications are inconsistent as to when the dry period began (AD 950, 1000, 1030, and 1100 are all given) and when the collapse began (AD 900, 1000, 1100, and 1150 are all given). Thus it is difficult to correlate the dry period with the collapse (Erickson 1999: 635). Furthermore, as mentioned above, the Puno area in Peru experienced the same general dry period without raised-field abandonment. Gartner (1996) raises other objections to this collapse theory.

The Quelccaya ice core also shows unusually high dust deposits derived from the Titicaca lake plain from *c.* AD 830 to AD 960, a wet period during which raised-field activity was probably intense. Thompson *et al.* (1988: 765) attribute this dust increase, in part, to the cultivation of raised fields. Subsequent to AD 960, there was a long dry period to *c.* AD 1500, with a marked decline in dust deposition. This would support Kolata's thesis. However, it can be questioned whether there was sufficient soil exposed by raised fields to contribute significant amounts of dust particles to the atmosphere over a very large region.

Graffam (1989: 43; 1992: 893–4, 899; also Erickson 1999: 641), believes that raised-field construction and use continued after the collapse of the Tiwanaku state until about AD 1480 when climatic change resulted in colder temperatures causing more frost and greater precipitation causing flooding of fields and sedimentation of canals.[9] He also recognizes the disruptive effects on all indigenous agriculture by the Spanish occupation a few decades later.

The raised fields in the Juli-Pomata area along the southern shore of Titicaca in Peru have been studied by archaeologist Charles Stanish (1994). He dates the earliest raised fields there to 800–200 BC, with 41 per cent of the regional population living in the field area. The population portion in the raised-field areas increased to 69 per cent during 200 BC–AD 400 and was 57 per cent during the Tiwanaku period, AD 400–1100. (Most of the remaining population was located in areas with rainfed terraces.) The population living in or near the raised-field areas then decreased to about 40 per cent of the total, and to less than 15 per cent during the Inca period, AD 1450–1532. By the early colonial period few if any raised fields seemed to be in use. Stanish believes the demise of raised-field cultivation can be attributed to long-term drought and colder temperatures after AD 1000, while maximum construction and use were associated with pre-Inca periods of political complexity and large populations.

I question, as does Erickson (1999: 637–41), whether either drought *or* increased rainfall could cause major raised-field abandonment, since the system is adapted to both, the two conditions having alternated throughout human times. When these conditions were extreme, farmers could extend raised-field cultivation either toward a receding lake shore or away from an advancing lake shore. Erickson (1999) argues that changing political dominance can best account for the expansion and contraction over time of raised-field cultivation and associated settlement in different sectors of the Titicaca region, without necessarily complete abandonment.

Raised-field Restoration (Fig. 13.7)

An exciting spin-off of the Huatta and Pampa Koani archaeological projects has been the construction of new raised fields and the restoration of relic fields. This work was initiated by Erickson at Huatta in Peru in 1981, initially for experimental purposes to obtain information on field construction, functions, and productivity (Erickson 1985; 1988*b*; Erickson and Candler 1989; and Garaycochea 1987). These fields were built by local farmers using traditional tools and crops. As reported above, the yields were spectacular. The resulting attention was both local and international (Mullen in the *Chicago Tribune* 1986; Stevens in the *New York Times* 1988). Impressed by the high yields obtained, Indians near Huatta began constructing raised fields independently, with encouragement and support from Erickson and the Ministry of Agriculture. In 1986–7, Kolata and Oswaldo Rivera inspired Aymara farmers at Lakaya at the edge of Pampa Koani in Bolivia to restore several hectares of relic raised fields (Kolata 1991; 1993: 190–8; 1996*a*: 211–59; Kolata *et al.* 1996). By 1992, 840 ha of prehistoric raised fields had been restored by fifty indige-

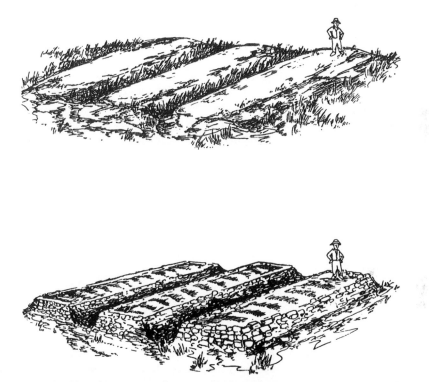

Fig. 13.7. Prehistoric eroded (top) and recently rehabilitated (bottom) platform raised fields at Huatta, Lake Titicaca, Peru. The fields are on flat terrain, contrary to this perspective. By D. A. Brinkmeier, in Erickson (1996: 161)

nous communities in Peru (Erickson 1994*a*). By 1994, raised-field cultivation had been introduced to fifty-two communities and 2,500 people on the Bolivian side of Titicaca (Kolata *et al.* 1996: 107).

A lost, ancient technology may be on the verge of a significant revival, an example of applied archaeology (Erickson 1998; Erickson and Brinkmeier 1991). As Kolata (1991: 108; 1996*a*: 267) points out:

Continuous cultivation on fixed, permanent fields, short or no fallow periods, and two episodes of sowing and harvesting within the same agricultural year [had previously been] inconceivable in the contemporary landscape of the high plateau. Yet these may have been standard features of Tiwanaku agricultural practice . . . There is no environmental reason why the altiplano should not bloom again . . . By our projections the altiplano with raised-field technology is capable of producing enough wheat to supply Bolivia's entire internal market.

Significantly, 'Both the Peruvian and Bolivian governments, in addition to several governmental and nongovernmental organizations, have added raised-field technology to their development programs for the rural altiplano' (Erickson 1994*a*).

The raised fields on the shore plain around Lake Titicaca are only one part of a fossil cultural landscape massively altered by earth- and stoneworks: settlements, causeways, aqueducts, canals, the heaping up of earth to create cultivable land in wet places, and the making of stone-faced terraces which rise like staircases almost to the elevational limit of cultivation. It is fairly certain that methods of expanding cultivable land enabled a dense population to live in the exacting environment of the Titicaca region and made possible the rise of the Tiwanaku state. Even in the late Inca period, after the preceding people had been decimated by the Inca wars of conquest and subjected to Inca rule for eighty or ninety years, there were still nearly as many people in one part of the raised-field area, the province of Chucuito, as there were in 1940 (C. Smith 1970).

Now, raised fields are being restored or created anew around Lake Titicaca, the most successful such effort to date in the Americas. It is entirely possible that this ancient system of reclaiming marginal lands may once again provide a secure, highly productive, sustainable means of feeding large numbers of people.

Notes

[1] Raised fields were examined by us in 1966 at Huatta, Juliaca, Paucarcolla, Lake Umayo and its surroundings, Capachica, and Pomata in Peru and at Aygachi (Pampa Koani) in Bolivia.

[2] Climatic data are for variable years between 1943 and 1965 (Monheim 1956: 44; ONERN 1965: 1: 156–63).

[3] The only other known embanked raised fields elsewhere in the Americas are in the Llanos de Mojos of Bolivia near the Río Apere (Plafker 1963).

[4] In 1986, most prehistoric raised fields and all modern non-raised fields were inundated follow-

ing unusually heavy rains, but many of the restored raised fields remained productive (Erickson 1992: 294).

[5] Active raised fields were reported in the 1960s by David Preston (pers. comm.) in Bolivia near Cerro Cohana *c*. 15 km northeast of Aygachi and near Batallas along the Pan American Highway. Archaeologist Carlos Ponce Sanginés (pers. comm.) saw apparent raised fields in cultivation in the Chipaya area north of the Salar de Coipasa in the Bolivian altiplano.

[6] Terminology persists. In Quechua, large raised fields are called *waru-waru*, and in Aymara they are called *suka-kollus* (Erickson 1996: 30).

[7] Estimates of the portion of the Titicaca ridges and ditches that is cultivated surface are 45–50 per cent (C. Smith *et al.* 1968: 355), 46 per cent (Denevan 1982: 190), and 30–60 per cent (Biesboer *et al.* 1999: 255); 50 per cent (Erickson n.d.) seems a good compromise.

[8] Kolata (1996*b*: 277) believes that: (1) the initial construction of raised fields was by autonomous farmers but later involved state management, and that (2) both autonomous and centralized management of intensive agriculture coexisted in 'time and space' in most agrarian states (Kolata 1996*b*: 266).

[9] For a counter-argument, see Kolata (1996*b*: 21).

14

Ditched Fields, Drainage Canals, and River Canalization

Most raised fields provide for drainage by means of elevated surfaces in association with surrounding ditches which catch water, drain subsoils, and in some instances remove water from the field vicinity, both on flats and on slopes. In addition, there are examples of drainage by ditching alone without raised planting surfaces or with only minimal raising from earth removed from narrow, shallow ditches. Some of the prehistoric Mayan drained fields in Belize are believed to be primarily ditched or channelized fields (Turner 1983; Pohl and Bloom 1996).

In South America, prehistoric ditched fields occur in Amazonia in the Llanos de Mojos and on Andean slopes in Ecuador and Colombia. Present-day Karinya Indians ditch fields to improve drainage in the eastern Orinoco Llanos in Venezuela. Also, there are some instances of ditched fields on flats in the Lake Titicaca Basin and in other Andean basins. Another means of assisting drainage was the Inca technique of straightening and confining river channels to facilitate water flow and thereby reduce overflow.

Llanos de Mojos

In Mojos in Bolivia (Chapter 12), there are a large number of drained fields in which the drainage was accomplished by digging closely spaced ditches (see Fig. 11.3). These fields are very common along both sides of the Río Apere from 50 km north of San Ignacio to La Esperanza (see Fig. 12.1). They also occur west of San Ignacio near the Río Apere. Air photos show about 12,000 ditched fields almost covering an area of about 20 km^2 in the vicinity of La Esperanza (Fig. 14.1). I saw at least this number elsewhere along the Río Apere. The ditches are difficult to distinguish from the air and on the ground unless the grass cover has been burned off. Some of the ditches are clustered in rectangular patterns.

The ditches at La Esperanza are 15 to 30 cm deep and range in length from 8 to 150 m, but average about 120 m. They are 0.6 to 1.2 m wide and are spaced 1.5 to 7.5 m apart. The close, furrow-like spacing of the ditches gives the appearance of ploughed fields.

Fig. 14.1. Pre-European ditched fields at La Esperanza in the Llanos de Mojos, Bolivian Amazonia. From Denevan (1966*a*: 182; also see photos on pp. 180 and 181)

At La Esperanza, there are only a few scattered trees, such as *Acrocomia* and *Curatella* in the area of the fields. The soils of both fields and ditches are alluvial with no mottling within the upper 25 to 30 cm. The topsoil of one of the fields was a grayish-brown loam with a pH of 6.0, while that of an adjacent ditch was a light-gray loam with a pH of 4.9, so the ditch soils are much more acid. The entire area is flooded when the Río Apere overflows, but most of the water drains off rapidly into *bajios* (depressions) some distance from the river. Most of the fields are on the gentle slope between the natural levees and the *bajios*. The heavier silt is deposited on the fields and the finer clays in the *bajios*. There are a few large ditches leading to the *bajios*, which may have been intentional drainage canals, but many of the ditches run parallel to these canals rather than leading to them.

Northern Andes

The ditched fields in Ecuador primarily occur west and south of Pimampiro in the northern highlands (Knapp and Preston 1987). These are straight, linear features running downslope and are quite prominent on air photos. Some extend from ridge crest to slope bottom for as much as 900 m. Most are 10 to 20 m apart, and are 0.5 to 1.0 m deep and about 1.5 m wide at the ditch tops. Most occur on slopes of 15 to 20 per cent at elevations between 2,700 and 2,900 m but as high as 3,000 m. Some ditches may have been enlarged by erosion, but it is unlikely that these regular ditch patterns are natural gullies.

Knapp and Preston argue that the primary functions of the ditches was to facilitate runoff on heavy, clay-rich soils which drain poorly, with a possible secondary function being field bounding. Other suggestions of erosion control and irrigation seem unlikely, except in some flatter areas where ditches may have served for both drainage and irrigation. In some places faint lines horizontal to the slope, between the ditches, may have been low ridges to divert runoff into the ditches. Also, in places there are long horizontal ditches between the up- and downslope ditches that may have been channels which served to divert water from the ditches into depressions or into nearby ravines. The ditches are self-maintaining once constructed and therefore not labor-intensive.

Knapp and Preston believe that the ditches are pre-Columbian in origin, given an association with other prehistoric features. The region was considerably depopulated in colonial times and fields on slopes were abandoned. Many of the ditches appeared when settlers removed brush or forest in the twentieth century. Tractor plowing has since destroyed some ditches. There is no evidence that the ditched fields are colonial or recent in origin; they are not mentioned by the Spaniards. Current agriculture on slopes, however, often involves ditching, but in a box lattice pattern around fields; stone diversion walls direct water into the ditches (Knapp 1991: 67).

Ditches vertical to slopes also occur in the Colombian Andes. One of the first reports was by J. J. Parsons (1949: 34–5) in Quindío. He mentioned 'extensive, ridged and furrowed Indian old fields (*surcos de los indios*)' between 1,400 and 2,400 m. These were up- and downslope suggesting a drainage function. The ridges (*eras*) were about 0.3 m high and 1.5 m from crest to crest, mainly located on volcanic ash soils. Possible crops were arracacha plus manioc at lower elevations and potatoes higher up. Bray *et al.* (1987) and Eidt (1984: 67–72) describe similar ditches as 'field lines' or 'ditched fields' in Calima, some of which Bray dated to the period AD 1200–1600. Possible functions are: (1) drainage (but the ditches occur on different types of soil and highly variable degrees of slope), (2) prevention of landslips where volcanic ash overlies heavy clay, and (3) field boundaries. Other ditches on slopes have been reported elsewhere in the Colombian Andes (Bray *et al.* 1987: 457). The low frequency of ditches on slopes in Peru and Bolivia probably reflects lower rainfall there than in the northern Andes, which supports the argument for a primary drainage function for such ditches.

For Peru, in the Province of Espinar in the southern part of the Department of Cuzco, Orlove (1977: 96, 98) describes contemporary furrows vertical to the slope that serve for both drainage and water retention. On steeper slopes, over 30 degrees, diagonal furrows serve similar functions in the provinces of Canas and Canchis, also in Cuzco. Ditched fields also occur on the Lake Titicaca floodplain. Schaedel (1986: 324) mentions a combination of irrigation canals and drainage canals in wetlands in Cuzco.

The Karinya, Orinoco Llanos

Karinya field systems in the eastern Llanos have been described in Chapter 5. The most interesting system, unique in tropical South America, is the drainage of swamps by means of ditching. These ditches were studied in 1972 by myself and one of my students, Roland Bergman, who did most of the data-gathering. Information here is drawn from Denevan and Schwerin (1978: 23–31) and from Denevan and Bergman (1975).

The Orinoco savannas seem never to have been very attractive to early farmers because there was little game and few crops could survive the long dry season and low soil fertility. Most of the low plains in the west were even more unattractive because of seasonal flooding, with only limited strips along the middle courses of streams being suitable for swidden cultivation (Leeds 1961: 14–15). In the eastern Llanos (see Fig. 5.12), however, the higher *mesas* are well drained, but the valleys are swampy. By using ditching to drain saturated soils, the Karinya were able to cultivate the cool, moist, and fertile soils of the *morichales* (moriche palm groves) in the stream channels.

The antiquity of Karinya swamp drainage is unknown, but is considerable according to oral tradition. Archaeological investigation of the *morichales* could confirm this. In any case, Carib Indians in the region probably had mastered these techniques and moved into the *mesas* some time before AD 1600. Berrío and Ralegh (Ralegh 1997: 179) in the 1590s reported a well-established town of 'Canibals' at Acamacari between the Río Caris and the Río Limo. The Karinya descendants of this group live in the same region today in the village of Tabaro. The Carib also settled at various sites along the Río Pao. At Santa Clara de Aribí, Fuchs (1964) differentiated between crops planted on *tierra alta* and those planted on *tierra baja*. The latter probably refers to *morichal* fields (which were also observed by Schwerin at nearby Cabeceras del Pao). There is no mention of ditching, however, in the historical accounts, in the ethnographic literature, or elsewhere until it was briefly noted by Schwerin (1966: 32).

From the lower Orinoco the Carib spread to rivers such as the Guanipa, the Amana, and the Guarapiche, which empty into the Atlantic. It was mastery of

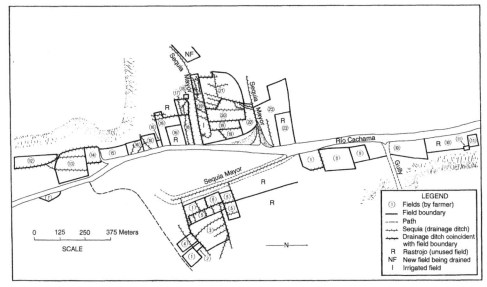

Fig. 14.2. Map of Karinya ditched fields at Bajo Hondo, Cachama, Orinoco Llanos, Venezuela. Fields are numbered by owner. Based on rough field measurements by Roland Bergman (March 1972). From Denevan and Schwerin (1978: 30)

the technique of draining the *morichal* swamps which border these rivers that made possible an adaptation to the particular conditions of the *mesas* and facilitated widespread occupation of the area. This system of cultivation is clearly more intensive, in terms of both cropping frequency and labor inputs, than swidden or annual floodplain agriculture. Once perfected, it continued with little change until recently.

The description here is primarily based on the *morichal* fields (*conucos*) of Bajo Hondo at Cachama (Fig. 14.2). However, ditched fields were also briefly examined at San Joaquín de Mapiricure, along the Río Morichal Largo, at Mamo and Nuevo Mamo (a nearby *mestizo* town), at Marcano on the Río Yabo, at Cabeceras del Pao, and at the highly acculturated community of Tácata (see Fig. 5.12). In 1972, the ditched fields at Cachama were particularly numerous, large, and complex, and they were the main means of cultivation by the community. This traditional agriculture survived here despite the immediate proximity of the major highway from the north coast to El Tigre and Ciudad Bolívar, the oil industry on Cachama lands, the availability of jobs in *criollo* cities, and the efforts of the Ministry of Agriculture to encourage cultivation of the upland savannas.

In reclaiming a *morichal* swamp, sites were chosen where flooding was not excessive, where tall vegetation and hence better soil existed, and where leaf cutter ants were minimal. The first and most important task in preparing a field was the digging of *sequias* or ditches to provide adequate drainage for the plot which had been chosen. This was difficult and strenuous work due both to the wetness of the soil and the density of the undergrowth. The tools used to dig the drainage ditches were the shovel, machete, and *chicora* (hardwood digging stick). Ditches were made by cutting down through the grass sod with a machete. Then the earth was thrown out by hand or with a shovel. Most of the ditches in 1972 were 0.3 to 0.7 m deep, 0.7 to 1.0 m wide, and 10 to 100 m long, with spacing 10 to 20 m apart.

Initially, one or more large canals (*sequias mayores*) were excavated through a section. The largest one at Bajo Hondo followed a natural channel and drained into the Río Cachama; it was 1.3 to 1.7 m wide, 1.0 to 1.3 m deep, and the maintained portion was about 0.5 km long. It was kept meticulously clean, and at the height of the dry season it was observed full of fast-flowing water that had seeped in from the saturated subsoil. Other shallow tributary channels of the Río Cachama were kept clean of vegetation and silt and served as additional *sequias mayores*. After substantial drainage by means of the main ditches or improved channels, smaller ditches were dug around each field, and often very shallow ditches were dug across fields as well. These secondary ditches drained into a *sequia mayor* or stream channel. Although most *morichal* subsoils are underlain with clay, which impedes drainage, there was lateral movement of soil moisture toward and into the ditches. Complete drainage of a new or long-abandoned section usually took several years.

A *morichal* contains a variable density of undergrowth, but once drainage had been initiated and the soil was dried out somewhat, it became relatively easy to clear the undergrowth and many of the trees with axes and machetes. The debris was left to dry and later was burned. Some moriche palms were always left standing, however, for the fruit provided a much appreciated supplement to the diet; the fronds were used for roof thatching and hammock fiber, and the branches were used for wall framing. Also, it was believed that a field would become excessively dry when deprived of moriche shade. Many of the palms were so damaged by fire that they eventually died anyway. To control burns, debris was often gathered and piled for burning. Ash concentrations were also the focus for planting.

When a field was newly cleared there was a period of intensive planting in May, June, and July during the early rainy season. The major crop grown in the *morichal* fields was manioc. The banana was second in importance. Manioc was harvested eight to ten months after planting. Most vegetatively reproduced crops can, however, be left in the ground for many months after they are ready for harvest, for they will increase in size for a long time before they become unfit to eat. As they were harvested, appropriate slips or cuttings were removed in the field and replanted soon thereafter. Manioc was planted in single cuttings or pairs, using a machete or the *chicora* to break the ground. Where soil was sufficiently moist, manioc and banana would sprout in the dry season without rain. In this way both harvest and planting could be spread out over the whole year, thus ensuring a food supply in all seasons.

Manioc, the local staple, is considered to require good drainage, but in some *morichal* fields it was observed growing in boggy soils, suggesting that some Karinya varieties are much more tolerant of poor drainage than is generally believed possible (Bolian 1971). At Marcano, manioc 2.5 m high was growing in standing water during the 1972 dry season. It was choked by weeds and doing poorly, but the farmer pointed out that some but not all of the crop would rot and be lost.[1] The poor drainage was explained as resulting from failure to clean out the ditches that year. Secondary tuber crops included yautia and taro, which are more tolerant of poor drainage than is manioc, and also yampee, yam, and sweet potato. Seed crops, requiring good drainage, included maize, beans, squash, and gourd.

Without ditching, most of the ground in the *morichales* is not only saturated, but carries standing water much or all of the year, so that cropping is difficult. Even with ditching, the subsoil is usually damp at a shallow depth long after the rains have ended. Consequently a second crop could be planted and harvested during the dry season. On the other hand, if the artificial drainage was excessive the soil became too dry for a dry season crop, as had happened on several sections at Bajo Hondo (e.g., Field 19, Fig. 14.2). This was countered by simple removable dams (sheets of metal) on the *sequia mayor* which backed water up into the field ditches. These were also dammed, forcing the overflow onto the manioc fields. This crude system was one of the few known instances of Indian irrigation in the tropical lowlands of South America (Fig. 14.3). It was of minor extent at Cachama and was not practiced, apparently, in other Karinya areas. It may well be a recent innovation. In any event, there was clearly a delicate balance between excessive and inadequate soil moisture, which was regulated by the depth and spacing of the ditches and the degree to which they were kept clean. Some land may have been sufficiently high (such as Fields 9, 10, 11, and 23 at Bajo Hondo, Fig. 14.2) so that bordering ditches were not required. In March 1972, during the dry season, soil moisture variation in the *morichales* was well exemplified by the presence of dry soil to 30 or 40 cm depth in many *conucos* at Cachama, by moist top soil at Mamo, and by manioc growing in standing water at Marcano.

In Cachama, constant reference in 1972 to the land being 'dryer now than formerly' and there being 'less water in the river now' suggests that extensive drainage, by channeling water directly into the Río Cachama, may have encouraged deeper cutting of the channel. Nonetheless, this was not serious enough within the period of human memory to make these river-bottom fields too dry for successful cultivation.

Once a field had been drained and cleared it did not require much more labor to maintain than is the case with plots cleared in heavy forest and only cultivated for two or three years. In order to keep the fields relatively free of unwanted herbaceous growth they had to be weeded every two or three months. As long as the weeds remained fairly small and consisted mostly of grass they were fairly easily removed with the machete and *garabato* (hooked

Fig. 14.3. Karinya drainage/irrigation ditch at Bajo Hondo, Cachama, Orinoco Llanos, Venezuela. From Denevan and Schwerin (1978: 81)

stick). Weeds and grass cut by machete or cleared from ditches were often spread on the ground between the rows of manioc as a mulch, rather than being burned. This may have been as much for preserving soil moisture during the dry season as for improving fertility. Once the manioc outstripped the weeds, however, a field could be left unattended until it was harvested.

Ditches were cleaned once each year before planting. No effort was made to pile up the excavated earth into mounds or ridges for planting surfaces, possibly because such surfaces might become excessively dry in the dry season.

According to the Cachamans it took two to three years for a newly cleared field to build up fertility from the brush which was burned or mulched and turned into the soil. In fact, the Karinya believed that drainage and cultivation made the soil more fertile. However, natural soil fertility was sufficiently high to permit continuous cultivation for many years with only short periods of fallow. Farmers from both Cachama and Mamo reported that fields were usually cultivated for two to four years before being fallowed for one to two years, but some individuals in Mamo cultivated particular fields for twenty years and longer. There was also an erratic cycle of long fallowing lasting ten years or more. One field (NF on Fig. 14.2) being ditched and cleared at Bajo Hondo in 1972 had not been cropped for forty years and another not for fifteen to twenty years. These longer fallow periods may have resulted from non-ecological considerations. For example, when a man or woman died their house was usually abandoned, and frequently this applied to their fields as well. It was also possible for the owner to move away for some reason while still retaining rights to the land.

The ecological reasons for fallowing were not completely clear, but excessive grass and weed invasion seemed more important than fertility decline. During fallow, brush was allowed to grow up to 2.0 to 3.0 m in height before being cut and burned. Fields were also long fallowed if they became excessively dry, or if they were buried by a thick layer of sand left behind by high water. Within a few years after abandonment the ditches became choked with silt and weeds and drainage deteriorated again. Groves of mangos were often planted during long fallowing.

It appeared to be more difficult to reclear an abandoned field than to clear a new field from *morichal* swamp. When left uncultivated, fields became overgrown with a tangled mass of grasses, vines, shrubs, and brush. Nevertheless, various overriding considerations encouraged the trend towards intensive rather than extensive cultivation. Most important, the necessity of draining an area before it could be planted, a task usually requiring the cooperation of several men, inhibited movement to new areas. The presence of already drained river bottoms close by counter balanced, in part at least, the additional labor of clearing second growth. Also, the distance to another suitable streambed was at times considerable, and one might be reluctant to leave kin in order to establish a new field elsewhere. Consequently, the cycle of cultivation tended towards long-term and more or less continuous cultivation.

Actually, there is evidence of some movement to new areas in the past, and this was still going on to a minor extent in 1972. Nonetheless, the limited amount of arable land and the long period of occupation of most extant Karinya villages were indicative of long dependence on continuous cultivation. Indigenous swidden cultivation in the infertile savannas, if present, would almost certainly have necessitated periodic village shifting. However, the inhabitants of Tabaro apparently have occupied the same general site for better than 400 years. *Morichal* agriculture may have made possible such long-term occupation of the same site.

At Bajo Hondo in 1972, twenty-three individual owners from sixteen households were working thirty-four drained fields. Figure 14.2 shows those under cultivation, plus sections in short fallow (*rastrojo*), but not necessarily actual land-holding boundaries; each number represents a different owner. In several instances husband and wife or father and son from the same household had separate fields, which may or may not have been adjacent. For example, owner 14 was the wife of owner 23, and owner 11 was the son of owner 10. Most holdings were contiguous units, but in two instances (1 and 7), owners had two separate fields. The total land cropped on Fig. 14.2 was 6.24 ha, with individual fields ranging in size from 0.065 ha to 0.5 ha. No data were obtained on productivity, but the average of only 0.39 ha of cultivated land (not counting house gardens) per household (six to eight persons for most) was indicative of highly productive agriculture.

Although this is a description of recent agricultural practices, it probably was similar to prehistoric Carib agriculture as it developed in adaptation to the conditions of the *mesas* in the early years after the Caribs occupied the region. As the Caribs became the modern Karinya, there were some modifications of detail, yet the basic pattern probably did not change. Several new Old World cultigens that have been introduced (banana, plantain, sugar cane) may have been acquired even before expansion into the *mesas*. As the major male occupations of trading and warfare declined or were restricted in the eighteenth and nineteenth centuries, more and more of the field work was taken over by men, until today only the harvest of crops is still primarily a female task, but the nature of the work has remained the same. Introduction of the machete greatly facilitated work done previously with other tools, but it has not changed the tasks or even basic skills in many cases. Moriche palms (softwood) can be readily cleared with stone axes, so the introduction of steel axes was not critical.

The picture we get then for river-bottom cultivation in communities like Cachama is one of initial intensive labor for clearing and planting a field, followed by many years (ten, twenty, or more) of relatively low labor inputs which sufficed to maintain a field at a satisfactory level of production to meet food needs.

In addition to the ditching of stream bottoms within the *mesas*, there was also ditching of *morichales* on river terraces at the edge of the Orinoco floodplain (see Fig. 5.13). These were farmed by people from Mamo several kilometers away. These swampy terraces are fed by seepage from adjacent *mesa* sides, making possible year-round cultivation. Otherwise, techniques and crops were similar to those at Cachama.

We can only be impressed by the Karinya's remarkable system of channelized swamp drainage, of unknown antiquity, persisting in close proximity to gas and oil activity and the major ports and industrial cities of Ciudad Bolívar and Ciudad Guyana. Apparently, however, these external pressures have prevailed, and the drained fields we observed in 1972 at Cachama and other Karinya villages were mostly abandoned a decade or so later (Karl Schwerin, pers. comm.).

La Tigra, Orinoco Llanos

During the regional archaeological project in the Río Canagua drainage of western Barinas northwest of La Calzada (Fig. 11.5), directed by Charles Spencer and Elsa Redmond, drained fields were found at La Tigra next to a causeway (Redmond and Spencer 1994; C. Spencer *et al.* 1994). These fields consist of a network of canals covering an area of 35 ha (Fig. 14.4). The canals or ditches range from 1.0 to 8.0 m in width. They are more similar in form to the Karinya drained fields of the eastern Llanos than to the Ventosidad ridged fields in Barinas (Chapter 11). They have been dated to the Late Gaván period, AD 550–1000, based on association with surrounding habitation sites.

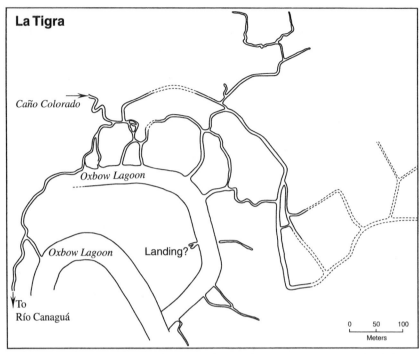

Fig. 14.4. Map of ditched fields at La Tigra, Orinoco Llanos, Venezuela. Adapted from C. Spencer *et al.* (1994: 129)

Pollen analyses for the La Tigra fields are interesting in that the maize pollen frequency was very high suggesting a maize staple. Secondary crops included arrowroot, ají, and tomato, plus cocopalm, palomero (fruit; *Myrica pubescens*), and yopo (C. Spencer *et al.* 1994: 131–2).

During Late Gaván times, C. Spencer *et al.* (1994) found no evidence of demographic pressure, in contrast to what we postulated for Ventosidad. Their estimated population density of 20 to 30/per km^2, high for the tropical lowlands, could have been supported by a short crop/fallow cycle system using only available alluvial land, without savanna drainage canals. They suggest that the intensive drainage cultivation was carried out to produce a surplus production which was transported via causeway to the regional center at Gaván 3 km distant. This could well have been the situation. However, it seems to me that the intensified crop production at La Tigra still would have been responding to a food demand at the nearby regional center.

River Channel Canalization

There are several known prehistoric instances of river channels having been artificially canalized, and/or straightened, and sometimes incised in order to

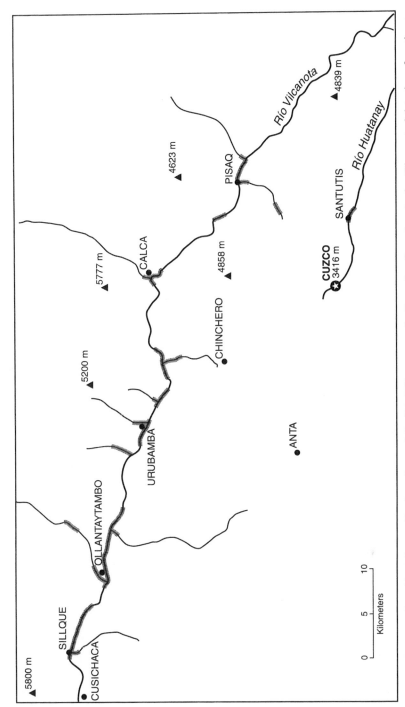

Fig. 14.5. Map of prehistoric canalized sections of the **Río** Vilcanota and tributaries, Peruvian Andes. These straightenings function to reduce flooding of adjacent fields. Adapted from Farrington (1983*b*: 223)

facilitate water flow (Farrington 1983*b*; Regal 1970: 91–3). This is a means of improving drainage without ditching. The best examples are in the Cuzco region. An early reference to river management appears in a land claim dated 1555 for Coya near Pisac on the Río Vilcanota (Fig. 14.5). Four chiefs testified that the land concerned was unused until the Inca (probably Huascar) had the land reclaimed by diverting water to it (Rostworowski 1962: 135–6, 142; Farrington 1983*b*: 232). In the city of Cuzco, the Río Huatanay and the Río Tullumayo are canalized (Rowe 1946: 233; Regal 1970: 91–2), and the artificial sections can be readily observed today by visitors.

Farrington (1983*b*) studied canalization for the Vilcanota and its tributaries from Cuzco to just below Ollantaytambo (Fig. 14.5).[2] Over a distance of about 200 km, there are at least thirteen canalized segments totaling 23 km. Retaining walls of boulders and gravel were constructed 2.0 to 8.0 m wide, with a height sufficient to prevent normal flooding on to floodplains. Canalization occurs both where *quebradas* join the Vilcanota and upstream in the *quebradas*. Meandering was reduced, and to counter a river's normal tendency to meander, strong walls were built, many of which have survived intact to the present. The main function of this canalization was drainage for land reclamation on floodplains, in addition to flood control and erosion protection. A straighter channel moves water faster and drains off water from surrounding land more quickly.

Associated with some of the Titicaca raised fields are canalized channels which are confined by paralleling artificial levees or walls (Lennon 1982: 190–4; Kolata 1993: 224–8). Some of these channels are connected to canal bypass systems which shunted excess river water, prone to overflow, away from raised fields. Finally, there are canalized (stone side walls) sections of the Río Colca and its tributaries near Chivay (Brooks 1998: 463–9).

Notes

[1] Manioc that will mature with a short growing period (four to six months) on the Amazon floodplain is poorer in quality and quantity than is manioc grown on *terra firme* (Ohly and Junk 1999: 288).

[2] Squier (1877: 507) at Ollantaytambo made one of the first modern observations of canalized river channels.

Part V

Conclusion

It is clear that we have been incredibly naive in our previous assessments of agricultural systems of the past.

(Ray Matheny and Deanne Gurr 1983: 99)

15

Implications of Indigenous Agricultural Technology

> An environment can only be described in terms of the knowledge and preferences of the occupying persons: 'natural resources' are in fact cultural appraisals.
>
> (Carl O. Sauer 1952: 2–3)

As Sauer indicated, environments are interpreted and utilized in terms of the cultures which occupy them. The technology available, desirable, and ecologically, economically, and socially feasible determines how different habitats can be used, for how long, and how productively. The environment is changed in the process of utilization and by intentional modification to make possible or to improve crop productivity. Such change may make sustained output possible, or it may lead to slow or rapid environmental degradation with production declining or terminating as result. In examining indigenous field systems in South America, what is remarkable is their diversity, the diversity of the habitats they occur in, their longevity, and their productivity given limited external inputs.

In this volume I have tried first to review the substantial literature on indigenous fields in the Andes and in Amazonia, and second to provide specific examples, both cultural and regional, based in large part on my own field research. Third, I have given some consideration to field abandonment, but only briefly to origins and diffusions. Fourth, several conceptual issues, controversies, and viewpoints have been briefly introduced, some receiving further consideration in this summation. I have not been consistent in the amount of attention I have given to the various field types, case studies, processes, and issues, reflecting space constraints, my own background and interests, and my judgement as to what should be emphasized, condensed, or eliminated. To compensate, I have made a considerable effort to lead the reader to relevant research.

As a further apologia, I should point out that this treatment of Indian field technology has neglected the Indian farmers who developed and applied that technology. These farmers were and are characterized by ingenuity, persistence, hard work, flexibility, and possession of detailed environmental and crop knowledge. The human processes involving environment, technology, decision-making, and behavior at the field level cannot be resurrected very well from the past, but they are certainly accessible today. However, we know more about current indigenous farming at the cultural, community, and household levels than we do at the field level.

Changing Perceptions of Indian Agriculture

In 1961, when I began field work in eastern Bolivia on prehistoric cultivation, there was little appreciation for the complexity, diversity, and productivity of indigenous agricultural fields. That year the window on agrarian prehistory began to open up with several projects and publications by geographers and anthropologists (Chapter 1). This was followed by four decades of both surveys of field systems and local studies by both individual scholars and by large interdisciplinary teams. Previously, while terracing and irrigation systems were well known, they had seldom been the subjects of systematic field research. Simple shifting cultivation seemed to dominate in the tropical lowlands. The new research, however, revealed a great diversity of types of fields nearly everywhere and an indigenous agricultural technical knowledge that is or was the equal of traditional agriculture anywhere in the world (Chapter 2).

Clearly, since the early 1960s there has been an elevation of the status of agricultural research within indigenous studies—prehistoric, ethnohistoric, and ethnographic. Indian societies, economies, demography, and environmental impacts could not be adequately analyzed otherwise. In addition, agricultural scholars and practitioners now recognize: (1) the indigenous roots of much of current peasant and small family-farm agriculture, and (2) the contribution that elements of traditional agriculture can make to agricultural development, sustainable production, and the viable utilization of marginal habitats.

This new and still continuing interest in Indian fields, crops, and technology reflects a parallel concern with the connections between food production, population growth, cultural evolution and collapse, and environmental change. These interests are tied to theoretical issues about people and environment with ramifications far beyond indigenous people, but which may have their roots in studies of such people in the Americas and elsewhere. During this same period since 1961 there have been significant new techniques for mapping, measuring, dating, identifying, and analyzing relic agricultural features, as well as an explosion of technical research on existing Indian fields, much of it collaborative and interdisciplinary. While we have come to know much more about the forms of Indian agriculture, there is considerable debate on the significance of this agriculture to both past and future societies, as well as on how and why different forms originated and have changed.

In retrospect, having long been engaged in studying Indian fields, I find that some of my own beliefs have changed over time. I no longer think that shifting cultivation was the dominant form of pre-Columbian agriculture in the tropics (Chapter 7); that the Amazon floodplains were uniformly intensively cultivated

(Chapter 6), or that raised fields were unique to Amazonia (Chapter 11). On the other hand, I remain firmly convinced that environmental causation is an inadequate or incomplete explanation of agro-based settlement expansion or collapse.

The People without a Prehistory

Eric Wolf in his treatment of 'the people without history' (1982) argued that historical study has focused on the elites and has underplayed the history of 'primitives', 'peasantries', and 'minorities'. The same argument can be made for prehistory, even more so since prehistorians have concentrated on the visible buildings and monuments of urban civilizations. The rural farmers who fed the Andean civilizations, as well as the people of other regions, such as Amazonia, the Caribbean fringe, and southern South America, have received much less attention. Of course, prehistoric rural folk left much less behind to study; however, we now know that large numbers of remnants of their field systems still survive. That pre-Columbian farmers have been a 'people without a prehistory' reflects the academic glitter of the Indian elites, inadequate funding for rural archaeology, the sparsity of early descriptions of the hinterlands, and rapid depopulation, even extinction, of rural people following the European conquest. Our knowledge of native farmers of the past has been further confounded by the tendency of both scholars and the public to incorrectly project characteristics of surviving Indians of the nineteenth and twentieth centuries back to the conquest and before.

Origins of Field Types

The varied types of indigenous field technologies found in Amazonia and the Andes had their origins in the distant past, well before the Incas, and we know little about them. We have indicated the earliest known dates for some of them—pre 1000 BC for canals, terraces, and raised fields. However, dating of agricultural fields is difficult, and the antiquity of some incipient types may be considerable. Some forms evolved locally over time as adaptations to local conditions and then were dispersed and modified in other locations. Some may have diffused from Mesoamerica, but there is no confirmation of this—possibly basic ideas but not specific forms. The fact that tools are distinctive in Mexico from those in the Andes, and the fact that agricultural terms are distinctive in both regions suggest mostly independent evolution of specific field types. We do know that larger and more complex features, such as bench terraces and megacanals, appear later than smaller and simpler features, such as sloping-field terraces and small canals (Chapter 8). Larger systems reflect more people to be fed and more control over the environment. Even large complex systems, however, may have been locally constructed and managed, although at some point in time state supervision may have taken over (Chapter 13).

We are not always certain that some traditional field features in South America had an indigenous origin. The *qanats* (underground canals) in Peru and Chile (also Mexico) were originally believed to have been of native American origin, but were probably introduced from Spain (Chapter 8). Also, some traditional canal features are of Spanish origin (William Doolittle, pers. comm.).

Changes in field systems from one type to another reflect adaptations to changing environments and socio-economic–demographic conditions, based on either innovation or diffusion. Regardless of origin, knowledge of applicable techniques is available to some farmers in a community *before* those techniques became widely adopted. Introduction of a new, useful technique does not result in general use unless there is a perceived need for it (Denevan 1983).

Agrarian Collapse

The presence of large numbers of abandoned fields in current landscapes is both impressive and disturbing. In South America, almost all the raised fields (Chapter 11), probably over 50 per cent of the terraces (Chapter 9), and many of the mountain and coastal canals (Chapter 8) are abandoned. Why fields were abandoned and when are questions not easily answered. We know that large sectors were no longer in cultivation in 1492, probably as result of population shifts and depopulation, and possibly natural or induced environmental change. Examples are the Titicaca plain (Chapter 13) and the Chicama and Moche valleys (Chapter 8). We also know that massive field abandonment occurred following the epidemics and forced population shifts after 1492. Empty farmlands were reclaimed by wild vegetation or, often, they were used for cattle pasture. Some abandoned fields have been restored or replaced by European field systems, but large areas of relic prehistoric fields remain abandoned and not reclaimed for reasons that are not completely clear (Chapter 10). In addition, further field abandonment has occurred in the Andes as a result of rural to urban migration, not just because of perceived land scarcity but because of the attraction of perceived opportunities in the cities. In many Andean communities the modern exodus has resulted in a reduction in rural populations since about 1955 (Preston 1996: 175–6).

In much of South America today food supplies are inadequate; seemingly there is insufficient land for additional food production; and people are migrating to the cities where few opportunities exist for them. Development of marginal lands in the deserts and rainforests is promoted, but both environments present major problems. It is ironic, then, that large areas of once cultivated lands in the Andes and the tropical lowlands are now abandoned—mostly depopulated, with cultivation terminated or reduced in area and in intensity.

Agrarian collapse, local or regional, is a major issue about which little is known. The magnitude of abandonment seems to be sufficient to justify a careful examination of the natural and human circumstances involved. What is the extent and distribution of land abandonment; what were the causes of specific abandonments; and why haven't such lands been brought back into production? These are questions we raised in the Colca Valley Project (Chapter 10). Possibly some of these lands can be farmed once again.

Colonial Changes

During the colonial period new crops, animals, tools, and field technologies were introduced from Europe and replaced or were merged with indigenous systems of land use. This process has not been given attention here. (See Gade 1992 on Andean introductions and Crosby 1972 on biological introductions.) Elements of Spanish irrigation technology, including arched stone aqueducts, were introduced (CEHOPU 1993). The Spanish ard or scratch plow was a major change. It contributed to greater erosion, the destruction of wetland raised fields in the Andes, and in some areas terrace walls were torn out in order to use animal-drawn plows (Brooks 1998: 170–6); livestock hooves also damaged terrace walls and continue to do so. There were reports in 1567 and 1613 of Spanish livestock destroying Indian irrigation canals (Antúnez de Mayolo 1986*b*: 49). Livestock occupied abandoned fields, and many former crop fields have remained in pasture to this day, such as around Lake Titicaca. Uncontrolled livestock spread on to still used but unfenced Indian fields leading to abandonment or disruption. Where allowed to overgraze, livestock contributed to soil erosion. European commercial crops, such as wheat, barley, sugar cane, coffee, and pineapple, required new forms of agricultural fields and management.

Persistence of Indian Field Technology

Despite a depopulation of native people amounting to possibly 90 per cent overall, the cultural disruptions of the conquest period, and the introduction of colonial technologies and crops, many elements of indigenous agriculture have persisted to the present, although often in altered form. Much of this survival has been with the emerging mestizo population or with deculturated Indians, as well as with remaining Indian communities, but also as elements in modern agriculture.

Almost all the terraces in the Andes are of prehistoric origin; although repairs have been made subsequently, few new terraces have been built. New, cement-lined canals (to reduce filtration) have replaced the native canals on the Peruvian coast, but the old routes are often unchanged. Even though some tractors, animal-drawn ploughs, and modern hand tools are now common in the Andes, the footplow and other traditional tools are still widely used, especially on steep slopes and in remote areas. Hatch (1976) in his study of the Motupe Valley of northern coastal Peru showed that small farmers retain many aspects of indigenous irrigation technology in a region with large, modern irrigated farms on which many of the small farmers have worked.

The old raised fields were all abandoned but successful attempts are now being made to rehabilitate them or to construct new ones in Mojos (Chapter 12) and at Lake Titicaca (Chapter 13). In Amazonia, where so many Indian cultures have become extinct, many techniques for managing tropical soils, pests, and plants have been lost, but much also still persists (Chapter 4). Indeed, agricultural field technology and crops have been among the most persistent components of Native American cultures, despite many losses and changes.

Regarding agricultural survival, we must emphasize that much of the indigenous technology found today, while seemingly ancient in form and use, has changed since 1492 because of new needs and adaptations, modifications by local people themselves, and because of contact with and borrowing from European technology. Agriculture, whether traditional or modern, is constantly undergoing change, rapid or slow, as people adapt their food production to changes in the environment and in their demographic and socioeconomic circumstances. Considering present Indian agricultural fields, even in the backlands, to be analogs of prehistoric fields may be justified in general, but less so in detail.

Survival of Relic Fields and Landscapes

To a much greater degree than realized a few decades ago, remnants of abandoned prehistoric fields, and field features such as canals, survive in enormous numbers, probably in the millions, forming entire and distinctive landscapes. New discoveries have been the result of the greater availability of aerial photography, combined with a greater awareness of such fields by geographers, archaeologists, and others and hence a recognition of them and search for them by air and ground survey. Raised fields and certain other field types were barely known to exist in 1960, and the extent and diversity of canals and terraces was unappreciated.

What we can still see must be only a fraction of the Indian fields that were on the landscape in 1492, given the field destruction that has since occurred. Many raised fields have been buried under sediment or have been plowed up. Old terrace walls and canals have been removed or buried. The process of intentional destruction continues. Many of the raised fields in Puno in Peru and in Guayas in Ecuador are being destroyed by modern agriculture. Thus, potential local models for reviving traditional technologies appropriate for local conditions are being lost.

While many Indian fields of the past can still be observed after centuries of non-use, most did not rearrange the microlandscape. Rather they were ephemeral in that they quickly vanished from view after just a few years without cultivation. However, such fields may be detected by soil techniques such as concentrated phosphorus (Eidt 1977; 1984) and micromorphology (Gartner 1992), presence of crop pollen, remnants of field boundary markers or walls, and by soil coloring such as *terra preta* (Chapter 6). Future research and new techniques should tell us much more about the distribution, character, and functions of former agricultural fields.

Field Form, Function, and Risk

Many agricultural fields are simple plots with adequate soil and moisture on which crop production takes place. Many of the indigenous fields described here, however, are much more in that they are designed to improve the crop environment, ensure labor efficiency, enhance sustainability, maximize productivity, and minimize risk from changing environments, particularly unpredictable climatic conditions. Attention to these different functions has been the concern of the new discipline of agroecology (Altieri 1995), and field functions have been consistently discussed in this study. Not all these functions may be equally achieved, of course. Minimizing risk often conflicts with maximizing production and efficiency. Maximizing production usually conflicts with sustainability. 'The principal economic organizing strategy of Andean arid land producers is risk management' (Browman 1987*b*: 2).

Prominent field functions described here include water management, frost reduction, erosion control, and soil accumulation. Managing a variety of types of fields in a variety of types of habitats is a common way that communities reduce risk and also diversify production. We saw this with plot scattering in the Colca Valley (Chapter 10) and with the practice of altitudinal zonation of crops (verticality) in the Andes (Chapter 3).

Agricultural Intensification

A major theoretical issue regarding indigenous agriculture in the New World is that of agricultural intensification. Under what circumstances do less intensive systems (in terms of frequency of cultivation, labor inputs, and productivity— all related) become more intensive? Or, why do some intensive systems become less intensive? These questions are fundamental to explaining agricultural change over time. As we have seen in South America, there is a great range in agricultural intensity from long-fallow shifting cultivation (Chapter 4) to permanent systems with several crops each year and high overall productivity, with major landscape modifications, soil amendments, and pest control (Chapters 3, 8, 11). Is this variation simply a matter of technological development and knowledge? Is it related to environmental variation? Or to demography?

In 1965, at about the time interest in indigenous agriculture began to accelerate, the Danish agricultural economist Ester Boserup argued that the key to understanding agricultural change was relative population pressure. Intensification, she reasoned (in a non-industrial economy), requires progressively higher labor inputs, but the resulting productivity increases at a lower rate. In other words, intensification is relatively inefficient in terms of inputs and outputs, and thus is resisted unless there is pressure from population growth. A vigorous discussion of Boserup's thesis ensued over the next two decades. It was pointed out that there are exceptions in which intensive systems are more, not less, efficient than extensive systems; that there are other pressures to increase production (economic, social) besides population; and that risk minimization may supersede labor efficiency in cultivation strategies (e.g. Grigg 1979). Thus, the consensus now seems to be that while Boserup's thesis is valid for explaining some changes in agriculture, other factors must also be considered.

In South America there does seem to be a general relationship between degree of agricultural intensity and population density. We see this in Amazonia with low densities and in the Andes where densities are high.[1] However, for pre-European Amazonia I have argued for clustered populations with intensive agriculture, both riverine and interfluve, surrounded by near-empty lands *with the same* natural resources and few obstacles to population dispersal (Chapter 7). Carneiro's (1970*b*) circumscription theory, in contrast, argues that concentrations of natural resources led to population pressure, agricultural intensification, conflict over the best land, and greater cultural complexity and political evolution.[2]

Also, in looking at intensive systems of raised fields, terraces, and irrigation, we have seen that these may require large labor inputs to construct, and thus are inefficient over the short term, whereas once fields are constructed labor inputs for maintenance may be relatively low over the long term. Thus, such systems are more labor-efficient, given high productivity, than lower input systems. This has been demonstrated for the Lake Titicaca raised fields by Clark Erickson (Chapter 13).

In addition, the change from one type of field system to another may be an adaptive (but optional), change reflecting changes in the environment which may be either natural or human induced (Denevan 1983). We have suggested for the Colca Valley that the conversion of rainfed, sloping-field terraces to irrigated bench terraces may have been related to a decrease in precipitation and possibly colder temperatures (Chapter 10). At the same time, however, population growth may have encouraged construction of less risky irrigated terracing, given seasonal rainfall uncertainty, without a decrease in total rainfall.

Finally, is it possible that reoccupation and reintensification of farm lands abandoned or deintensified, because of depopulation, are inhibited by social institutions and technologies which become adaptive to low production demands? And, to what extent have there been cycles of intensification and deintensification which have involved similar or different field technologies?

Carrying Capacity

The concept of agricultural intensification is very elastic regarding production of food from the environment. Human beings can change the environment in order to make it more productive. Limits are thus determined by the costs involved (labor, capital, external energy) in terms of technology and the nature of the environment concerned, and in terms of what people are willing to do as perceived needs change. The concept of carrying capacity, in contrast, tends to be more rigid. Carrying capacity (for people) is the density that can be supported at an acceptable standard of living for a given habitat on a sustainable basis. While a specific form of land-use technology must exist as part of the calculation of productivity, limits are nevertheless assumed to be determined by the environment.

The concept of carrying capacity is useful in wildlife ecology and range management where animal populations are involved. However, it is less appropriate for humans because they have varied technologies and hence options for deriving food from a particular environment (Denevan 1987b). And humans also can change the environment by adding fertility and building terraces, etc. Furthermore, the environment is not uniform, even over small distances, nor stable, changing seasonally and from year to year. All the variables for determining carrying capacity are highly dynamic (the environment itself, types of crops, standard of living, technology, costs). A farmer varies cultivation practices both from year to year and from field to field for reasons that may be spatial, environmental, economic, informational, demographic, or idiosyncratic. Carrying capacity thus can only be calculated at a given point in time for a specific piece of uniform land and with a specific technology and a specific standard of living.

Environmental Limitations, Agriculture, and Culture

When people argue that certain environments set limits on the amount of food that can be produced and hence on population density and on size and permanence of settlements, and by extension on level of culture, without considering the production technology options available, they become environmental determinists who ignore the capabilities of humans for managing their environments. For example, some prehistorians believe that permanent agriculture is impossible in Amazonia because of environmental limitations, or low agricultural potential, or low carrying capacity (Chapter 7). We have seen, however, that indigenous people in Amazonia were and are capable of controlling low soil fertility, poor drainage, and pests by means of permanent gardens and raised fields (Chapters 4, 11, 12). And we have seen how seemingly marginal agricultural habitats today—steep slopes, drylands, wetlands—have been made highly productive by native people by means of terracing, irrigation, raised fields, and other landscape-altering techniques. We have seen three examples of persistence of improved soil fertility by long-term human activity: *terra preta* soils in Amazonia (Chapter 6), terrace soils in the Colca Valley (Chapter 10), and raised-field soils at Mojos, Titicaca, and elsewhere (Chapters 11–13).

Arguments for the rise and fall of societies because of environmental conditions or changes thereof (e.g., Meggers 1954; 1996; Ortloff and Kolata 1993) need to be tempered with the realization that such conditions and changes in them are culturally and socially interpreted and responded to. Climatic change explanations for cultural collapse have once again become popular with anthropologists, historians, and in the popular press (e.g. Fagan 1999). War-wick Bray, an archaeologist working primarily in Colombia, has commented on environmental stress and the difficulty of determining causality of cultural and demographic change. He points out that '[t]he tendency among archae-ologists has been to assume that climate (i.e. the environment) is the prime mover'. He wisely states that we must distinguish between 'changes in the global climate, . . . [p]urely local events, . . . and [c]hanges resulting from human activities'; that we should 'look at the interplay of *all* factors, and . . . not assume that the global climate is the only driving force'; and that coinci-dence in time is 'a correlation, not an explanation' (Bray 1995: 97, 112). Fur-thermore, people not only respond culturally to natural fluctuations 'in their environments, but [they] also create, shape and transform those very environ-ments' (Erickson 1999: 641).

Environments are neutral. Their potential for human use is determined by both their characteristics (which people can change) *and* by the technology available for exploiting them combined with the will to do so. 'Agricultural potential' is not something inherent in nature. The concept contains the word 'culture'.

Sustainability and Degradation

It is apparent that many indigenous field systems have continued to produce in the same place nearly continuously for centuries. We have seen this for several *terra preta* sites in Amazonia (Chapter 6), in the Colca Valley terraces (Chap-ter 10), in the Titicaca raised fields (Chapter 13), and in the Moyobamba house gardens in northern Peru (Chapter 4). Generally, however, we cannot be certain about continuity or whether long periods of use were separated by long or short periods of abandonment or fallow, even for historical times. For irrigation, drained fields, and terraces, at least, we can probably assume long periods of production in order to justify large labor inputs for construction.

For long-term cultivation to be possible, soil maintenance and pest control techniques must be employed (Chapter 3). Prehistoric terraced, irrigated, and drained fields appear to have been sustainable because of the endurance of their landscaped components; actually they were sustainable because of effective soil and pest management. The presence of such fields alone implies such management, but we know little about it. For prehistory we know most about intensive forms of landscape modification (enduring fields) and least about shifting cultivation, agroforestry, rainfed cultivation, and floodwater farming (ephemeral fields). The latter tended to be on superior soils, but they were still dependent on soil and pest management for longevity. In time, even fertile soils decrease in tilth, increase in pest levels, and decrease in productivity. On marginal lands, however, cultivation may not even be possible without landscape modification.

For sustained production to occur, resource degradation had to be preventable or reversible. We can define 'degradation' as human-induced environmental change in which biological productivity is decreased in terms of either natural ecosystems or agroecosystems. Otherwise, 'degradation' is a subjective concept; deforestation can be considered degradation, but a forest may be replaced by a highly productive and sustainable agroecosystem which could be considered constructive. Erosion on an upper slope can be considered degradation, but deposit of that same soil on a field at the bottom of the slope might be a positive result. D. Johnson and Lewis (1995: 228) make a distinction between 'land degradation' (or 'destructive creation') and 'creative destruction'.

While we know that many indigenous field systems were and are sustainable, we also know of instances where excessive pressure on the environment brought about degradation and failure of the agroecosystem. For example, severe erosion is common today on many unterraced Indian lands in the Andes (Godoy 1984). Prehistoric canals were abandoned because of sediment or salt accumulation; terraces were abandoned because of inadequate maintenance. Throughout the Andes and also Amazonia, land cover, wildlife, soils, and drainage were modified, if not actually degraded, wherever people were present. Even in sparsely populated regions, significant impacts accumulated over time (Denevan 1992b).

Environmental disturbance is the inevitable result of any form of agriculture. Deforestation is required and soils are modified. Fertility declines, erosion occurs, pests increase, and water regimes are altered. It can be hypothesized that techniques of sustainability evolved in response to these habitat changes in order to continue an adequate level of productivity, given limited arable land. Of course, such adaptation is not inevitable, and field abandonment and agricultural collapse may occur as a result of excessive pressure on the land due to population growth, inappropriate or inadequate technology, or environmental change (natural or human-caused). Furthermore, knowledge of sustainable technology does not suffice if sources of fertilizer are depleted, if infestations of pests are overwhelming, if long-term climatic change occurs, or if costs are too high.[3] On the other hand, landscaped cultivation systems are fully dependent on maintenance of the physical infrastructure (terraces, etc.). If this is neglected, the system will collapse in full or in part. The reasons may be environmental (floods, drought, tectonic, volcanic), demographic decline (the most labor-intensive fields will generally be the ones abandoned), or social (warfare, managerial, costs). Once abandoned, failure to restore the physical infrastructure may reflect other food-obtaining options or migration, as we found in the Colca Valley (Chapter 10). Ecological and social factors, of course, may interact to bring about agrarian collapse.

Andean–Amazonian Comparisons

When I began writing this volume some years ago, I assumed that certain major contrasts between Andean and Amazonian indigenous field systems would emerge. Indeed, there are significant differences, given basic environmental differences. With adequate rainfall and relatively gentle slopes, irrigation and terracing are rare in lowland Amazonia. Erosion is much more of a management problem in the Andes, whereas management of weed and pest invasion is much more of a problem in Amazonia. Crop diversity and use of tree crops is greater in Amazonia. Landscape engineering is more widespread in the Andes. There is a greater diversity of field systems and resource management techniques in the Andes, but probably a greater microdiversity within fields in Amazonia. There has been a greater enduring impact (deforestation, erosion) at the hands of native cultivators in the Andes, whereas in Amazonia the human impacts have been more subtle and more ephemeral. (The expansion and creation of some savannas by clearing and burning is an exception to this generalization.)

However, I am also struck by some general similarities in cultivation practices in these two very different biomes. Raised fields were constructed for wetland agriculture in both Amazonia and the Andes (including the desert Pacific Coast). Soil fertility improvement occurred in both areas, both intentionally and inadvertently. Intensive, long-lasting cultivation was made possible in marginal habitats in both regions by means of sophisticated resource management techniques and landscape modification. Intensive cultivation and the densest populations were clustered in both the Andes and Amazonia, as result of both natural mosaic patterns of uneven resource distribution and variable social factors, in contrast to more even spatial distributions elsewhere. These characteristics of traditional cultivation, of course, are not unique to South America.

Traditional Knowledge

Several scholars have argued recently that agricultural methods used in ancient times could serve as models, or at least provide insights, for a modern sustainable agriculture (e.g., R. Smith 1987; Erickson 1992; Denevan 1995). Many others have made the same case for surviving traditional farmers. Hence, every effort should be made to reconstruct agroecological knowledge of the past and to record the still existing knowledge of indigenous and peasant farmers. Given that much of traditional agriculture has been sustainable, productive, controlling of risk, and elastic and responsive to changing conditions, there is no question that: (1) such agriculture needs to be allowed to continue, and (2) modern agriculture can benefit from it, both in agroecological details and in terms of 'principles of permanence' (Clarke 1977: 372–7). Prehistoric methods are particularly instructive because they provide a long-term perspective on successful as well as unsuccessful crop management (Denevan 1995). Restoration of past field technologies has been feasible in an appropriate context. Also, regardless of what we can learn from traditional knowledge, if we are interested in changing, by 'improving', traditional agriculture, we must first understand, appreciate, and build on that agriculture that is to be changed, rather than simply replace it (Sillitoe 1996: 403). However, '[t]he environmental image constructed by the development specialists still remains uninformed by long-term historical and cultural perspectives on human use' (Kolata 1996*a*: 266; also see Sluyter 1999: 394–6).

We must not be carried away, however. As we have seen, pre-European and traditional agriculture can fail and can bring about land degradation, just as can modern agriculture. The 'ecological Indian' is often a myth, and we should choose our models carefully. Indians act in their own social interests and have goals of efficiency, productivity (including surplus production), and risk avoidance, which may take precedence over sustainability and good stewardship, especially where land is in abundance.

Furthermore, most traditional agriculture is place specific, evolving over time in a particular habitat and culture. And this is where and why it tends to be successful. Transfers to other places and contexts thus may fail if soils, crops, tools and social and economic organization are different. Raised fields have been successfully revived at Lake Titicaca because the people there are already familiar with working communally with traditional hand tools, such as the foot plow, to move earth to make small lazy beds for potatoes and other crops (Chapter 13). Efforts to establish tropical raised fields (*chinampas*) in Tabasco, Mexico failed, however, because they did not fit readily into the local social and ecological systems (Chapin 1988).

Out of pre-European agricultural methods emerged two patterns: (1) continuity (with continuing modification) maintained by traditional farmers, and (2) evolution into modern agriculture, with increasing emphasis on machinery, non-renewable energy, and use of chemicals. Now, given the high cost of fossil fuels and environmental decline, we must consider integrating the two approaches, seeking a compromise between the general sustainability of the former and the high productivity objective of the latter (Denevan 1995: 39).

I am convinced that indigenous agricultural fields in South America once supported very large numbers of people on a largely sustainable basis (Denevan 1992*a*). In terms of agroecological understanding and complex management of landscapes and environmental variables, these fields systems are the equal of modern 'scientific' agriculture. The frequent failure of the latter in Third World development in part reflects attitudes of a superiority of Western agricultural technology and a reluctance to consider alternatives from other traditions. The belief that Indian agriculture is inferior dates to the European conquest and still persists.[4]

Finally, I think that we can do a much better job of discovering, mapping, and describing Indian fields and of explaining the agricultural changes that these fields bare witness to. These changes bring or brought about greater or lower food production and thus the feeding of more or fewer people.

My primary concern here has been with agricultural fields themselves, less with their origins and impact on environment and relation to cultural history, more as agroecological landscape features created by people for the purpose of food production, means of adapting to changing environments, demography, and technology. This information should have utility for scholars and others with broader agendas.

Notes

[1] As early as *c*. 1615, Guamán Poma de Ayala (1980: 885) stated that in pre-Inca times in Peru population pressure ('tanta suma de yndios') resulted in the construction of terraces and irrigation canals.

[2] For a recent discussion of Carneiro's theory as applied to Amazonia and coastal Peru, see D. J. Wilson (1999: 180–5, 356–71).

[3] '[T]echnologically sustainable agriculture theoretically can be carried out anywhere, in the Amazon, in the desert, or on a spacecraft. However, . . . the technology must be economically feasible' (Jordan 1987: 102). This includes labor inputs as well as material inputs. However, the costs that will be tolerated will vary with circumstances.

[4] '[H]ow common it is among some Mexican writers and historians to describe ancient indigenous societies as primitive, technologically backward, Neolithic, or in similar terms' (T. Rojas 1987: 199).

APPENDIX 1A. CULTIVATED PLANTS OF SOUTH AMERICA[a]

Common name[b]	Other names	Botanical classification[c]	Comments (location, use)
Achiote (*see* annatto)			
Achira, canna	Capacho	*Canna edulis* Ker-Gawl. (**Cannaceae**)	Lower Andes, root[d]
Achocha	Cáygua, achokcha	*Cyclanthera pedata* (L.) Schrad. (**Cucurbitaceae**)	Andes, vegetable
Agave, maguey	Pakpa, cabuya	*Agave americana* L. (**Amaryllidaceae**)	Andes, fiber
Aguaje (*see* moriche)			
Ahipa	Ajipa, huitoto	*Pachyrhizus ahipa* (Wedd.) Parodi (**Fabaceae**)	Lower Andes, coast, root
Ají (*see* chili pepper)			
Algarrobo	—	*Prosopis* spp. (**Fabaceae**)	Pacific Coast, fruit
Almendra	Almendro, sawari, piqui, piquiá	*Caryocar* spp. (**Caryocaraceae**)	Tropics, fruit, oil, nut, fish poison
Amaranth	Kiwicha, achita, bledo, millmi, chaquillón	*Amaranthus caudatus* L. (= *A. edulis* Speg.: *A. mantegazzianus* Passer) (**Amaranthaceae**)	Andes, Pacific Coast, cereal
Amaranth	Bledo, jataco	*Amaranthus cruentus* L. (**Amaranthaceae**)	Andes, cereal
Amaranth	Kiwillo, sangorache	*Amaranthus quitensis* Kunth (**Amaranthaceae**)	Andes, cereal
American oil palm	Corozo, puma yarina, caiaué, noli	*Elaeis oleifera* Kunth (**Arecaceae**)	N Amazonia, fruit, oil
American yam (*see* yampee)			
Anchovy pear	Sacha mango	*Grias peruviana* Miers; *Grias* spp. (**Lecythidaceae**)	W Amazonia, fruit
Angel's trumpet (*see* floripondo)			
Annatto, achiote	Mantur, urucú	*Bixa orellana* L. (**Bixaceae**)	Tropics, red dye
Annona, sweetsop, sugar apple	Annona, fruta de conde	*Annona squamosa* L. (**Annonaceae**)	Tropics, fruit

Common name[b]	Other names	Botanical classification[c]	Comments (location, use)
Añu	Isaño, mashua	*Tropaeolum tuberosum* R. & P. (**Tropaeolaceae**)	Andes, root
Araźá	Guayaba brasileira, araçá-boi	*Eugenia stipitata* McVaugh (**Myrtaceae**)	W Amazonia, fruit
Arracacha	Apio, raccacha, lakacha	*Arracacia xanthorrhiza* Bancr. (**Umbelliferae**)	Andes, root
Arrayan	—	*Myrcianthes foliosa* (Kunth) McVaugh (= *Eugenia foliosa* (Kunth) DC) (**Myrtaceae**)	Andes, fruit
Arrow cane	Flecha, lata, caña brava, giant cane	*Gynerium sagittatum* (Aubl.) Beauv. (**Poaceae**)	Tropics, arrows, construction
Arrowroot, maranta	Guapo, sagu, ereu	*Maranta arundinacea* L. (**Marantaceae**)	Tropics, root
Assai	Açaí, palmito, chonta, murrapo	*Euterpe oleracea* Mart. (**Arecaceae**)	N tropics, fruit
Avocado	Palta, aguacate, abacate	*Persea americana* Mill. (**Lauraceae**)	Tropics, fruit
Ayahuasca	Caapi, yagé, pindé, timbo branco, napi	*Banisteriopsis caapi* (Spruce ex Griesb.) Morton (= *Banisteria caapi* Spruce ex Griesb.) (**Malpighiaceae**)	Amazonia, hallucinogen, anesthetic
Ayapana	—	*Eupatorium ayapana* Vent. (**Asteraceae**)	Amazonia, condiment
Babaco	Papayo calentano	*Carica pentagona* Heilborn (**Caricaceae**)	Andes, fruit
Babassu	Babaçu	*Orbignya phalerata* Mart. (**Arecaceae**)	S Amazonia, fruit, oil
Bacuri (see Guiana orange)			
Bacuripari	Baçu, pacuri	*Rheedia brasiliensis* Planch. & Triana (**Clusiaceae**)	E Amazonia, fruit
Banana passion fruit	Curuba, tasco	*Passiflora mollissima* (Kunth) Bailey (= *Tacsonia mollissima* Kunth) (**Passifloraceae**)	Andes, fruit
Barbados cherry	Cerezo, acerola, cerejeira do Pará	*Malpighia punicifolia* L., *malpighhia* spp. (**Malpighiaceae**)	Tropics, fruit
Barbasco, rotenone	Cube, timbó, nicou	*Lonchocarpus nicou* (Aubl.) DC; *Lonchocarpus* spp. (**Fabaceae**)	Tropics, fish poison, arrow poison
Barbasco	Ciruelito, para-para, timbó	*Phyllanthus piscatorum* Kunth; *Phyllanthus* spp. (**Euphorbiaceae**)	N Amazon, fish poison, medicinal
Barbasco	Cunumi, huaco	*Clibadium asperum* (Aubl.) DC; *clibadium* spp. (**Asteraceae**)	Amazon, fish poison

Barbasco	*Tephrosia sinapou* (Buch.) A. Chev. (= *T. toxicaria* Sw. Pers.); *Tephrosia* spp. (**Fabaceae**)	Cube, kumu, timbó	Amazonia, fish poison, medicinal
Basul	*Erythrina edulis* Triana; *Erythrina* spp. (**Fabaceae**)	Sachaporoto, pajuro	Andes, edible seed, living fences
Begonia	*Begonia* spp. (**Begoniaceae**)	—	Amazonia, medicinal, ornamental
Biribá	*Rollinia mucosa* (Jacq.) Baill (= *Annona mucosa* Jacq.); (**Annonaceae**)	Anona	Tropics, fruit
Blackberry	*Rubus* spp. (**Rosaceae**)	Zarzamora, mora de Castilla	Andes, fruit
Borojoa	*Borojoa sorbilis* (Ducke) Cuatr. (= *Aliberta sorbilis* Ducke = *Thieleodoxa sorbilis* Ducke) (**Rubiaceae**)	Purui grande, borjó	Amazonia, fruit
Bottle gourd (see gourd)			
Brazil nut	*Bertholletia excelsa* H. & B. (**Lecythidaceae**)	Castaña, castanha do Pará	Tropics, nut, medicinal
Breadnut, ramón	*Brosimum alicastrum* Sw. (**Moraceae**)	Machinga	Tropics (Brazil, Colombia, Peru), fruit, edible seeds
Bullock's heart, custard apple	*Annona reticulata* L. (**Annonaceae**)	Sacha anona	Tropics, fruit
Cabuya	*Furcraea cabuya* Trel. (**Amaryllidaceae**)	Pia floja	Andes, fiber
Cacao, cocoa	*Theobroma cacao* L. (**Sterculiaceae**)	Cacau	Tropics, fruit, medicinal, stimulant
Cacaui	*Theobroma speciosum* Willd. (**Sterculiaceae**)	Cacau sacha, choclatillo	Amazonia, fruit
Cache	*Phaseolus polyanthus* Greenman (**Fabaceae**)	Petaco, matatropa, toda la vida	N Andes, legume, seed
Caimito	*Pouteria caimito* (R. & P.) Radlk. (= *Lucuma caimito* Roem. Schult.) (**Sapotaceae**)	Abiu, caimo, cauje, abieiro, sapote	Tropics, fruit
Cainito (see star apple)			
Calabash, tree gourd	*Crescentia cujete* L. (**Bignoniaceae**)	Calabaza, higuera, jicaro, pate, pati, pamuco, tutuma, cuia, ancara	Tropics, fruit, gourd

Common name[b]	Other names	Botanical classification[c]	Comments (location, use)
Canarana	Mishquipanga, huitillo	*Renealmia* spp. (**Zingiberaceae**)	Amazonia, fruit, dye (Vickers and Plowman 1984: 33)
Cañihua	Kaniwa, cañahua	*Chenopodium pallidicaule* Aellen (**Chenopodiaceae**)	Andes, cereal
Canistel	Lucma, cutité	*Pouteria macrophylla* (Lam.) Eyma. (= *Lucuma* spp.; *Aridella* spp.; *Zitellaria* spp.) (**Sapotaceae**)	Amazonia, Andes, fruit
Canna (see achira)			
Cape gooseberry (see goldenberry)			
Casabanana	Secana, curua, cajuba, curuba	*Sicana odorifera* (Vell.) Naud. (**Cucurbitaceae**)	Tropics, vegetable
Cashew	Caju, merey, marañon	*Anacardium occidentale* L. (**Anacardiaceae**)	Tropics, nut, fruit
Cayaponia	Munteká	*Cayaponia ophthalmica* R. E. Schultes (**Cucurbitaceae**)	Colombian Amazonia, vine, medicinal
Chambira	Fiber palm, tucum, cumare	*Astrocaryum chambira* Burret.; *Astrocaryum* spp. (**Arecaceae**)	NW Amazonia, fruit, medicinal, edible seeds, fiber
Chayote	Christophine, chuchu	*Sechium edule* (Jacq.) Sw. (**Cucurbitaceae**)	Tropics fruit, vine
Cherimoya, custard apple	Chirimoya, masa, cherimolia	*Annona cherimola* Mill. (**Annonaceae**)	Tropics and Andes, fruit
Chicle (see sapodilla)			
Chilean mango, mangu	Tecu	*Bromus mango* E. Desv. (**Poaceae**)	Central Chile, Argentina, rare cereal, edible seed
Chilean strawberry	Quellen, frutilla, morango de Chile	*Fragaria chiloensis* Georgi (**Rosaceae**)	S Andes, fruit
Chili pepper; pepper, aji	Chile, rocoto, pimenta, uchu, tabasco, panka, maiagueta	*Capsicum annuum*. L.; *C. baccatum* L.; *C. chinense* Jacq,; *C. frutescens* L.; *C. pubescens* R. & P.; *Capsicum* spp. (**Solanaceae**)	Andes and tropics, condiment

Common name	Vernacular name	Scientific name (Family)	Region, uses
Ciruela (see green plum)			
Coca	—	*Erythroxylon coca* Lam. var. *coca* (**Erythroxylaceae**)	Lower Andes, coastal Peru, narcotic, medicinal
Coca Colombina	—	*Erythroxylon novogranatense* (Morris) Hieron. (**Erythroxylaceae**)	N Andes, coastal Peru, narcotic, medicinal
Coca ipadu	Ipadú	*Erythroxylon coca* Lam. var. *ipadu* Plowman (**Erythroxylaceae**)	Amazonia, narcotic, medicinal
Cocoa (see cacao)			
Cocona (see peach tomato)			
Coconilla	Coconilla colorado	*Solanum coconilla* Huber (**Solanaceae**)	Tropics, fruit
Coconut	Côco, coqueiro	*Cocos nucifera* L. (**Arecaceae**)	Tropics, fruit, oil
Cocopalm	Mucajá, macaúba	*Acrocomia sclerocarpa* Mart. (**Arecaceae**)	Central Amazon, fruit, oil, palm heart
Cocoyam, yautia	Huitina, mangareto, otó, malanga, onko, uncucha, ocumo, tannia, taioba, tegüe	*Xanthosoma sagittifolium* Schott; *Xanthosoma* spp. (**Araceae**)	Tropics, root
Colombian passion fruit	—	*Passiflora antioquiensis* Karst. (**Passifloraceae**)	N Andes, fruit
Common bean (see kidney bean)			
Coquito (see Quito palm)			
Coriander	Culantro de monte	*Eryngium foetidum* L. (**Apiaceae**)	Tropics, spice, medicinal
Corn (see maize)			
Cotton	Algodón, algodão	*Gossypium barbadense* L.; *G. hirsutum* L. (**Malvaceae**)	Tropics and subtropics, fiber
Cow tree	Sorveira, sorva	*Couma utilis* (Mart.) Muel. Arg. (**Apocynaceae**)	Amazonia, fruit, latex, gum
Cupá	Kupá, cipó-babão	*Cissus gongyloides* Burch. (**Vitaceae**)	Brazilian Amazonia, vegetable
Cupuassu	Cupuaçu	*Theobroma grandiflorum* (Willd. ex Spreng) Schum. (**Sterculiaceae**)	E Amazonia, fruit
Custard apple (see cherimoya and bullock's heart)			

Common name[b]	Other names	Botanical classification[c]	Comments (location, use)
Datura	Nongué	*Datura* spp. (**Solanaceae**)	Tropics, medicinal, narcotic
Fame flower	—	*Talinum triangulare* Willd. (**Portulacaceae**)	N South America, vegetable
Flat sedge	Piripiri, pripreoca	*Cyperus* spp. (**Cyperaceae**)	W Amazonia, sedge, condiment, medicinal
Floripondo, angels trumpet	Floripondio, toé, toatoé, maricahua	*Brugmansia aurea* Langerhein; *B. insignis* (Barb. Rodr.) Lockwood; *B. suaveolens* (H. & B. ex Willd.) (**Solanaceae**)	Amazonia, lower Andes, medicinal leaves, hallucinogen
Fragrant granadilla	Maracuja grande	*Passiflora alata* Curtis (**Passifloraceae**)	Tropics, fruit
Galupa	Bejuco, chulupa	*Passiflora pinnatistipula* Cav. (**Passifloraceae**)	Andes, fruit
Genipa	Genipap, caruto, huito	*Genipa americana* L. (**Rubiaceae**)	Tropics, black dye, fruit
Gherkin	—	*Cucumis anguria* (**Cucurbitaceae**)	Tropics, fruit
Giant granadilla, passion fruit	Granadilla, tumbo, maracujá-açu, badea	*Passiflora quadrangularis* L. (**Passifloraceae**)	Tropics, fruit
Golden bell	Campanilla de oro, canaria	*Allamanda cathartica* L. (**Apocynaceae**)	Amazonia, vine, ornamental, medicinal
Goldenberry, cape gooseberry	Capulí, camapun, aguaymanto, topotopo	*Physalis peruviana* Mill. (**Solanaceae**)	Andes, Amazonia, fruit
Gourd, bottle gourd	Poro, maté, tapara, camasa, totuma, calabaza, cabaça, cuiera, cuia	*Lagenaria siceraria* (Molina) Standley (**Cucurbitaceae**)	Tropics and temperate Andes, multi uses
Granadilla, passion fruit (also see fragrant granadilla, sweet granadilla, giant granadilla, water lemon)	—	*Passiflora* spp. (**Passifloraceae**)	Tropics, fruit
Granadilla de Quijos	Curubejo	*Passiflora popenovii* Killip (**Passifloraceae**)	Ecuadorian Amazonia, fruit
Grape-leaf passion fruit	Granadillo del monte, guate, tumbo	*Passiflora vitifolia* Kunth (**Passifloraceae**)	Tropics, fruit
Grape tree (see uvilla)			

Green plum, ciruela	Ciruela de fraile, ciruelo, usum	*Bunchosia armeniaca* (Cav.) Rich. (**Malpighiaceae**)	Andes, Peruvian coast, Amazonia, fruit
Ground cherry (see husk tomato)			
Guaba, pacay	Guabilla, guabo, guamo, guama, inga sipo, pacae, shimbillo	*Inga edulis* Mart.; *Inga* spp. (**Fabaceae**)	Tropics, fruit
Guanábana, soursop	Quabana, graviola, masasamba	*Annona muricata* L. (**Annonaceae**)	Tropics and temperate Andes, fruit
Guaraná	Cupana	*Paullinia cupana* Kunth (= *P. sorbilis* Mart.) (**Sapindaceae**)	Amazonia, stimulant, beverage
Guava, guayaba	Guayabo, goiaba, sawintu, araça	*Psidium guajava* L.; *Psidium* spp. (**Myrtaceae**)	Tropics, fruit
Guayaba (see guava)			
Guayusa	—	*Ilex guayusa* Loes. (**Aquifoliaceae**)	W Amazonia, lower Andes, medicinal, stimulant
Guiana orange, bacuri	Bacuri do Paraná	*Platonia esculenta* (Arruda) Rickett & Stafleu. (= *P. insignis* Mart.) (**Clusiaceae**)	E Amazonia, Paraguay, fruit
Heliconia	Millue	*Heliconia hirsuta* L.f.; *Heliconia* spp. (**Heliconiaceae**)	Tropics, root, medicinal, beverage
Hog plum, mombin	Yellow mombin, ciruela, cajá, taperebá, jobo, ubos, taperibá	*Spondias mombin* L. (= *S. lutea* L.) (**Anacardiaceae**)	Tropics, fruit
Husk tomato, ground cherry	Capuli	*Physalis pubescens* L. (**Solanaceae**)	Lower Andes, Amazonia, fruit, medicinal
Indigo	Añil, mutui, velvet bean	*Indigofera suffruticosa* Mill. (= *I. anil* L.) (**Fabaceae**)	Lower Andes, purple dye
Ipecac	Ipecacuana, raicilla, golden root	*Psychotria ipecacuana* (Brot.) Stokes (= *Cephaelis ipecacuanha* (Brot.) Tussac) (**Rubiaceae**)	Amazonia, root, medicinal
Ivory palm	Tagua, yarina, vegetable ivory, ivory nut	*Phytelephas macrocarpa* R. & P. (**Arecaceae**)	W Amazonia, nut, crafts
Jaboticaba	Camboim	*Myrciaria cauliflora* McVaugh (**Myrtaceae**)	Brazil, Paraguay, fruit

Common name[b]	Other names	Botanical classification[c]	Comments (location, use)
Jacaranda	Ashpingo, paraparaúba	*Jacaranda copaia* (Aubl.) (**Bignoniaceae**)	Amazonia, medicinal
Jackbean	Pallar de gentiles, nescafé	*Canavalia ensiformis* (L.) DC.; *C. plagiosperma* Piper (**Fabaceae**)	Tropics, Pacific Coast, legume, seed
Jesus' heart	—	*Caladium bicolor* (Ait.) (**Araceae**)	Amazonia, ornamental, potions
Jicama (see potato bean)			
Kaahée	Sweet herb, stevia	*Stevia rebaudiana* Bertoni (**Asteraceae**)	Paraguay, sweetener, medicinal
Kidney bean, common bean	Frijol, poroto, nuñas, feijão	*Phaseolus vulgaris* L. (**Fabaceae**)	Tropics and Andes, legume, seed
Lerén, sweet cornroot	Lirén, alluia, topinampur, dale dale, arruruz, ariã	*Calathea allouia* (Aubl.) Lindl. (= *Maranta lutea* Aubl.) (**Marantaceae**)	Tropics, root
Lima bean	Pallar, haba, feijão de Lima	*Phaseolus lunatus* L. (**Fabaceae**)	Tropics, lower Andes, legume, seed
Lucuma	Lucmo, lucma	*Pouteria lucuma* (R. & P.) Ktze.; *Pouteria* spp. (= *Lucuma* spp.) (**Sapotaceae**)	Amazonia, Andes, fruit
Lupine (see tarwi)			
Maca	Maka	*Lepidium meyenii* Walp. (**Cruciferae**)	Andes, root
Macambo	Patashte, pataste macambillo, cacau do Perú	*Theobroma bicolor* H. & B. (**Sterculiaceae**)	W Amazonia, fruit, seed
Macoubea	Amapa	*Macoubea witotorum* R. E. Schultes (**Apocynaceae**)	Colombian Amazonia, fruit
Madi	Madivilcun, tarweed	*Madia sativa* Molina (**Asteraceae**)	Central Chile, oil plant, edible seeds
Madrono	Naranjito, bacuripari	*Rheedia* spp. (**Clusiaceae**)	Amazonia, fruit
Maguey (see agave)			
Maize, corn	Maíz, sara, milho	*Zea mays* L. (**Poaceae**)	Andes, tropics, cereal
Mamey	Mammey, mammee, abricó	*Mammea americana* L. (**Clusiaceae**)	Tropics, fruit
Mamón	Mamoncillo, maca, pitomba, genip	*Melicocca bijuga* L. (= *M. bijugatus* Jacq.) (**Sapindaceae**)	N Amazonia, fruit

Mangu (see Chilean mango) Manioc	Yuca, cassava, casabe, mandioca, ruma	*Manihot esculenta* Crantz (**Euphorbiaceae**)	Tropics, root
Mapuey (see yampee)			
Maranta (see arrowroot)			
Marimari	—	*Cassia leiandra* Benth.; *Cassia* spp. (**Fabaceae**)	Amazonia, fruit
Maté	Yierba maté	*Ilex paraguariensis* St. Hil. (**Aquilfoliaceae**)	Subtropics, beverage
Mauka	Miso	*Mirabilis expansa* R. & P. Standl. (**Nyctaginaceae**)	Temperate Andes, root
Melon pear, pepino	Pepino, cachun, kachan	*Solanum muricatum* Ait. (**Solanaceae**)	Temperate Andes, fruit
Mito (see papaya, dwarf)			
Molle	Arbol de pimienta, mulli	*Schinus molle* L. (**Anacardiaceae**)	Andes, medicinal, beverage, ornamental
Mombin (see hog plum, Spanish plum)			
Moriche, aguaje	Buriti, miriti	*Mauritia flexuosa* L.f. (**Arecaceae**)	Amazonia, fruit, starch, fiber, construction
Mullaca	Bolsa mullaca	*Physalis angulata* L. (**Solanaceae**)	Amazonia, medicinal, narcotic (Davis and Yost 1983: 167, 203)
Nambicuara peanut	Amendoim indigena	*Arachis nambyquarae* Hoehne (**Fabaceae**)	S Amazonia, legume, seed
Nance	Muruçi, chaparro	*Byrsonima crassifolia* (L.) Kunth (**Malpighiaceae**)	Tropics, fruit
Naranjilla, Quito orange	Lulo, toronja	*Solanum quitoense* Lam. (**Solanaceae**)	Tropics, fruit
Oca	Cavi	*Oxalis tuberosa* Molina (=*Xanthoxolis tuberosa* (Molina) Holub.) (**Oxalidaceae**)	Andes, root
Oleander (see yellow oleander)			
Orinoco nut	Inchi, cacay, palo de nuez, tacay nut	*Caryodendron orinocense* Karst. (**Euphorbiaceae**)	N Amazonia, nut
Pacay (see guaba)			
Paco	Membrillo	*Gustavia superba* (Kunth) Berg. (**Lecythidaceae**)	Tropics, fruit

Appendix 1A

Common name[b]	Other names	Botanical classification[c]	Comments (location, use)
Palillo	Guaribo	*Campomanesia lineatifolia* R. & P. (**Myrtaceae**)	W Amazonia, lower Andes, fruit, medicinal
Panama hat palm	Bonbonaje	*Carludovica palmata* R. & P. (**Cyclanthaceae**)	Tropics, fiber
Papaya	Lechosa, mamão, pawpaw	*Carica papaya* L.; *Carica* spp. (**Caricaceae**)	Tropics, lower Andes, fruit
Papaya (dwarf), mito	Chamburu	*Carica candicans* Gray (**Caricaceae**)	Andes, fruit
Papaya (mountain)	Chamburo	*Carica pubescens* Solms (**Caricaceae**)	Lower Andes, fruit
Pará cress	Berro, botoncillo	*Spilanthes acmella* (L.) Murr. (**Asteraceae**)	Amazonia, condiment
Parinari	Supay ocote, umarirana	*Couepia subcordata* Benth.; *Couepia* spp. (**Chrysobalanaceae**)	Amazonia, fruit, nut
Passion fruit, granadilla (see grape-leaf passion fruit, Colombian passion fruit, banana passion fruit, purple passion fruit, rosy passion fruit, yellow passion fruit)	—	*Passiflora* spp. (**Passifloraceae**)	Tropics, fruit
Patauá, ungurahui	Bataua, batana, seje	*Jessenia bataua* (Mart.) Burret (= *Oenocarpus bataua* Mart.) (**Arecaceae**)	Tropics, fruit
Peach palm	Pejibaye, chonta duro, pijuayo, pupunha	*Bactris gasipaes* Kunth. (= *Guilielma speciosa* Mart., *G. gasipaes* Kunth Bailey) (**Arecaceae**)	Tropics, fruit, palm heart
Peach tomato, cocona	Topiro, lulo, cubiu	*Solanum sessiliflorum* Dunal (= *S. topiro* H. B. ex Dunal) (**Solanaceae**)	Tropics, fruit
Peanut	Mani, inchis, amendoim	*Arachis hypogaea* L. (**Fabaceae**)	Amazonia, Pacific Coast, legume, seed
Penca Pepino (see melon pear) Pepper (see chili pepper)	Pakpa, chuchao	*Furcraea andina* Trel. (**Amaryllidaceae**)	Andes, fiber, ornamental
Pineapple	Piña, abacaxi, ananás, curauá, achupalla	*Ananas comosus* (L.) Merr. (= *A. sativus* Schult.; *A. ananas* (L.) Voss.) (**Bromeliaceae**)	Tropics, fruit

Common name	Local names	Scientific name (Family)	Distribution / Use
Pineapple (wild)	—	*Ananas erectifolius* L. B. Smith (**Bromeliaceae**)	Amazonia, Orinoco, fiber
Pitomba	—	*Talisia esculenta* Radlk. (= *Sapindus edulis* St. Hil.) (**Sapindaceae**)	Tropics, fruit
Potato	Papa, apichu, batata	*Solanum tuberosum* L.; *Solanum* spp. (**Solanaceae**)	Andes, root
Potato (tropical)	Moshaki, amista, kurahji	*Solanum hygrothermicum* Ochoa (**Solanaceae**)	Peruvian Amazonia (Ochoa 1984), root
Potato bean, jícama	Macucu, nupe	*Pachyrhizus tuberosus* (Lam.) Spreng. (**Fabaceae**)	Amazonia, Paraguay, root
Prickly pear	Pata kiska, espino, cacto	*Opuntia exaltata* Berger; *Opuntia* spp. (**Cactaceae**)	Andes, Pacific Coast, fencing, misc.
Pumpkin (see squash)			
Purple passion fruit (see yellow passion fruit)			
Quinoa	Kiuna, trigo Inca	*Chenopodium quinoa* Willd. (**Chenopodiaceae**)	Andes, cereal grain
Quishuar	—	*Buddleja incana*; *Buddleja* spp. (**Buddlejaceae**)	Andes, small tree, cane, tools, carvings, fuel, leaves for fertilizer and medicine (Gade 1999: 62–3)
Quito orange (see naranjilla)			
Quito palm, coquito	—	*Parajubaea cocoides* Burr. (**Arecaceae**)	Ecuadorian Andes, nut, ornamental
Ramón (see breadnut)			
Rope plant	Caraua, carúa, caroá, curratow	*Neoglaziovia variegata* (Arruda) Mez. (= *Bromelia variegata* Arruda) (**Bromeliaceae**)	Tropics, fiber
Rosy passion fruit	Curuba bogotana, red banana passion fruit	*Passiflora cumbalensis* Harms (**Passifloraceae**)	N Andes, fruit
Rotenone (see barbasco)			
Rubber	Shiringa, seringa, jebe, goma, caucho, seringuiera	*Hevea brasiliensis* (Willd.) Muell. Arg.; *Hevea* spp. (**Euphorbiaceae**)	Amazonia, latex, edible seeds
Sacaca	Cascarilla	*Croton cajucara* Benth. (**Euphorbiaceae**)	Amazonia, medicinal

Common name[b]	Other names	Botanical classification[c]	Comments (location, use)
San Pedro cactus	Gigantón, jahuackollai	*Trichocereus cuzcoensis* Britton and Rose; *Trichocereus.* spp. (**Cactaceae**)	Andean and coastal Peru, hallucinogen, edible fruit
Sapodilla, chiclé	Níspero, zapote, sapoti, naseberry, caimito brasilero	*Manilkara achras* (Mill.) Fosberg (= *M. zapota* (L.) Van Royen; *Achras zapota* L.) (**Sapotaceae**)	N South America, fruit, latex
Sapote	Sapote de monte, chupa-chupa, sapota	*Quararibea cordata* (H. & B.) Vischer (= *Matisia cordata* H. & B.) (**Bombacaceae**)	W Amazonia, fruit
Sapucaia	Castaña de monte	*Lecythis pisonis* Caub. (**Lecythidaceae**)	Amazonia, nut
Scarlet runner bean	Ayocote, fava	*Phaseolus coccineus* (L.) DC (= *P. multiflorus* Willd.) (**Fabaceae**)	Subtropics, rare in South America, legume, seed
Soursop (see guanábana)			
Spanish plum, mombin	Cajazeiro, red mombin	*Spondias purpurea* L. (**Anacardiaceae**)	N South America, tropics, fruit
Squash, pumpkin	Zapallo, auyama, jerimum, achocha, abóbora	*Cucurbita maxima* Duchesne; *C. ficifolia* Bouché (= *Pepo ficifolia* (Bouche) Britton); *C. moschata* (Duchesne) Poir.; *C. ecuadorensis* Cutler and Whitaker (**Cucurbitaceae**)	Temperate Andes and tropics, vine, vegetable
			Ecuador coast
Star apple, cainito	Caimito	*Chrysophyllum cainito* L. (**Sapotaceae**)	Tropics, fruit
Strawberry (see Chilean strawberry)			
Strawberry guava	—	*Psidium cattleianum* Sabine (**Myrtaceae**)	Coastal Brazil, fruit
Sugar apple (see anona)			
Sunsapote	—	*Licania platypus* (Helmsley) Fritsch (**Chrysobalanaceae**)	N South America, fruit
Surinam cherry	Pitanga, cereza	*Eugenia uniflora* L. (= *Stenocalyx uniflorus* (L.) Kausel) (**Myrtaceae**)	Tropics, fruit
Sweet calabash	Chulupa, curuba	*Passiflora maliformis* L. (**Passifloraceae**)	Colombian Andes, fruit
Sweet cornroot (see lerén)			
Sweet granadilla	Sweet passion fruit, granadilla, tumbo, badea	*Passiflora ligularis* Juss. (**Passifloraceae**)	Temperate Andes, fruit

Common name	Scientific name (Family)	Local names	Region, use
Sweet potato	*Ipomoea batatas* (L.) Lam. (**Convolvulaceae**)	Batata, camote, cumara, aje	Tropics and lower Andes, root
Sweetsop (see anona)			
Tarwi, lupine	*Lupinus mutabilis* Sweet (**Fabaceae**)	Tarhui, chocho	Andes, edible seeds
Tobacco	*Nicotiana rustica* L.; *N. tabacum* L. (**Solanaceae**)	Tobaco, tabaco	Tropics, lower Andes, narcotic, medicinal
Tomato	*Lycopersicum esculentum* Mill. (**Solanaceae**)	Tomate, paconca	Tropics, lower Andes, fruit
Totaí	*Acrocomia totai* Mart. (**Arecaceae**)	Mbocayá, coco paraguayo	E Bolivia, Paraguay, fruit, oil, palm heart
Tree gourd (see calabash)			
Tree tomato	*Cyphomandra betacea* (Cav.) Sendt.; *Cyphomandra* spp. (**Solanaceae**)	Sacha tomate, tomatillo, tamarillo	Temperate Andes, fruit
Trompillo	*Alibertia edulis* A. Rich. ex DC (**Rubiaceae**)	Guabillo, huitillo, purui	Tropics, fruit
Tucumá	*Astrocaryum aculeatum* G. Meyer (= *A. tucuma* Mart.; *A. princeps* Barb. Rodr.) (**Arecaceae**)	Tucumã, chonta	Amazonia, fruit, fiber
Tucumão	*Astrocaryum vulgare* Mart. (**Arecaceae**)	Curua, cumare, aoura	Brazilian Amazon, fruit, fiber
Ugni	*Myrtus ugni* Moll. (= *Ugni ugni* Macloskie) (**Myrtaceae**)	Murta	Chilean Andes, fruit
Ulluco Umari	*Ullucus tuberosus* Caldas (**Basellaceae**) *Poraqueiba sericea* Tul.; *P. paraensis* Ducke (**Icacinaceae**)	Papa lisa, ullucu, melloco Mari	Andes, root W Amazonia, fruit, oil E Amazonia, fruit, oil
Ungurahui (see pataua) Uvilla, grape tree	*Pourouma cecropiaefolia* Mart. (**Moraceae**)	Ubilla, puruma	Tropics, fruit
Vanilla	*Vanilla planifolia* Andr.; *Vanilla* spp. (**Orchidaceae**)	—	Tropics, Andes, vine, flavoring
Water lemon	*Passiflora laurifolia* L. (**Passifloraceae**)	Bell apple, yellow granadilla, maracujá, Jamaica honeysuckle	N tropics, fruit
Yacón	*Polymnia sonchifolia* Poeppig & Endlicher (**Asteraceae**)	Llacón, aricona, aricuma, jiquima	Temperate Andes, root

Common name[b]	Other names	Botanical classification[c]	Comments (location, use)
Yampee, American yam, mapuey	Cará, sacha papa, cush-cush, ñame	*Dioscorea trifida* L.; *Dioscorea* spp. (**Dioscoreaceae**)	Tropics, root
Yautia (see cocoyam)			
Yellow oleander	Flor amarilla	*Thevetia peruviana* (Pers.) Schumann (= *Cerbera peruviana* Pers.; *C. thevetia* L.) (**Apocynaceae**)	Tropics, medicinal, poison
Yellow passion fruit, purple passion fruit	Yellow granadilla, purple granadilla, maracujá	*Passiflora edulis* Sims (**Passifloraceae**)	Tropics, S Brazil, fruit
Yoco	—	*Paullinia yoco* Schultes & Killip (**Sapindaceae**)	W Amazonia, medicinal, hallucinogen
Yopo	Cohoba, paricá	*Anadenanthera peregrina* (L.) Speg. (**Fabaceae**)	Amazonia, medicinal bark, hallucinogen

[a] Domesticates, semi-domesticates, and other selected planted and protected (cultivated) plants, native to the Americas, and prehistoric or probably prehistoric in South America. Listed by at least two sources as domesticated, semi-domesticated, or cultivated.

[b] Alphabetized by common name used in text, as given in Schery (1972) or elsewhere.

[c] Alternate family names: **Amaryllidaceae** (Agavaceae), **Arecaceae** (Palmae), **Asteraceae** (Compositae), **Clusiaceae** (Guttiferae), **Fabaceae** (Leguminoseae), **Poaceae** (Gramineae).

[d] Root crops here include tubers, rhizomes, and corms.

Sources: For comprehensive lists and descriptions see: Balée (1994); Brücher (1989); Cárdenas (1989); Clement (1999); Duke and Vásquez (1994); Gade (1975); Le Cointe (1947); León (1994); NRC (1989); Patiño (1963–9); Piperno and Pearsall (1998); C. Sauer (1950); J. Sauer (1993); Schery (1972); Schultes and Raffauf (1990); N. Smith *et al.* (1992); Towle (1961); Vásquez and Gentry (1989); Zeven and de Wet (1982). For additional semi-domesticates and incipient domesticates in the Brazilian Amazon, see Balée and Moore (1991: 232–7) and Clement (1999). Most scientific names have been checked in botanical data bases, including the *Index Kewensis* (updated 1991–5) and the *Gray Card Index* of New World plants (updated 1995). Inconsistencies are frequent. The listing here has been checked and added to by ethobotanists William Balée and (especially) Joseph McCann. However decisions on what information to provide and which plants to include were made by the author.

Appendix 1B. Roster of Cultivated Plants by Species Name[a]

Achras zapota (sapodilla)
Acrocomia sclerocarpa (cocopalm)
Acrocomia totai (totaí)
Agave americana (agave)
Aliberta edulis (trompillo)
Aliberta sorbilis (borojoa)
Allamanda cathartica (golden bell)
Amaranthus caudatus (amaranth)
Amaranthus cruentus (amaranth)
Amaranthus edulis (amaranth)
Amaranthus mantegazzianus (amaranth)
Amaranthus quitensis (amaranth)
Anacardium occidentale (cashew)
Anadenanthera peregrina (yopo)
Ananas ananas (pineapple)
Ananas comosus (pineapple)
Ananas erectifolius (pineapple, wild)
Ananas sativas (pineapple)
Annona cherimola (cherimoya)
Annona mucosa (biribá)
Annona muricata (guanábana)
Annona reticulata (bullock's heart)
Annona squamosa (anona)
Arachis hypogaea (peanut)
Arachis nambyquarae (Nambicuara peanut)
Aridella spp. (canistel)
Arracacia xanthorrhiza (arracacha)
Astrocaryum aculeatum (tucumá)
Astrocaryum chambira (chambira)
Astrocaryum princeps (tucumá)
Astrocaryum tucuma (tucumá)
Astrocaryum vulgare (tucumão)
Bactris gasipaes (peach palm)
Banisteria caapi (ayahuasca)
Banisteriopsis caapi (ayahuasca)
Begonia spp. (begonia)
Bertholletia excelsa (Brazil nut)
Bixa orellana (annatto)
Borojoa sorbilis (borojoa)
Bromelia variegata (rope plant)
Bromus mango (Chilean mango)

Brosimum alicastrum (breadnut)
Brugmansia aurea (floripondo)
Brugmansia insignis (floripondo)
Brugmansia suaveolens (floripondo)
Buddleja incana (quishuar)
Bunchosia armeniaca (green plum)
Byrsonima crassifolia (nance)
Caladium bicolor (Jesus' heart)
Calathea allouia (lerén)
Campomanesia lineatifolia (palillo)
Canavalia ensiformis (jackbean)
Canavalia plagiosperma (jackbean)
Canna edulis (achira)
Capsicum annuum (chili pepper)
Capsicum baccatum (chili pepper)
Capsicum chinense (chili pepper)
Capsicum frutescens (chili pepper)
Capsicum pubescens (chili pepper)
Carica candicans (papaya, dwarf)
Carica papaya (papaya)
Carica pentagona (babaco)
Carica pubescens (papaya, highland)
Carludovica palmata (Panama hat palm)
Caryocar spp. (almendra)
Caryodendron orinocense (Orinoco nut)
Cassia leiandra (marimari)
Cayaponia ophthalmica (cayaponia)
Cephaelis ipecacuanha (ipecac)
Cerbera peruviana (yellow oleander)
Cerbera thevetia (yellow oleander)
Chenopodium pallidicaule (cañihua)
Chenopodium quinoa (quinoa)
Chrysophyllum cainito (star apple)
Cissus gongyloides (cupá)
Clibadium asperum (barbasco)
Clibadium sylvestre (barbasco)
Cocos nucifera (coconut)
Couepia subcordata (parinari)
Couma utilis (cow tree)
Crescentia cujete (calabash)
Croton cajucara (sacaca)
Cucumis anguria (gherkin)

Cucurbita argyrosperma (squash)
Cucurbita ecuadorensis (squash)
Cucurbita ficifolia (squash)
Cucurbita maxima (squash)
Cucurbita mixta (squash)
Cucurbita moschata (squash)
Cyclanthera pedata (achocha)
Cyperus spp. (flat sedge)
Cyphomandra betacea (tree tomato)
Datura spp. (datura)
Dioscorea trifida (yampee)
Elaeis oleifera (American oil palm)
Eryngium foetidum (coriander)
Erythrina edulis (basul)
Erythroxylon coca var. *coca* (coca)
Erythroxylon coca var. *ipadu* (coca ipadú)
Erythroxylon novogranatense (coca
  Colombina)
Eugenia foliosa (arrayan)
Eugenia stipitata (arazá)
Eugenia uniflora (Surinam cherry)
Eupatorium ayapana (ayapana)
Euterpe oleracea (assai)
Fragaria chiloensis (Chilean strawberry)
Furcraea andina (penca)
Furcraea cabuya (cabuya)
Genipa americana (genipa)
Gossypium barbadense (cotton)
Gossypium hirsutum (cotton)
Grias peruviana (anchovy pear)
Guilielma gasipaes (peach palm)
Guilielma speciosa (peach palm)
Gustavia superba (paco)
Gynerium sagittatum (arrow cane)
Heliconia hirsuta (heliconia)
Hevea brasiliensis (rubber)
Ilex paraguariensis (maté)
Ilex guayusa (guayusa)
Indigofera suffruticosa (indigo)
Inga edulis (guaba)
Ipomoea batatas (sweet potato)
Jacaranda copaia (jacaranda)
Jessenia bataua (patauá)
Lagenaria siceraria (gourd)
Lecythis pisonis (sapucaia)
Lepidium meyenii (maca)
Licania platypus (sunsapote)
Lonchocarpus nicou (barbasco)

Lucuma caimito (caimito)
Lucuma spp. (lucuma)
Lupinus mutabilis (tarwi)
Lycopersicum esculentum (tomato)
Macoubea witotorum (macoubea)
Madia sativa (madi)
Malpighia punicifolia (Barbados cherry)
Mammea americana (mamey)
Manihot esculenta (manioc)
Manilkara achras (sapodilla)
Manilkara zapota (sapodilla)
Maranta arundinacea (arrowroot)
Maranta lutea (lerén)
Matisia cordata (sapote)
Mauritia flexuosa (moriche)
Melicocca bijuga (mamón)
Melicoccus bijugatus (mamón)
Mirabilis expansa (mauka)
Myrcianthes foliosa (arrayan)
Myrciaria cauliflora (Jaboticaba)
Myrtus ugni (ugni)
Neoglaziovia variegata (rope plant)
Nicotiana rustica (tobacco)
Nicotiana tabacum (tobacco)
Oenocarpus bataua (patauá)
Opuntia exaltata (prickly pear)
Orbignya phalerata (babassu)
Oxalis tuberosa (oca)
Pachyrhizus ahipa (ahipa)
Pachyrhizus tuberosus (potato bean)
Parajubaea cocoides (Quito palm)
Passiflora alata (fragrant granadilla)
Passiflora antioquiensis (Colombian
  passion fruit)
Passiflora cumbalensis (rosy passion
  fruit)
Passiflora edulis (purple passion fruit,
  yellow passion fruit)
Passiflora laurifolia (water lemon)
Passiflora ligularis (sweet granadilla)
Passiflora maliformis (sweet calabash)
Passiflora mollissima (banana passion
  fruit)
Passiflora pinnatistipula (galupa)
Passiflora popenovii (granadilla de
  Quijos)
Passiflora quadrangularis (giant
  granadilla)

Passiflora vitifolia (grape-leaf passion fruit)
Paullinia cupana (guaraná)
Paullinia sorbilis (guaraná)
Paullinia yoco (yoco)
Pepo ficifolia (squash)
Persea americana (avocado)
Phaseolus coccineus (scarlet runner bean)
Phaseolus lunatus (lima bean)
Phaseolus multiflorus (scarlet runner bean)
Phaseolus polyanthus (cache)
Phaseolus vulgaris (kidney bean)
Phyllanthus piscatorum (barbasco)
Physalis angulata (mullaca)
Physalis peruviana (goldenberry)
Physalis pubescens (husk tomato)
Phytelephas macrocarpa (ivory palm)
Platonia esculenta (Guiana orange)
Platonia insignis (Guiana orange)
Polymnia sonchifolia (yacón)
Poraqueiba paraensis (umarí)
Poraqueiba sericea (umarí)
Pourouma cecropiaefolia (uvilla)
Pouteria caimito (caimito)
Pouteria lucuma (lucuma)
Pouteria macrophylla (canistel)
Prosopis spp. (algarrobo)
Psidium cattleianum (strawberry guava)
Psidium guajava (guava)
Psychotria ipecacuana (ipecac)
Quararibea cordata (sapote)
Renealmia spp. (canarana)
Rheedia spp. (madrono)
Rheedia brasiliensis (bacuripari)
Rollinia mucosa (biribá)
Rubus spp. (blackberry)
Sapindus edulis (pitomba)

Schinus molle (molle)
Sechium edule (chayote)
Sicana odorifera (casabanana)
Solanum coconilla (coconilla)
Solanum hygrothermicum (potato, tropical)
Solanum muricatum (melon pear)
Solanum quitoense (naranjilla)
Solanum sessiliflorum (peach tomato)
Solanum topiro (peach tomato)
Solanum tuberosum (potato)
Spilanthes acmella (Pará cress)
Spondias lutea (hog plum)
Spondias mombin (hog plum)
Spondias purpurea (Spanish plum)
Stenocalyx uniflorus (Surinam cherry)
Stevia rebaudiana (kaahée)
Tacsonia mollissima (banana passion fruit)
Talinum triangulare (fame flower)
Talisia esculenta (pitomba)
Tephrosia sinapou (barbasco)
Tephrosia toxicaria (barbasco)
Theobroma bicolor (macambo)
Theobroma cacao (cacao)
Theobroma grandiflorum (cupuassu)
Theobroma speciosum (cacauí)
Thevetia peruviana (yellow oleander)
Thieleodoxa sorbilis (borojoa)
Trichocereus cuzcoensis (San Pedro cactus)
Tropaeolum tuberosum (añu)
Ugni ugni (ugni)
Ullucus tuberosus (ulluco)
Vanilla planifolia (vanilla)
Xanthosoma sagittifolium (cocoyam)
Xanthoxolis tuberosa (oca)
Zea mays (maize)
Zitellaria spp. (canistel)

[a] See Appendix 1A for additional information alphabetized by common name indicated here and used in text.

REFERENCES

ACUÑA, CRISTÓBAL DE (1942) [1641]. *Nuevo descubrimiento del gran río de las Amazonas*, Buenos Aires: Emecé Editores.

AGUADO, PEDRO DE (1931) [1578]. *Primera parte de la recopilación historial resolutoria de Sancta Marta y Nuevo Reino de Granada de las Indias del Mar Oceano*, 3 vols, Madrid: Espasa-Calpe.

ALCORN, JANIS B. (1981). 'Huastec Noncrop Resource Management: Implications for Prehistoric Rain Forest Management', *Human Ecology*, 9: 395–417.

——(1984). *Huastec Mayan Ethnobotany*, Austin: University of Texas Press.

ALLEN, WILLIAM L. and JUDY HOLSHOUSER DE TIZÓN (1973). 'Land Use Patterns among the Campa of the Alto Pachitea, Peru', in Donald W. Lathrap and Jody Douglas (eds.), *Variation in Anthropology: Essays in Honor of John C. McGregor*, Urbana: Illinois Archaeological Survey, 137–53.

ALTAMIRANO, DIEGO FRANCISCO (1891) [*c*. 1710]. *Documentos históricos de Bolivia: Historia de la mision de los Mojos*, La Paz: Imprenta de El Comercio.

ALTIERI, MIGUEL A. (1995). *Agroecology: The Science of Sustainable Agriculture*, Boulder, CO: Westview Press.

ANDERSON, ANTHONY B. (1990). 'Extraction and Forest Management by Rural Inhabitants in the Amazon Estuary', in A. B. Anderson (ed.), *Alternatives to Deforestation: Steps Toward Sustainable Use of the Amazon Rain Forest*, New York: Columbia University Press, 65–85.

——and DARRELL A. POSEY (1989). 'Management of a Tropical Scrub Savanna by the Gorotire Kayapó of Brazil', *Advances in Economic Botany*, 7: 159–73.

ANDRADE, ANGELA (1986). *Investigación arqueológica de los antrosoles de Araracuara*, Fundación de Investigaciones Arqueológicas Nacionales, Bogotá: Banco de la República.

ANDREWS, JEAN (1984). *Peppers: The Domesticated Capsicums*, Austin: University of Texas Press.

ANONYMOUS (1985). 'Centuries-Old Fertilizer Formula Survives in Peru', *Ceres*, 107: 8.

ANTÚNEZ DE MAYOLO R., SANTIAGO ERIK (1980). 'Fertilizantes agrícolas en el antiguo Perú', *Boletín de la Sociedad Geográfica de Lima*, 99: 31–40.

——(1986*a*). 'El riego en Aija', *Allpanchis*, 28: 47–71.

——(1986*b*). 'Ciencia agrícola en el Perú precolombino', in *Estudios de historia de la ciencia en el Perú 1: Ciencias básicas e technológicas*, Lima: CONCYTEC: 152–91.

APARICIO, FRANCISCO DE (1946). 'The Comechingón and Their Neighbors of the Sierras de Córdoba', in Steward (1946–1959: 2: 673–85).

ARDISSONE, R. (1944). 'Andenes en la cuenca del torrente de Las Trancas', *Relaciones de la Sociedad Argentina de Antropología*, 4: 93–109.

——(1945). 'Las pircas de Ancasti: Contribución al conocimiento de los restos de andenes en el noroeste de la Argentina', *Gaea: Boletín de la Sociedad Argentina de Estudios Geográficos*, 7: 383–416.

ARMILLAS, PEDRO (1961). 'Land Use in Pre-Columbian America', in L. D. Stamp (ed.), *Arid Zone Research*, Paris: UNESCO, 17: 255–76.

ARRIAGA, PABLO JOSEPH DE (1968) [1621]. *The Extirpation of Idolatry in Peru*, L. Clark Keating (trans. and ed.), Lexington: University of Kentucky Press.

ATRAN, SCOTT (1993). 'Itza Maya Tropical Agro-Forestry', *Current Anthropology*, 34: 633–700.

BAKSH, MICHAEL (1985). 'Faunal Food as a Limiting Factor on Amazonian Cultural Behavior: A Machiguenga Example', in Barry L. Isaac (ed.), *Research in Economic Anthropology*, Greenwich, CT: JAI Press, 7: 145–75.

——and ALLEN JOHNSON (1990). 'Insurance Policies among the Machiguenga: An Ethnographic Analysis of Risk Management in a Non-Western Society', in Elizabeth Cashdan (ed.), *Risk and Uncertainty in Tribal and Peasant Economies*, Boulder, CO: Westview Press, 193–227.

BALÉE, WILLIAM (1987). 'Cultural Forests of the Amazon', *Garden*, 11/6: 12–14, 32.

——(1989). 'The Culture of Amazonian Forests', *Advances in Economic Botany*, 7: 1–21.

——(1992). 'People of the Fallow: A Historical Ecology of Foraging in Lowland South America', in Kent H. Redford and Christine Padoch (eds.), *Conservation of Neotropical Forests: Working from Traditional Resource Use*, New York: Columbia University Press, 35–57.

——(1994). *Footprints of the Forest: Ka'apor Ethnobotany: The Historical Ecology of Plant Utilization by an Amazonian People*, New York: Columbia University Press.

——(1995). 'Historical Ecology of Amazonia', in Sponsel (1995: 97–110).

——(ed.) (1998). *Advances in Historical Ecology*, New York: Columbia University Press.

——and ANNE GÉLY (1989). 'Managed Forest Succession in Amazonia: The Ka'apor Case', *Advances in Economic Botany*, 7: 129–58.

——and DENNY MOORE (1991). 'Similarity and Variation in Plant Names in Five Tupi-Guarani Languages (Eastern Amazonia)', *Bulletin of the Florida Museum of Natural History, Biological Sciences*, 35: 209–62.

BARANDIARAN, DANIEL DE (1967). 'Agricultura y recolección entre los indios Sanema-Yanoama, o el hacha de piedra y la psicología paleolítica de los mismos', *Antropológica*, 19: 24–50.

BARHAM, A. J. and D. R. HARRIS (1985). 'Relict Field Systems in the Torres Strait Region', *British Archaeological Reports, International Series*, 232: 247–83.

BARNES, MONICA and DAVID FLEMING (1991). 'Filtration-Gallery Irrigation in the Spanish New World', *Latin American Antiquity*, 2: 48–68.

BASILE, DAVID G. (1974). *Tillers of the Andes: Farmers and Farming in the Quito Basin*, Chapel Hill: Department of Geography, University of North Carolina.

BASSO, ELLEN B. (1973). *The Kalapalo Indians of Central Brazil*, New York: Holt, Rinehart, and Winston.

BATCHELOR, BRUCE (1980). 'Los camellones de Cayambe de la Sierra de Ecuador', *América Indígena*, 40: 671–89.

BECKERMAN, STEPHEN (1979). 'The Abundance of Protein in Amazonia: A Reply to Gross', *American Anthropologist*, 81: 533–60.

——(1983a). 'Does the Swidden Ape the Jungle?', *Human Ecology*, 11: 1–12.

——(1983b). 'Barí Swidden Gardens: Crop Segregation Patterns', *Human Ecology*, 11: 85–101.

——(1983c). 'Carpe Diem: An Optimal Foraging Approach to Bari Fishing and Hunting', in Hames and Vickers (1983a: 269–99).

BECKERMAN, STEPHEN (cont.) (1987). 'Swidden in Amazonia and the Amazon Rim', in B. L. Turner II and Steven B. Brush (eds.), *Comparative Farming Systems*, New York: Guilford Press, 55–94.

BENAVIDES, MARIA A. (1997). 'Cambios en la organización social, en la gestión de los recursos naturales y en la producción agropecuaria andina durante el período colonial', Lima, manuscript.

BENFER, ROBERT A., GLENDON H. WEIR, and BERNARDINO OJEDA ENRIQUEZ (1987). 'Early Water Management Technology on the Peruvian Coast', in Browman (1987*a*: 195–206).

BENNETT, WENDELL C. (1946). 'The Archaeology of the Central Andes', in Steward (1946–1959: 2: 61–147).

BERGMAN, ROLAND W. (1980). *Amazon Economics: The Simplicity of Shipibo Indian Wealth*, Dellplain Latin American Studies 6, Ann Arbor, MI: University Microfilms International.

BERNEDO MÁLAGA, LEÓNIDAS (1949). *La cultura puquina o prehistoria de la provincia de Arequipa*, Lima: Dirección de Educación y Artistica y Extensión Cultural.

BIESBOER, DAVID D., MICHAEL W. BINFORD, and ALAN KOLATA (1999). 'Nitrogen Fixation in Soils and Canals of Rehabilitated Raised-Fields of the Bolivian Altiplano', *Biotropica*, 31: 255–67.

BINFORD, MICHAEL W., ALAN L. KOLATA, MARK BRENNER, JOHN W. JANUSEK, MATTHEW T. SEDDON, MARK ABBOTT, and JASON H. CURTIS (1997). 'Climate Variation and the Rise and Fall of an Andean Civilization', *Quaternary Research*, 47: 235–48.

BLANK, PAUL W. (1976). 'Macusi Indian Subsistence, Northern Amazonia', Master's thesis, University of Wisconsin–Madison.

——(1981). 'Wet Season Vegetable Protein Use among Riverine Tropical American Cultures: A Neglected Adaptation?', *Social Science and Medicine*, 15D: 463–9.

BLOCK, DAVID (1994). *Mission Culture on the Upper Amazon: Native Tradition, Jesuit Enterprise, and Secular Policy in Moxos, 1660–1880*, Lincoln: University of Nebraska Press.

BODLEY, JOHN H. (1970). 'Campa Socio-Economic Adaptation', Ph. D. dissertation, University of Oregon.

BOGIN, BARRY (1979). 'Climatic Change and Human Behavior on the Southwest Coast of Ecuador', *Central Issues in Anthropology*, 1/2: 21–31.

BOLIAN, CHARLES E. (1971). 'Manioc Cultivation in Periodically Flooded Areas', unpublished paper presented at the American Anthropological Association Meeting, New York.

——(1975). 'Archaeological Excavations in the Trapecio of Amazonas: The Polychrome Tradition', Ph. D. dissertation, University of Illinois.

BONAVIA, DUCCIO (1967–8). 'Investigaciones arqueológicas en el Mantaro medio', *Revista del Museo Nacional*, Lima, 35: 211–94.

——(1985). *Mural Painting in Ancient Peru*, Patricia J. Lyon (trans.), Bloomington: Indiana University Press.

——and RAMIRO MATOS M. (1990). 'La recuperación de los terrenos agrícolas: Realidad o utopia?, *Revista Peruana de Ciencias Sociales*, 2/2: 61–72.

BOOMERT, AAD (1976). 'Pre-Columbian Raised Fields in Coastal Surinam', *Proceedings of the Sixth International Congress for the Study of Pre-Columbian Cultures of the Lesser Antilles*, Guadeloupe: Pointe a Pitre, 134–44.

BORAH, WOODROW (1964). 'America as Model: The Demographic Impact of European Expansion upon the Non-European World', *Actas y Memorias, XXXV Congreso Internacional de Americanistas*, 3: 379–87.

BOSERUP, ESTER (1965). *The Conditions of Agricultural Growth: The Economics of Agrarian Change under Population Pressure*, Chicago: Aldine.

BOSTER, JAMES (1983). 'A Comparison of the Diversity of Jivaroan Gardens with that of the Tropical Forest', *Human Ecology*, 11: 47–68.

BRAY, WARWICK (1995). 'Searching for Environmental Stress: Climate and Anthropogenic Influences on the Landscape of Colombia', in Stahl (1995: 96–112).

——LEONOR HERRERA, MARIANNE CARDALE SCHRIMPFF, PEDRO BOTERO, and JOSÉ G. MONSALVE (1987). 'The Ancient Agricultural Landscape of Calima, Colombia', *British Archaeological Reports, International Series*, 359: 443–81.

BROADBENT, SYLVIA M. (1964). 'Agricultural Terraces in Chibcha Territory, Colombia', *American Antiquity*, 29: 501–4.

——(1968). 'A Prehistoric Field System in Chibcha Territory, Colombia', *Ñawpa Pacha*, 6: 135–47.

——(1987). 'The Chibcha Raised-Field System in the Sabana de Bogotá, Colombia: Further Investigations', *British Archaeological Reports, International Series*, 359: 425–42.

BROOKFIELD, HAROLD C. (1962). 'Local Study and Comparative Method: An Example from Central New Guinea', *Annals of the Association of American Geographers*, 52: 242–54.

——(1968). 'New Directions in the Study of Agricultural Systems in Tropical Areas', in Ellen T. Drake (ed.), *Evolution and Environment*, New Haven, CT: Yale University Press, 413–39.

BROOKS, SARAH O. (1998). 'Prehistoric Agricultural Terraces in the Río Japo Basin, Colca Valley, Peru', Ph. D. dissertation, University of Wisconsin–Madison.

BROWMAN, DAVID L. (ed.) (1987a). *Arid Land Use Strategies and Risk Management in the Andes: A Regional Anthropological Perspective*, Boulder, CO: Westview Press.

——(1987b). 'Risk Management in Andean Arid Lands', in Browman (1987a: 1–23).

BRÜCHER, HEINZ (1989). *Useful Plants of Neotropical Origin and their Wild Relatives*, Berlin: Springer.

BRUHNS, KAREN (1981). 'Prehispanic Ridged Fields of Central Colombia', *Journal of Field Archaeology*, 8: 1–8.

BRUSH, STEPHEN B. (1976). 'Man's Use of an Andean Ecosystem', *Human Ecology*, 4: 147–66.

——(1977). *Mountain, Field, and Family: The Economy and Human Ecology of an Andean Valley*, Philadelphia: University of Pennsylvania Press.

——(1980a). 'Potato Taxonomies in Andean Agriculture', in D. Brokensha, M. Warren, and O. Werner (eds.), *Indigenous Knowledge Systems and Development*, New York: University Press of America, 37–47.

——(1980b). 'The Environment and Native Andean Agriculture', *América Indígena*, 40: 161–72.

——(1992). 'Ethnoecology, Biodiversity, and Modernization in Andean Potato Agriculture', *Journal of Ethnobiology*, 12: 161–85.

BULLOCK, D. S. (1958). 'La agricultura de los Mapuches en tiempos pre-hispánicos', *Boletín Sociedad Biológico Concepción*, Chile, 33: 141–54.

BURGOA, FRANCISCO DE (1934) [1674]. *Geográfica descripción*, México: Publicaciones del Archivo General, 25–6.

BUSTOS, VICTOR (1976). *Investigaciones arqueológicas en Trinidad, Departmento del Beni*, La Paz: Instituto Nacional de Arqueología, Publicación 22.

BUYS, JOZEF E. and MICHAEL MUSE (1987). 'Arqueología de asentamientos asociados a los campos elevados de Peñón del Río, Guayas, Ecuador', *British Archaeological Reports, International Series*, 359: 225–48.

CAILLAVET, CHANTAL (1983). 'Toponimia histórica, arqueología y formas prehispánicas de agricultura en la region de Otavalo—Ecuador', *Boletín del Instituto Francés de Estudios Andinos*, Lima, 12: 1–21.

CAMINHA, PEDRO VAZ DE (1938). 'Letter of Pedro Vaz de Caminha to King Manuel Written from Porto Seguro of Vera Cruz the 1st of May 1500', in William B. Greenlee (trans. and ed.), *The Voyage of Pedro Alvares Cabral to Brazil and India*, London: Hakluyt Society, 2nd Series, 81: 3–33.

CANZIANI AMICO, JOSÉ (1995). 'Las lomas de Atiquipa: Arqueología y problemas de desarrollo regional', *Gaceta Arqueológica Andina*, 24: 113–33.

CÁRDENAS, MARTÍN (1989). *Manual de plantas económicas de Bolivia*, La Paz: Los Amigos del Libro.

CARDICH, AUGUSTO (1985). 'The Fluctuating Upper Limits of Cultivation in the Central Andes and Their Impact on Peruvian Prehistory', *Advances in World Archaeology*, New York: Academic Press, 4: 293–333.

CARNEIRO, ROBERT L. (1957). 'Subsistence and Social Structure: An Ecological Study of the Kuikuro Indians', Ph. D. dissertation, University of Michigan.

——(1960). 'Slash-and-Burn Agriculture: A Closer Look at Its Implications for Settlement Patterns', in Anthony F. C. Wallace (ed.), *Men and Cultures*, Philadelphia: University of Pennsylvania Press, 229–34.

——(1961). 'Slash-and-Burn Cultivation Among the Kuikuru and Its Implications for Cultural Development in the Amazon Basin', in Wilbert (1961: 47–67).

——(1964). 'Shifting Cultivation among the Amahuaca of Eastern Peru', in Hans Becher (ed.), *Beiträge zur Völkerkunde Südamerikas: Festgabe für Herbert Baldas zum 65. Geburtstag*, Völkerkundliche Abhandlungen, Hannover: Kommissionsverlag Münstermann–Druck GMBH, 1: 9–18.

——(1970a). 'The Transition from Hunting to Horticulture in the Amazon Basin', in *Proceedings, Eighth Congress of International Anthropological and Ethnological Sciences, Tokyo, 1968*, Tokyo: Science Council of Japan, 3: 244–8.

——(1970b). 'A Theory of the Origin of the State', *Science*, 169: 733–8.

——(1974). 'On the Use of the Stone Axe by the Amahuaca Indians of Eastern Peru', *Ethnologische Zeitschrift Zürich*, 1: 107–22.

——(1979a). 'Tree Felling with the Stone Axe: An Experiment Carried Out Among the Yanomamö Indians of Southern Venezuela', in Carol Kramer (ed.), *Ethnoarcheology: Implications of Ethnography for Archeology*, New York: Columbia University Press, 21–58.

——(1979b). 'Forest Clearance Among the Yanomamö, Observations and Implications', *Antropológica*, 52: 39–76.

——(1983). 'The Cultivation of Manioc among the Kuikuru of the Upper Xingú', in Hames and Vickers (1983a: 65–111).

——(1995). 'The History of Ecological Interpretations of Amazonia: Does Roosevelt Have it Right?', in Sponsel (1995: 45–70).

CARNEY, HEATH J., MICHAEL W. BINFORD, ALAN L. KOLATA, RUBEN R. MARIN, and CHARLES R. GOLDMAN (1993). 'Nutrient and Sediment Retention in Andean Raised-Field Agriculture', *Nature*, 364: 131–3.

CARVAJAL, GASPAR DE (1934*a*) [1542]. 'Carvajal's Account', in José Toribio Medina (compiler), H. C. Heaton (ed.), *The Discovery of the Amazon According to the Account of Friar Gaspar de Carvajal and Other Documents*, Special Publication 17, New York: American Geographical Society, 167–242.

——(1934*b*) [1542]. 'The Version of Carvajal's Account in Oviedo's "Historia"', in H. C. Heaton (ed.), *The Discovery of the Amazon According to the Account of Friar Gaspar de Carvajal and Other Documents*, Special Publication 17, New York: American Geographical Society, 405–48.

CASTELLANOS, JUAN DE (1955) [1589]. *Elegías de varones ilustres de Indias*, 4 vols, Bogotá: Editorial ABC.

CASTILLO, JOSEPH DE (1906) [*c.* 1676]. 'Relación de la provincia de Mojos', in M. V. Ballivián (ed.), *Documentos para la historia geográfica de la República de Bolivia, Serie primera: Epoca colonial 1: Las provincias de Mojos y Chiquitos*, La Paz: J. M. Gamarra, 294–395.

CAVELIER DE FERRERO, INÉS, SANTIAGO MORA CAMARGO, and LUISA FERNANDA HERRERA DE TURBAY (1990). 'Estabilidad y dinámica agrícola: Las transformaciones de una sociedad agrícola', in S. Mora Camargo (ed.), *Ingenierias prehispánicas*, Bogotá: Fondo FEN, 73–109.

CAVIEDES, CÉSAR and GREGORY KNAPP (1995). 'Strategies for Agricultural Subsistence', in C. Caviedes and G. Knapp, *South America*, Englewood Cliffs, NJ: Prentice Hall, 162–83.

CEHOPU (El Centro de Estudios Historicos de Obras Públicas y Urbanismo) (1993). *Obras hidráulicas en América colonial*, Madrid: Ministerio de Obras Públicas, Transportes, y Medio Ambiente.

CHAGNON, NAPOLEON A. (1992). *Yánomamö* (4th edn), Fort Worth, TX: Harcourt Brace Jovanovich.

——and RAYMOND B. HAMES (1979). 'Protein Deficiency and Tribal Warfare in Amazonia: New Data', *Science*, 203: 910–13.

————(1980). 'La "hipótesis proteica" y la adaptación indígena a la cuenca del Amazonas: Una revisión crítica de los datos y de la teoría', *Interciencia*, 5: 346–58.

CHAPIN, MAC (1988). 'The Seduction of Models: Chinampa Agriculture in Mexico', *Grassroots Development*, 12/1: 8–17.

CHERNELA, JANET M. (1989). 'Managing Rivers of Hunger: The Tukano of Brazil', *Advances in Economic Botany*, 7: 238–48.

——(1994). 'Tukanoan Know-How: The Importance of the Forested River Margin to Neotropical Fishing Populations', *Research and Exploration*, 10: 440–57.

CHEVALIER, FRANÇOIS (1966). 'L'expansion de la grande propriété dans le Haut-Pérou au XXᵉ siècle', *Annales, Economies, Sociétés, Civilisations*, 21: 815–31.

CHRISTIANSEN G., JORGE (1967). *El cultivo de la papa en el Perú*, Lima: Centro Internacional de la Papa.

CIEZA DE LEÓN, PEDRO DE (1959) [1553]. *The Incas of Pedro Cieza de León*, Harriet de Onis (trans.), Norman: University of Oklahoma Press.

CLARK, KATHLEEN and CHRISTOPHER UHL (1987). 'Farming, Fishing, and Fire in the History of the Upper Río Negro Region of Venezuela', *Human Ecology*, 15: 1–26.

CLARKE, WILLIAM C. (1977). 'The Structure of Permanence: The Relevance of Self-

Subsistence Communities for World Ecosystem Management', in T. P. Bayliss-Smith and R. G. Feachem (eds.), *Subsistence and Survival: Rural Ecology in the Pacific*, London: Academic Press, 363–84.

CLARKSON, PERSIS and RONALD I. DORN (1995). 'New Chronometric Dates for the Puquios of Nasca, Peru', *Latin American Antiquity*, 6: 56–69.

CLEMENT, CHARLES R. (1999). '1492 and the Loss of Amazonian Crop Genetic Resources. I. The Relation Between Domestication and Human Population Decline', *Economic Botany*, 53: 188–202.

COBO, BERNABÉ (1956) [1653]. *Obras del P. Bernabé Cobo*, Biblioteca de Autores Españoles, Madrid: Ediciones Atlas, 91, 92.

——(1979) [1653]. *History of the Inca Empire*, Roland Hamilton (trans. and ed.), Austin: University of Texas Press.

——(1990) [1653]. *Inca Religion and Customs*, Roland Hamilton (trans. and ed.), Austin: University of Texas Press.

COCK, JAMES H. (1982). 'Cassava: A Basic Energy Source in the Tropics', *Science*, 218: 755–62.

COE, MICHAEL D. (1964). 'The Chinampas of Mexico', *Scientific American*, 211/1: 90–8.

COE, WILLIAM R. (1957). 'Environmental Limitation on Maya Culture: A Reexamination', *American Anthropologist*, 59: 328–35.

COHEN, MARK NATHAN (1977). *The Food Crisis in Prehistory: Overpopulation and the Origins of Agriculture*, New Haven, CT: Yale University Press.

COLCHESTER, MARCUS (1984). 'Rethinking Stone Age Economics: Some Speculations Concerning the Pre-Columbian Yanoama Economy', *Human Ecology*, 12: 291–314.

CONKLIN, HAROLD C. (1980). *Ethnographic Atlas of Ifugao: A Study of Environment, Culture, and Society in Northern Luzon*, New Haven, CT: Yale University Press.

COOK, NOBLE DAVID (1982). *The People of the Colca Valley: A Population Study*, Dellplain Latin American Studies 9, Boulder, CO: Westview Press.

COOK, O. F. (1916). 'Staircase Farms of the Ancients', *National Geographic Magazine*, 29: 474–534.

——(1920). 'Footplough Agriculture in Peru', *Annual Report of the Smithsonian Institution for 1918*, Washington, DC, 487–91.

COOMES, OLIVER T. (1992a). 'Blackwater Rivers, Adaptation, and Environmental Heterogeneity in Amazonia', *American Anthropologist*, 94: 699–701.

——(1992b). 'Making a Living in the Amazon Rain Forest: Peasants, Land, and Economy in the Tahuayo River Basin of Northeastern Peru', Ph.D. dissertation, University of Wisconsin–Madison.

——(1998). 'Traditional Peasant Agriculture along a Blackwater River of the Peruvian Amazon', *Revista Geográfica*, 124: 33–55.

——and G. J. BURT (1997). 'Indigenous Market-Oriented Agroforestry: Dissecting Local Diversity in Western Amazonia', *Agroforestry Systems*, 37: 27–44.

COVICH, ALAN P. and NORTON H. NICKERSON (1966). 'Studies in Cultivated Plants in Choco Dwelling Clearings, Darien, Panama', *Economic Botany*, 20: 285–301.

COWGILL, URSULA M. (1961). 'Soil Fertility and the Ancient Maya', *Transactions of the Connecticut Academy of Arts and Sciences*, 42: 1–56.

COX, GEORGE W. and MICHAEL D. ATKINS (1979). *Agricultural Ecology: An Analysis of World Food Production Systems*, San Francisco: W. H. Freeman.

CRÉPEAU, ROBERT R. (1990). 'L'écologie culturelle américaine et les sociétés amazoniennes', *Recherches Amérindiennes au Québec*, 20: 89–104.

CROSBY, ALFRED W. (1972). *The Columbian Exchange: Biological and Cultural Consequences of 1492*, Westport, CT: Greenwood Press.

CRUZ, LAUREANO DE LA (1885) [1653]. 'Nuevo descubrimiento del Marañon, año 1651', in Francisco Maria Compte (ed.), *Varones ilustres de la Orden Seráfica en el Ecuador desde la fundación de Quito hasta nuestros días*, Quito: Imprenta del Clero, 1: 148–205.

DAGODAG, TIM and GARY KLEE (1973). 'A Review of Some Analogies in Sunken Garden Agriculture', *Anthropological Journal of Canada*, 11: 10–15.

D'ALTROY, TERENCE N. (2000). 'Andean Land Use at the Cusp of History', in Lentz (2000): 357–90.

DARCH, J. P. (ed.) (1983). 'Drained Field Agriculture in Central and South America', *British Archaeological Reports, International Series*, 189.

DAVIS, E. WADE and JAMES A. YOST (1983). 'The Ethnobotany of the Waorani of Eastern Ecuador', *Botanical Museum Leaflets*, 29/3: 159–217.

DAY, KENT C. (1974). 'Walk-In-Wells and Water Management at Chanchan, Peru', in C. C. Lamberg-Karlovsky and J. A. Sabloff (eds.), *The Rise and Fall of Civilizations*, Menlo Park, CA: Cummings Publishing Company, 182–90.

DeBOER, WARREN R. (1981a). 'The Machete and the Cross: Conibo Trade in the Late Seventeenth Century', in Peter D. Francis, F. J. Kense, and P. G. Duke (eds.), *Networks of the Past: Regional Interaction in Archaeology, Proceedings of the Twelfth Annual Conference*, The Archaeological Association of the University of Calgary, 31–47.

——(1981b). 'Buffer Zones in the Cultural Ecology of Aboriginal Amazonia: An Ethnohistorical Approach', *American Antiquity*, 46: 364–77.

—— KEITH KINTIGH, and ARTHUR G. ROSTOKER (1996). 'Ceramic Seriation and Site Reoccupation in Lowland South America', *Latin American Antiquity*, 7: 263–78.

DENEVAN, WILLIAM M. (1962). 'Informe preliminar sobre la geografía de los Llanos de Mojos, noreste de Bolivia', *Boletín de la Sociedad Geográfica e Histórica Sucre*, 47: 91–113.

——(1963a). 'Additional Comments on the Earthworks of Mojos in Northeastern Bolivia', *American Antiquity*, 28: 540–5.

——(1963b). 'The Aboriginal Settlement of the Llanos de Mojos: A Seasonally Inundated Savanna in Northeastern Bolivia', Ph. D. dissertation, University of California, Berkeley.

——(1966a). *The Aboriginal Cultural Geography of the Llanos de Mojos of Bolivia*, Ibero-Americana 48, Berkeley: University of California Press.

——(1966b). 'A Cultural-Ecological View of the Former Aboriginal Settlement in the Amazon Basin', *The Professional Geographer*, 18: 346–51.

——(1970a). 'Aboriginal Drained-Field Cultivation in the Americas', *Science*, 169: 647–54.

——(1970b). 'The Aboriginal Population of Western Amazonia in Relation to Habitat and Subsistence', *Revista Geográfica*, Rio de Janeiro, 72: 61–86.

——(1971a). 'Campa Subsistence in the Gran Pajonal, Eastern Peru', *The Geographical Review*, 61: 496–518.

——(1971b). 'Prehistoric Cultural Change and Ecology in Latin America', *Proceedings of the Conference of Latin Americanist Geographers*, 1: 138–51.

——(1976). 'The Aboriginal Population of Amazonia', in W. M. Denevan (ed.), *The Native Population of the Americas in 1492*, Madison: University of Wisconsin Press, 205–34.

DENEVAN, WILLIAM M. (cont.) (ed.) (1980*a*). 'La agricultura intensiva prehispánica', *América Indígena*, 40: 613–815.

——(1980*b*). 'Latin America', in Gary A. Klee (ed.), *World Systems of Traditional Resource Management*, London: Edward Arnold, 217–44.

——(1980*c*). 'Tipología de configuraciones agrícolas prehispánicas', *América Indígena*, 40: 619–52.

——(1982). 'Hydraulic Agriculture in the American Tropics: Forms, Measures, and Recent Research', in Kent V. Flannery (ed.), *Maya Subsistence*, New York: Academic Press, 181–203.

——(1983). 'Adaptation, Variation, and Cultural Geography', *Professional Geographer*, 35: 399–407.

——(1984*a*). 'Ecological Heterogeneity and Horizontal Zonation of Agriculture in the Amazon Floodplain', in Marianne Schmink and Charles H. Wood (eds.), *Frontier Expansion in Amazonia*, Gainesville: University of Florida Press, 311–36.

——(1984*b*). Review of *Adaptive Responses of Native Amazonians*, Raymond B. Hames and William T. Vickers (eds.), *Economic Geography*, 60: 91–3.

——(1986). 'Abandono de terrazas en el Perú andino: Extensión, causas, y propuestas de restauración', in Carlos de la Torre and Manuel Burga (eds.), *Andenes y camellones en el Perú andino: Historia presente y futuro*, Lima: Consejo Nacional de Ciencia y Tecnología, Lima, 255–8.

——(ed.) (1986–8). *The Cultural Ecology, Archaeology, and History of Terracing and Terrace Abandonment in the Colca Valley of Southern Peru*, 2 vols, Technical Report to the National Science Foundation and the National Geographic Society, Department of Geography, University of Wisconsin–Madison.

——(1987*a*). 'Terrace Abandonment in the Colca Valley, Peru', *British Archaeological Reports, International Series*, 359: 1–43.

——(1987*b*). Review of *Human Carrying Capacity of the Brazilian Rainforest*, by Philip M. Fearnside, *The Geographical Review*, 77: 479–81.

——(1988). 'Measurement of Abandoned Terracing from Air Photos, Colca Valley, Peru', *Yearbook, Conference of Latin Americanist Geographers*, 14: 20–30.

——(1991). 'Prehistoric Roads and Causeways of Lowland Tropical America', in Charles D. Trombold (ed.), *Ancient Road Networks and Settlement Hierarchies in the New World*, Cambridge: Cambridge University Press, 230–42.

——(1992*a*). 'Native American Populations in 1492: Recent Research and a Revised Hemispheric Estimate', in W. M. Denevan (ed.), *The Native Population of the Americas in 1492* (2nd edn), Madison: University of Wisconsin Press, xvii–xxxviii.

——(1992*b*). 'The Pristine Myth: The Landscape of the Americas in 1492', *Annals of the Association of American Geographers*, 82: 369–85.

——(1992*c*). 'Stone vs Metal Axes: The Ambiguity of Shifting Cultivation in Prehistoric Amazonia', *Journal of the Steward Anthropological Society*, 20: 153–65.

——(1993). 'The 1931 Shippee–Johnson Aerial Photography Expedition to Peru', *The Geographical Review*, 83: 238–51.

——(1995). 'Prehistoric Agricultural Methods as Models for Sustainability', *Advances in Plant Pathology*, London: Academic Press, 11: 21–43.

——(1996). 'A Bluff Model of Riverine Settlement in Prehistoric Amazonia', *Annals of the Association of American Geographers*, 86: 654–81.

DENEVAN, WILLIAM M. (cont.) (1998). 'Comments on Prehistoric Agriculture in Amazonia', *Culture and Agriculture*, 20: 54–9.

—— (n.d.). 'The Native Population of Amazonia in 1492 Reconsidered', unpublished manuscript.

—— and ROLAND W. BERGMAN (1975). 'Karinya Indian Swamp Cultivation in the Venezuelan Llanos', *Yearbook of the Association of Pacific Coast Geographers*, 37: 23–37.

—— and KENT MATHEWSON (1983). 'Preliminary Results of the Samborondón Raised-Field Project, Guayas Basin, Ecuador', *British Archaeological Reports, International Series*, 189: 167–81.

—— and CHRISTINE PADOCH (eds.) (1988). 'Swidden-Fallow Agroforestry in the Peruvian Amazon', *Advances in Economic Botany*, 5.

—— and KARL H. SCHWERIN (1978). 'Adaptive Strategies in Karinya Subsistence, Venezuelan Llanos', *Antropológica*, 50: 3–91.

—— and JOHN M. TREACY (1988). 'Young Managed Fallows at Brillo Nuevo', *Advances in Economic Botany*, 5: 8–46.

—— and B. L. TURNER II (1974). 'Forms, Functions, and Associations of Raised Fields in the Old World Tropics', *Journal of Tropical Geography*, 39: 24–33.

——, J. M. TREACY, J. B. ALCORN, C. PADOCH, J. DENSLOW, and S. FLORES PAITÁN (1984). 'Indigenous Agroforestry in the Peruvian Amazon: Bora Indian Management of Swidden Fallows', *Interciencia*, 9: 346–57.

—— and ALBERTA ZUCCHI (1978). 'Ridged-Field Excavations in the Central Orinoco Llanos, Venezuela', in David L. Browman (ed.), *Advances in Andean Archaeology*, The Hague: Mouton, 235–45.

——, KENT MATHEWSON, and GREGORY KNAPP (eds.) (1987). 'Pre-Hispanic Agricultural Fields in the Andean Region', *British Archaeological Reports, International Series*, 359.

DENSLOW, JULIE SLOAN and GARY S. HARTSHORN (1994). 'Tree-Fall Gap Environments and Forest Dynamic Processes', in Lucinda A. McDade *et al.* (eds.), *La Selva: Ecology and Natural History of a Neotropical Rain Forest*, Chicago: University of Chicago Press, 120–7.

DESCOLA, PHILIPPE (1994). *In the Society of Nature: A Native Ecology in Amazonia*, Cambridge: Cambridge University Press.

DEUEL, LEO (1969). *Flights into Yesterday: The Story of Aerial Archaeology*, New York: St. Martin's Press.

DICK, R. P., J. A. SANDOR, and N. S. EASH (1994). 'Soil Enzyme Activities after 1500 Years of Terrace Agriculture in the Colca Valley, Peru', *Agriculture, Ecosystems and Environment*, 50: 123–31.

DIENER, PAUL, KURT MOORE, and ROBERT MUTAW (1980). 'Meat, Markets, and Mechanical Materialism: The Great Protein Fiasco in Anthropology', *Dialectial Anthropology*, 5: 171–92.

DIEZ DE SAN MIGUEL, GARCI (1964). *Visita hecha a los indios de la provincia de Chucuito, 1567*, Lima: Casa de la Cultura del Perú.

DILLEHAY, TOM D., JACK ROSSEN, and PATRICIA J. NETHERLY (1997). 'The Nanchoc Tradition: The Beginnings of Andean Civilization', *American Scientist*, 85: 46–56.

DOBYNS, HENRY F. (1966). 'Estimating Aboriginal American Population: An Appraisal of Techniques with a New Hemispheric Estimate', *Current Anthropology*, 7: 395–449.

—— and PAUL L. DOUGHTY (1976). *Peru*, New York: Oxford University Press.

DONKIN, R. A. (1970). 'Pre-Columbian Field Implements and their Distribution in the Highlands of Middle and South America', *Anthropos*, 65: 505–29.

——(1979). *Agricultural Terracing in the Aboriginal New World*, Viking Fund Publications in Anthropology 56, Tucson: University of Arizona Press.

DOOLITTLE, WILLIAM E. (1984). 'Cabeza de Vaca's Land of Maize: An Assessment of Its Agriculture', *Journal of Historical Geography*, 10: 246–62.

——(1990a). *Canal Irrigation in Prehistoric Mexico: The Sequence of Technological Change*, Austin: University of Texas Press.

——(1990b). 'Terrace Origins: Hypotheses and Research Strategies', *Yearbook, Conference of Latin Americanist Geographers*, 16: 94–7.

——(1992). 'Agriculture in North America on the Eve of Contact: A Reassessment', *Annals of the Association of American Geographers*, 82: 386–401.

——(2000). *Cultivated Landscapes of Native North America*, Oxford: Oxford University Press.

DOUGHERTY, BERNARDO and HORACIO CALANDRA (1981). 'Nota preliminar sobre investigaciones arqueológicas en los Llanos de Moxos, Departamento del Beni, República de Bolivia', *Revista del Museo de la Plata, Sección Antropología*, 53: 87–106.

————(1984). 'Prehispanic Human Settlement in the Llanos de Moxos, Bolivia', in Jorge Rabassa (ed.), *Quaternary of South America and Antarctic Peninsula*, Rotterdam: A. A. Balkema, 2: 163–99.

DOUROJEANNI, MARC J. (1983). 'Desarrollo y conservación en la Sierra del Perú', *Boletín de Lima*, 28: 63–72.

DUKE, JAMES A. and RODOLFO VÁSQUEZ (1994). *Amazonian Ethnobotanical Dictionary*, Boca Raton: CRC Press.

DUMOND, D. E. (1961). 'Swidden Agriculture and the Rise of Maya Civilization', *Southwestern Journal of Anthropology*, 17: 301–16.

DUMONT, JEAN FRANÇOIS, SANDRINE LAMOTTE, and FRANCIS KAHN (1990). 'Wetland and Upland Forest Ecosystems in Peruvian Amazonia: Plant Species Diversity in the Light of Some Geological and Botanical Evidence', *Forest Ecology and Management*, 33–4: 125–39.

EARLS, JOHN (1989). *Planificación agrícola andina*, Lima: COFIDE.

—— and IRENE SILVERBLATT (1981). 'Sobre la instrumentación de la cosmología inca en el sitio arqueológico de Moray', in Heather Lechtman and Ana María Soldi (eds.), *La tecnología en el mundo andino*, México: Universidad Nacional Autónoma de México, 443–73.

EASH, NEAL S. (1989). 'Natural and Ancient Agricultural Soils in the Colca Valley, Peru', Masters thesis, Iowa State University.

—— and J. A. SANDOR (1995). 'Soil Chronosequence and Geomorphology in a Semi-Arid Valley in the Andes of Southern Peru', *Geoderma*, 65: 59–79.

EDEN, MICHAEL J. (1980). 'A Traditional Agro-System in the Amazon Region of Colombia', in J. I. Furtado (ed.), *Tropical Ecology and Development*, Kuala Lumpur: International Society of Tropical Ecology, 1: 509–14.

——(1990). *Ecology and Land Management in Amazonia*, London: Belhaven Press.

——, WARWICK BRAY, LEONOR HERRERA, and COLIN MCEWAN (1984). 'Terra Preta Soils and their Archaeological Context in the Caquetá Basin of Southeast Colombia', *American Antiquity*, 49: 125–40.

EDER, FRANCISCO JAVIER (1985) [1791]. *Breve descripción de las reducciones de Mojos*, Josep M. Barnadas (trans.), Cochabamba: Historia Boliviana.

EIDT, ROBERT C. (1959). 'Aboriginal Chibcha Settlement in Colombia', *Annals of the Association of American Geographers*, 49: 374–92.

——(1977). 'Detection and Examination of Anthrosols by Phosphate Analysis', *Science*, 197: 1327–33.

——(1981). 'Rural Society and Land Use Change in the Highland Basins of Colombia', *Latin American Studies*, Ibaraki, Japan, 3: 25–45.

——(1984). *Advances in Abandoned Settlement Analysis: Application to Prehistoric Anthrosols in Colombia, South America*, Center for Latin America, University of Wisconsin–Milwaukee.

ELING, HERBERT H. (1986). 'Prehispanic Irrigation Sources and Systems in the Jequetepeque Valley, Northern Peru', in Ramiro Matos Mendieta, Solveig A. Turpin, and H. H. Eling (eds.), *Andean Archaeology*, Institute of Archaeology Monograph 27, University of California, Los Angeles, 130–49.

——(1987a). 'The Role of Irrigation Networks in Emerging Societal Complexity During Late Prehispanic Times: Jequetepeque Valley, North Coast, Peru', Ph. D. dissertation, University of Texas at Austin.

——(1987b). 'The Rustic Boca Toma: Traditional Hydraulic Technology for the Future', in Browman (1987a: 171–93).

ENGEL, FRÉDÉRIC A. (1973). 'New Facts About Pre-Columbian Life in the Andean Lomas', *Current Anthropology*, 14: 271–80.

——(1976). *An Ancient World Preserved: Relics and Records of Prehistory in the Andes*, New York: Crown Publishers.

ERICKSON, CLARK L. (1980). 'Sistemas agrícolas prehispánicos en los Llanos de Mojos', *América Indígena*, 40: 731–55.

——(1985). 'Applications of Prehistoric Andean Technology: Experiments in Raised Field Agriculture, Huatta, Lake Titicaca, 1981–82', *British Archaeological Reports, International Series*, 232: 209–32.

——(1987). 'The Dating of Raised-Field Agriculture in the Lake Titicaca Basin, Peru', *British Archaeological Reports, International Series*, 359: 373–84.

——(1988a). 'An Archaeological Investigation of Raised Field Agriculture in the Lake Titicaca Basin of Peru', Ph.D. dissertation, University of Illinois.

——(1988b). 'Raised Field Agriculture in the Lake Titicaca Basin: Putting Ancient Agriculture Back to Work', *Expedition*, 30/3: 8–16.

——(1992). 'Prehistoric Landscape Management in the Andean Highlands: Raised Field Agriculture and its Environmental Impact', *Population and Environment*, 13: 285–300.

——(1993). 'The Social Organization of Prehispanic Raised Field Agriculture in the Lake Titicaca Basin', in Vernon L. Scarborough and Barry L. Isaac (eds.), *Economic Aspects of Water Management in the Prehispanic New World*, Research in Economic Anthropology, Supplement 7, Greenwich, CT: JAI Press, 369–426.

——(1994a). 'Raised Fields as a Sustainable Agricultural System from Amazonia', unpublished paper presented at the Latin American Studies Association Meeting, Atlanta, Georgia.

——(1994b). 'Methodological Considerations in the Study of Ancient Andean Field Systems', in Miller and Gleason (1994a: 111–52).

——(1995). 'Archaeological Methods for the Study of Ancient Landscapes of the Llanos de Mojos in the Bolivian Amazon', in Stahl (1995: 66–95).

ERICKSON, ELARK L. (cont.) (1996). *Investigación arqueológica del sistema agrícola de los camellones en la cuenca del Lago Titicaca del Perú*, La Paz: PIWA.

——(1998). 'Applied Archaeology and Rural Development: Archaeology's Potential Contribution to the Future', in Michael B. Whiteford and Scott Whiteford (eds.), *Crossing Currents: Continuity and Change in Latin America*, Upper Saddle River, NJ: Prentice Hall, 34–45.

——(1999). 'Neo-Environmental Determinism and Agrarian "Collapse" in Andean Prehistory', *Antiquity*, 73: 634–42.

——(2000). 'The Lake Titicaca Basin: A Precolumbian Built Landscape', in Lentz (2000): 311–56.

——(n.d.). *Waru Waru: Ancient Andean Agriculture*, unpublished book manuscript.

——and DANIEL A. BRINKMEIER (1991). *Raised Field Rehabilitation Projects in the Northern Lake Titicaca Basin*, Report to the Interamerican Foundation, Department of Anthropology, Philadelphia: University of Pennsylvania.

——and KAY L. CANDLER (1989). 'Raised Fields and Sustainable Agriculture in the Lake Titicaca Basin of Peru', in John O. Browder (ed.), *Fragile Lands of Latin America: Strategies for Sustainable Development*, Boulder, CO: Westview Press, 230–48.

ESTRADA, EMILIO (1957). *Ultimas civilizaciones pre-históricas de la Cuenca del Río Guayas*, Guayaquil: Museo Víctor Emilio Estrada 2.

EVENARI, MICHAEL, LESLIE SHANAN, and NAPHTALI TADMOR (1971). *The Negev: The Challenge of a Desert*, Cambridge: Harvard University Press.

EVANS, TOM P., and BRUCE WINTERHALDER (2000). 'Modified Solar Insolation as an Agronomic Factor in Terraced Environments', *Land Degradation and Development*, 11: 273–87.

FAGAN, BRIAN (1995). 'Footplows and Raised Fields', in Brian Fagan, *Time Detectives: How Archaeologists Use Technology to Recapture the Past*, New York: Simon and Schuster, 210–26.

——(1999). *Floods, Famines, and Emperors: El Niño and the Fate of Civilizations*, New York: Basic Books.

FALESI, ITALO CLAUDIO (1974). 'Soils of the Brazilian Amazon', in Charles Wagley (ed.), *Man in the Amazon*, Gainesville: University Presses of Florida, 201–29.

FARRINGTON, IAN S. (1980a). 'The Archaeology of Irrigation Canals, with Special Reference to Peru', *World Archaeology*, 11: 287–305.

——(1980b). *Prehistoric Agricultural Facilities in the Cusichaca Valley: A Preliminary Study*, Cusichaca Project Land Use and Irrigation Report 1, Canberra: Australian National University.

——(1983a). 'The Design and Function of the Intervalley Canal: Comments on a Paper by Ortloff, Moseley, and Feldman', *American Antiquity*, 48: 360–75.

——(1983b). 'Prehistoric Intensive Agriculture: Preliminary Notes on River Canalization in the Sacred Valley of the Incas', *British Archaeological Reports, International Series*, 189: 221–35.

——(ed.) (1985). 'Prehistoric Intensive Agriculture in the Tropics', *British Archaeological Reports, International Series* 232.

——and C. C. PARK (1978). 'Hydraulic Engineering and Irrigation Agriculture in the Moche Valley, Peru: *c.* AD 1250–1532', *Journal of Archaeological Science*, 5: 255–68.

FEDICK, SCOTT L. (1995). 'Indigenous Agriculture in the Americas', *Journal of Archaeological Research*, 3: 257–303.

FEJOS, PAUL (1944). *Archeological Explorations in the Cordillera Vilcabamba, Southeastern Peru*, New York: Viking Fund Publications in Anthropology, 3.

FEMENIAS, BLENDA (1997). 'Ambiguous Emblems: Gender, Clothing, and Representation in Contemporary Peru', Ph.D. dissertation, University of Wisconsin–Madison.

FERDON, EDWIN N. (1959). 'Agricultural Potential and the Development of Cultures', *Southwestern Journal of Anthropology*, 15: 1–19.

FERGUSON, R. BRIAN (1995). *Yanomami Warfare: A Political History*, Santa Fe: School of American Research Press.

——(1998). 'Whatever Happened to the Stone Age? Steel Tools and Yanomami Historical Ecology', in Balée (1998: 287–312).

FIELD, CHRIS (1966). 'A Reconnaissance of Southern Andean Agricultural Terracing', Ph.D. dissertation, University of California, Los Angeles.

FLORES OCHOA, JORGE A. (1987). 'Cultivation in the Qocha of the South Andean Puna', in Browman (1987*a*: 271–96).

FRANK, ERWIN (1987). 'Delimitaciones al aumento poblacional y desarrollo cultural en las culturas indígenas de la Amazonía antes de 1492', *Tübinger Geographische Studien*, 95: 109–23.

FRECHIONE, JOHN (1982). 'Manioc Monozoning in Yekuana Agriculture', *Antropológica*, 58: 53–74.

FREZIER, A. (1717). *A Voyage to the South Sea along the Coast of Chile and Peru, in the Years 1712, 1713, and 1714*, London: J. Bowyer.

FRITZ, SAMUEL (1922) [1723]. *Journal of the Travels and Labours of Father Samuel Fritz in the River of the Amazons between 1686 and 1723*, George Edmundson (ed.), Hakluyt Society, 2nd Series 51, London: Cambridge University Press.

FUCHS, HELMUTH (1964). 'La agricultura en la comunidad indígena de Santa Clara de Aribí (Cariña), Estado Anzoátegui, Venezuela', *VIᵉ Congrès International des Sciences Anthropologiques et Ethnologiques*, Paris, 2: 27–32.

GADE, DANIEL W. (1970). 'Ecología del robo agrícola en las tierras altas de los Andes centrales', *América Indígena*, 30: 3–14.

——(1975). *Plants, Man and the Land in the Vilcanota Valley of Peru*, The Hague: W. Junk.

——(1992). 'Landscape, System, and Identity in the Post-Conquest Andes', *Annals of the Association of American Geographers*, 82: 460–77.

——(1999). *Nature and Culture in the Andes*, Madison: University of Wisconsin Press.

——and ROBERTO RIOS (1972). 'Chaquitaclla: The Native Footplough and Its Persistence in Central Andean Agriculture', *Tools and Tillage*, 2: 3–15.

GALLAGHER, JAMES P. (1989). 'Agricultural Intensification and Ridged-Field Cultivation in the Prehistoric Upper Midwest of North America', in David R. Harris and Gordon C. Hillman (eds.), *Foraging and Farming: The Evolution of Plant Exploitation*, London: Unwin Hyman, 572–84.

——(1992). 'Prehistoric Field Systems in the Upper Midwest', in William I. Woods (ed.), *Late Prehistoric Agriculture: Observations from the Midwest*, Studies in Illinois Archaeology 8, Springfield: Illinois Historic Preservation Agency, 95–135.

GARAYCOCHEA Z., IGNACIO (1987). 'Agricultural Experiments in Raised Fields in the Lake Titicaca Basin, Peru: Preliminary Considerations', *British Archaeological Reports, International Series*, 359: 385–98.

GARCILASO DE LA VEGA (1966) [1609, 1616–1617]. *Royal Commentaries of The Incas and General History of Peru*, 2 vols, Austin: University of Texas Press.

GARSON, ADAM G. (1980). 'Prehistory, Settlement and Food Production in the Savanna Region of La Calzada de Paez, Venezuela', Ph.D. dissertation, Yale University.

GARTNER, WILLIAM G. (1992). 'Soils and Sediments as Artifacts in Prehistoric Wisconsin', *Proceedings of the First International Conference on Pedo-Archaeology*, Knoxville: University of Tennessee Agricultural Experiment Station, 113–16.

——(1996). Review of *The Tiwanaku: Portrait of an Andean Civilization*, by Alan L. Kolata, *Annals of the Association of American Geographers*, 86: 153–6.

GEERTZ, CLIFFORD (1963). *Agricultural Involution: The Process of Ecological Change in Indonesia*, Berkeley: University of California Press.

GELLES, PAUL H. (1990). 'Channels of Power, Fields of Contention: The Politics and Ideology of Irrigation in an Andean Peasant Community', Ph.D. dissertation, Harvard University.

——(1994). 'Channels of Power, Fields of Contention: The Politics of Irrigation and Land Recovery in an Andean Peasant Community', in Mitchell and Guillet (1994: 233–73).

GIBBONS, ANN (1990). 'New View of Early Amazonia', *Science*, 248: 1488–90.

GILIJ, FELIPE SALVADOR (1965) [1782]. *Ensayo de historia americana*, 3 vols, Caracas: Biblioteca de la Academia Nacional de Historia.

GLASER, BRUNO, GEORG GUGGENBERGER, and WOLFGANG ZECH (n.d.). 'Black Carbon in Sustainable Soils of the Brazilian Amazon Region', *Proceedings of the Ninth Conference of the International Humic Substances Society*, Adelaide, 1998, forthcoming.

GLASSOW, MICHAEL A. (1980). *Prehistoric Agricultural Development in the Northern Southwest: A Study in Changing Patterns of Land Use*, Socorro, NM: Ballena Press Anthropological Papers 16.

GLEASON, KATHRYN L. (1994). 'To Bound and to Cultivate: An Introduction to the Archaeology of Garden and Field', in Miller and Gleason (1994*a*: 1–24).

GLIESSMAN, STEVEN R. (1982). 'Nitrogen Distribution in Several Traditional Agro-Ecosystems in the Humid Tropical Lowlands of South-Eastern Mexico', *Plant and Soil*, 67: 105–17.

GODOY, RICARDO A. (1984). 'Ecological Degradation and Agricultural Intensification in the Andean Highlands', *Human Ecology*, 12: 359–83.

——(1988). 'Survey of Andean Common Field Agriculture', *Culture and Agriculture*, 36: 16–18.

——(1991). 'The Evolution of Common-Field Agriculture in the Andes: A Hypothesis', *Comparative Studies in Society and History*, 33: 395–414.

GOLAND, CAROL (1993). 'Field Scattering as Agricultural Risk Management: A Case Study from Cuyo Cuyo, Department of Puno, Peru', *Mountain Research and Development*, 13: 317–38.

GOLOB, ANN (1982). 'The Upper Amazon in Historical Perspective', Ph.D. dissertation, City University of New York.

GOLOMB, BERL and HERBERT M. EDER (1964). 'Landforms Made by Man', *Landscape*, 14/1: 4–7.

GOLSON, JACK (1989). 'The Origins and Development of New Guinea Agriculture', in David R. Harris and G. C. Hillman (eds.), *Foraging and Farming: The Evolution of Plant Domestication*, London: Unwin Hyman, 678–87.

GÓMEZ-POMPA, ARTURO, HECTOR LUIS MORALES, EPIFANIO JIMENÉZ ÁVILA, and JULIO JIMÉNEZ ÁVILA (1982). 'Experiences in Traditional Hydraulic Agriculture', in Kent V.

Flannery (ed.), *Maya Subsistence: Studies in Memory of Dennis E. Puleston*, New York: Academic Press, 327–42.

——, J. SALVADOR FLORES, and V. SOSA (1987). 'The Pet Kot: A Man-Made Forest of the Maya', *Interciencia*, 12: 10–15.

GONDARD, PIERRE and FREDDY LÓPEZ (1983). *Inventario arqueológico preliminar de los Andes septentrionales del Ecuador*, Quito: MAG-PRONAREG-ORSTOM.

GONZÁLEZ, ALBERTO REX and VICTOR NÚÑEZ REGUEIRO (1962). 'Preliminary Report on Archaeological Research in Tafí del Valle, N. W. Argentina', *Akten des 34. Internationalen Amerikanistenkongresses*, 485–96.

GONZÁLEZ, FERNANDO LUIS (1995). 'Fences, Fields and Fodder: Enclosures in Lari, Valle del Colca, Southern Peru', Master's thesis, University of Wisconsin–Madison.

GOOD, KENNETH R. (1987). 'Limiting Factors in Amazonian Ecology', in Marvin Harris and Eric B. Ross (eds.), *Food and Evolution: Toward a Theory of Human Food Habits*, Philadelphia: Temple University Press, 407–21.

——(1995). 'Yanomami of Venezuela: Foragers or Farmers—Which Came First?', in Sponsel (1995: 113–20).

GORDON, BURTON L. (1982). *A Panama Forest and Shore: Natural History and Amerindian Culture in Bocas del Toro*, Pacific Grove, CA: The Boxwood Press.

GORECKI, PAWEL P. (1982). 'Ethnoarchaeology at Kuk', Ph.D. dissertation, University of Sydney.

GOUDIE, ANDREW S. (1993). 'Land Transformation', in Ronald J. Johnston (ed.), *The Challenge for Geography*, Cambridge, MA: Blackwell, 116–37.

GRAFFAM, GRAY C. (1989). 'Back Across the Great Divide: The Pakaq Señorío and Raised Field Agriculture', in Virginia J. Vitzthum (ed.), *Multidisciplinary Studies in Andean Anthropology*, Discussions in Anthropology 8, Ann Arbor: University of Michigan, 33–50.

——(1990). 'Raised Fields without Bureaucracy: An Archaeological Examination of Intensive Wetland Cultivation in the Pampa Koani Zone, Lake Titicaca, Bolivia', Ph.D. dissertation, University of Toronto.

——(1992). 'Beyond State Collapse: Rural History, Raised Fields, and Pastoralism in the South Andes', *American Anthropologist*, 94: 882–904.

Gray Card Index [of New World Plants] (updated 1995). CAMBRIDGE, MA: Gray Herbarium, Harvard University.

GRENAND, PIERRE (1981). 'Agriculture sur brûlis et changements culturels: Le cas des Indiens Wayãpi et Palikur de Guyane', *Journal d'Agriculture Traditionnelle et de Botanique Appliquée*, 28: 23–31.

GRIEDER, TERENCE, A. BUENO MENDOZA, C. EARLE SMITH, Jr., and R. M. MALINA (1988). *La Galgada, Peru: A Preceramic Culture in Transition*, Austin: University of Texas Press.

GRIGG, DAVID B. (1979). 'Ester Boserup's Theory of Agrarian Change: A Critical Review', *Progress in Human Geography*, 3: 64–84.

GROSS, DANIEL R. (1975). 'Protein Capture and Cultural Development in the Amazon Basin', *American Anthropologist*, 77: 526–49.

——(1983). 'Village Movement in Relation to Resources in Amazonia', in Hames and Vickers (1983a: 429–49).

GUAMÁN POMA DE AYALA, FELIPE (1980) [c. 1615]. *El primer nueva corónica y buen gobierno*, John V. Murra, Rolena Adorno, and Jorge L. Urioste (eds.), 3 vols, México D.F.: Siglo Veintiuno Editores.

GUILLET, DAVID W. (1987*a*). 'Contemporary Agricultural Terracing in Lari, Colca Valley, Peru: Implications for Theories of Terrace Abandonment and Programs of Terrace Restoration', *British Archaeological Reports, International Series*, 359: 193–206.

——(1987*b*). 'Terracing and Irrigation in the Peruvian Highlands', *Current Anthropology*, 28: 409–30.

——(1987*c*). 'Agricultural Intensification and Deintensification in Lari, Colca Valley, Southern Peru', *Research in Economic Anthropology*, Greenwich, CT: JAI Press, 8: 201–24.

——(1987*d*). 'On the Potential for Intensification of Agropastoralism in the Arid Zones of the Central Andes', in Browman (1987*a*: 81–98).

——(1992). *Covering Ground: Communal Water Management and the State in the Peruvian Highlands*, Ann Arbor: University of Michigan Press.

——and WILLIAM P. MITCHELL (1994). 'Introduction: High Altitude Irrigation', in Mitchell and Guillet (1994: 1–20).

GUMILLA, JOSÉ (1963) [1745]. *El Orinoco ilustrado y defendido*, Caracas: Biblioteca de la Academia Nacional de la Historia 68.

HAMES, RAYMOND B. (1983*a*). 'Monoculture, Polyculture, and Polyvariety in Tropical Forest Swidden Cultivation', *Human Ecology*, 11: 13–34.

——(1983*b*). 'The Settlement Pattern of a Yanomamö Population Bloc: A Behavioral Ecological Interpretation', in Hames and Vickers (1983*a*: 393–427).

——(1989). 'Time, Efficiency, and Fitness in the Amazonian Protein Quest', *Research in Economic Anthropology*, Greenwich, CT: JAI Press, 11: 43–85.

——and WILLIAM T. VICKERS (eds.) (1983*a*). *Adaptive Responses of Native Amazonians*, New York: Academic Press.

————(1983*b*). 'Introduction', in Hames and Vickers (1983*a*: 1–26).

HARNER, MICHAEL J. (1972). *The Jívaro: People of the Sacred Waterfalls*, New York: Doubleday/Natural History Press.

HARRIS, DAVID R. (1971). 'The Ecology of Swidden Cultivation in the Upper Orinoco Rain Forest, Venezuela', *Geographical Review*, 61: 475–95.

HARRIS, MARVIN and ERIC ROSS (1987). 'Population Regulation and Agricultural Modes of Production', in M. Harris and E. Ross (eds.), *Death, Sex, and Fertility*, New York: Columbia University Press, 37–71.

HART, ROBERT D. (1980). 'A Natural Ecosystem Analog Approach to the Design of a Successional Crop System for Tropical Forest Environments', *Biotropica*, Tropical Succession Supplement, 12: 73–82.

HASTORF, CHRISTINE A. (1993). *Agriculture and the Onset of Political Inequality before the Inca*, Cambridge: Cambridge University Press.

HATCH, JOHN K. (1976). *The Corn Farmers of Motupe: A Study of Traditional Farming Practices in Northern Coastal Peru*, Land Tenure Center Monographs 1, University of Wisconsin–Madison.

HEADLAND, THOMAS N. and ROBERT C. BAILEY (1991). 'Introduction: Have Hunter-Gatherers Ever Lived in Tropical Rain Forest Independently of Agriculture?', *Human Ecology*, 19: 115–22.

HECHT, SUSANNA B. (1982). 'Agroforestry in the Amazon Basin: Practice, Theory and Limits of a Promising Land Use', in S. B. Hecht (ed.), *Amazonia: Agriculture and Land Use Research*, Cali, Colombia: CIAT, 331–71.

——and DARRELL A. POSEY (1989). 'Preliminary Results on Soil Management Techniques of the Kayapó Indians', *Advances in Economic Botany*, 7: 174–88.

HECKENBERGER, MICHAEL J. (1996). 'War and Peace in the Shadow of Empire: Socio-political Change in the Upper Xingu of Southeastern Amazonia, AD 1400–2000', Ph.D. dissertation, University of Pittsburgh.

——(1998). 'Manioc Agriculture and Sedentism in Amazonia: The Upper Xingu Example', *Antiquity*, 72: 633–48.

——, JAMES B. PETERSEN, and EDUARDO GOÉS NEVES (1999). 'Village Size and Perma-nence in Amazonia: Two Archaeological Examples from Brazil', *Latin American Antiquity*, 10: 353–76.

HEISER JR., CHARLES B. (1979). *The Gourd Book*, Norman: University of Oklahoma Press.

HEMMING, JOHN (1978). *Red Gold: The Conquest of the Brazilian Indians, 1500–1760*, Cambridge: Harvard University Press.

HENLEY, PAUL, MARIE-CLAUDE MATTEI-MÜLLER, and HOWARD REID (1994–6). 'Cul-tural and Linguistic Affinities of the Foraging People of Northern Amazonia: A New Perspective', *Antropológica*, 83: 3–37.

HERIARTE, MAURICIO DE (1952) [1692]. 'The Province of the Tapajós', John H. Rowe (trans. and ed.), *Kroeber Anthropological Society Papers*, 6: 16–18.

HERNÁNDEZ DE ALBA, GREGORIO (1948). 'Tribes of Northwestern Venezuela', in Stew-ard (1946–1959: 4: 469–79).

HERRERA, LUISA FERNANDA, INÉS CAVELIER, CAMILIO RODRÍGUEZ, and SANTIAGO MORA (1992). 'The Technical Transformation of an Agricultural System in the Colombian Amazon', *World Archaeology*, 24: 98–113.

HERZOG, WERNER (1982). In *Burden of Dreams*, directed and produced by Les Blank, documentary film on the making of *Fitzcarraldo*.

HILL, KIM and HILLARD KAPLAN (1989). 'Population and Dry-Season Subsistence Strategies of the Recently Contacted Yora of Peru', *National Geographic Research*, 5: 317–34.

HILL, R. L. (1959). 'Inflow to Lake Titicaca', *Journal of Geophysical Research*, 64: 789–94.

HIRAOKA, MARIO (1985). 'Floodplain Farming in the Peruvian Amazon', *Geographical Review of Japan*, 58/Ser. B: 1–23.

——(1986). 'Zonation of Mestizo Riverine Farming Systems in Northeast Peru', *National Geographic Research*, 2: 354–71.

HODDER, IAN (1983). *The Present Past: An Introduction to Anthropology for Archaeolo-gists*, New York: Pica Press.

HOLMBERG, ALLAN R. (1969) [1950]. *Nomads of the Long Bow: The Sirionó of Eastern Bolivia*, New York: Natural History Press.

HORKHEIMER, HANS (1960). *Nahrung und Nahrungsgewinnung im vorspanischen Peru*, Berlin: Bibliotheca Ibero-Americana.

——(1990) [1960]. *Alimentación y obtención de alimentos en los Andes prehispánicos*, La Paz: Hisbol.

HUBBELL, STEPHEN P. and ROBIN B. FOSTER (1990). 'Structure, Dynamics, and Equilibrium Status of Old-Growth Forest on Barro Colorado Island', in Alwyn H. Gentry (ed.), *Four Neotropical Rainforests*, New Haven, CT: Yale University Press, 522–41.

HVALKOF, SOREN (1989). 'The Nature of Development: Native and Settler Views in Gran Pajonal, Peruvian Amazon', *Folk*, 21: 125–50.

HYSLOP, JOHN (1990). *Inka Settlement Planning*, Austin: University of Texas Press.

Index Kewenis (updated 1991–5, Supplement 20). Kew, UK: Royal Botanic Gardens.

INEI (Instituto Nacional de Estadística e Informática) (1993). *Compendio Estadístico 1992–93: Región Arequipa*, Lima.

INNIS, DONALD Q. (1997). *Intercropping and the Scientific Basis of Traditional Agriculture*, London: Intermediate Technology Publications.

IRVINE, DOMINIQUE (1989). 'Succession Management and Resource Distribution in an Amazonian Rain Forest', *Advances in Economic Botany*, 7: 223–37.

ISAAC, BARRY L. (1977). 'The Sirionó of Eastern Bolivia: A Reexamination', *Human Ecology*, 5: 137–54.

ISBELL, WILLIAM H. (1968). 'New Discoveries in the Montaña, Southeastern Peru', *Archaeology*, 21/2: 108–14.

JACKS, GRAHAM V., R. TRAVERNIER, and D. H. BOALCH (eds.) (1960). *Multilingual Vocabulary of Soil Science*, Rome: Land and Water Division, Food and Agricultural Organization of the United Nations.

JOHNSON, ALLEN (1977). 'The Energy Costs of Technology in a Changing Environment: A Machiguenga Case', in Heather Lechtman and R. Merrill (eds.), *Material Culture: Styles, Organization, and Dynamics of Technology*, St. Paul: West Publishing Co., 155–67.

——(1982). 'Reductionism in Cultural Ecology: The Amazon Case', *Current Anthropology*, 23: 413–28.

——(1983). 'Machiguenga Gardens', in Hames and Vickers (1983*a*: 29–63).

JOHNSON, DOUGLAS L. and LAURENCE A. LEWIS (1995). *Land Degradation: Creation and Destruction*, Oxford: Blackwell.

JOHNSON, GEORGE R. (1930). *Peru from the Air*, New York: American Geographic Society, Special Publication 12.

JORDAN, CARL E. (ed.) (1987). *Amazonian Rain Forests: Ecosystem Disturbance and Recovery*, New York: Springer.

JULIEN, CATHERINE J. (1985). 'Guano and Resource Control in Sixteenth-Century Arequipa', in Shozo Masuda, Izumi Shimada, and Craig Morris (eds.), *Andean Ecology and Civilization*, Tokyo: University of Tokyo Press, 185–231.

KAERGER, KARL (1901). *Landwirtschaft und Kolonisation im spanischen Amerika*, Leipzig: Verlag von Dunker & Humblot.

KAPLAN, HILLARD (1985). 'Tropical Forest Clearance with a Stone Axe: Experimental Data', unpublished paper presented at the American Anthropological Association Meeting, Washington, DC.

KATZER, F. (1944). 'A terra preta', *Boletim da Secção de Fomento Agrícola no Estado do Pará*, 3/2: 35–8.

KEELEY, HELEN C. M. (1984). 'Soil Studies in the Cusichaca Valley, Peru', *British Archaeological Reports, International Series*, 210: 323–43.

——(1985). 'Soils of Prehistoric Terrace Systems in the Cusichaca Valley, Peru', *British Archaeological Reports, International Series*, 232: 547–68.

——(1988). 'Soils of Pre-Hispanic Field Systems in the Río Salado Basin, Northern Chile—A Preliminary Report', *British Archaeological Reports, International Series*, 410: 183–206.

——and FRANK M. MEDDENS (1993). 'Prehispanic Agricultural Terrace Systems in the Chicha-Soras Valley Peru', *Bulletin of the Institute of Archaeology*, 29: 121–38.

KELLY, KENNETH (1965). 'Land-use Regions in the Central and Northern Portions of the Inca Empire', *Annals of the Association of American Geographers*, 55: 327–38.

KENDALL, ANN (1992). 'Infraestructura agrícola e hidráulica pre-hispánica: Presente e futuro', Lima: Cusichaca Trust.

——(ed.) (1997). *Restoration and Rehabilitation of Pre-Hispanic Agricultural Systems in the Southern Highlands of Peru*, Bellbroughton, UK: Cusichaca Trust.

KERR, WARWICK E. and DARRELL A. POSEY (1984). 'Informações adicionais sobre a agricultura dos Kayapó', *Interciencia*, 9: 392–400.

KEY, MARY (*c*. 1961). 'An Archaeologist's Mystery', unpublished manuscript, Riberalta, Bolivia: Summer Institute of Linguistics.

KILLION, THOMAS W. (1990). 'Cultivation Intensity and Residential Site Structure: An Ethnographic Examination of Peasant Agriculture in the Sierra de los Tuxtlas, Veracruz, Mexico', *Latin American Antiquity*, 1: 191–215.

KIRCHHOFF, PAUL (1948). 'The Native Tribes North of the Orinoco River', in Steward (1946–1959: 4: 481–93).

KIRKBY, ANNE V. T. (1973). *The Use of Land and Water Resources in the Past and Present Valley of Oaxaca, Mexico*, Memoirs of the Museum of Anthropology 5, University of Michigan, Ann Arbor.

KLEE, GARY A. (ed.) (1980). *World Systems of Traditional Resource Management*, London: Edward Arnold.

KNAPP, GREGORY (1979). 'The Sunken Fields of Chilca: Horticulture, Microenvironment, and History in the Peruvian Coastal Desert', Master's thesis, University of Wisconsin–Madison.

——(1982). 'Prehistoric Flood Management on the Peruvian Coast: Reinterpreting the "Sunken Fields" of Chilca', *American Antiquity*, 47: 144–54.

——(1983). 'Reply to Richard T. Smith', *American Antiquity*, 48: 150–1.

——(1984). 'Soil, Slope, and Water in the Equatorial Andes: A Study of Prehistoric Agricultural Adaptation', Ph.D. dissertation, University of Wisconsin–Madison.

——(1991). *Andean Ecology: Adaptive Dynamics in Ecuador*, Dellplain Latin American Studies 27, Boulder, CO: Westview Press.

——and WILLIAM M. DENEVAN (1985). 'The Use of Wetlands in the Prehistoric Economy of the Northern Ecuadorian Highlands', *British Archaeological Reports, International Series*, 232: 185–207.

——and PATRICIA MOTHES (1998). 'Quilotoa Ash and Human Settlements in the Equatorial Andes', in P. Mothes (ed.), *Actividad volcánica y pueblos precolombinos en el Ecuador*, Quito: Ediciones Abya-Yala.

——and DAVID A. PRESTON (1987). 'Evidence of Prehistoric Ditched Fields on Sloping Land in Northern Highland Ecuador', *British Archaeological Reports, International Series*, 359: 403–23.

——and ROY RYDER (1983). 'Aspects of the Origin, Morphology, and Function of Ridged Fields in the Quito Altiplano, Ecuador', *British Archaeological Reports, International Series*, 189: 201–20.

KOLATA, ALAN L. (1986). 'The Agricultural Foundations of the Tiwanaku State: A View from the Heartland', *American Antiquity*, 51: 748–62.

——(1991). 'The Technology and Organization of Agricultural Production in the Tiwanaku State', *Latin American Antiquity*, 2: 99–125.

KOLATA, ALAN L. (cont.) (1993). *The Tiwanaku: Portrait of an Andean Civilization*, Cambridge, MA: Blackwell.

—— (1996a). *Valley of the Spirits: A Journey into the Lost Realm of the Aymara*, New York: John Wiley and Sons.

—— (ed.) (1996b). *Tiwanaku and Its Hinterland: Archaeology and Paleoecology of an Andean Civilization 1, Agroecology*, Washington, DC: Smithsonian Institution Press.

—— and CHARLES ORTLOFF (1989). 'Thermal Analysis of Tiwanaku Raised Field Systems in the Lake Titicaca Basin of Bolivia', *Journal of Archaeological Science*, 16: 233–63.

———— (1996). 'Tiwanaku Raised-Field Agriculture in the Lake Titicaca Basin of Bolivia', in Kolata (1996b: 109–151).

——, OSWALDO RIVERA, JUAN CARLOS RAMÍREZ, and EVELYN GEMIO (1996). 'Rehabilitating Raised-Field Agriculture in the Southern Lake Titicaca Basin of Bolivia: Theory, Practice, and Results', in Kolata (1996b: 203–30).

KOSOK, PAUL (1965). *Life, Land and Water in Ancient Peru*, New York: Long Island University Press.

KOZÁK, VLADIMÍR (1972). 'Stone Age Revisited', *Natural History*, 81/8: 14–24.

——, D. BAXTER, L. WILLIAMSON, and R. L. CARNEIRO (1979). *The Héta Indians: Fish in a Dry Pond*, New York: Anthropological Papers of the American Museum of Natural History 55.

KRAMER, BETTY JO (1977). 'Las implicaciones ecológicas de la agricultura de los Urarina', *Amazonia Peruana*, 1: 75–86.

KRAMER, FRITZ L. (1966). *Breaking Ground: Notes on the Distributions of Some Simple Tillage Tools*, Sacramento Anthropological Society 5.

KROEBER, ALFRED L. (1946). 'The Chibcha', in Steward (1946–1959: 2: 887–909).

KUS, JAMES S. (1972). 'Selected Aspects of Irrigated Agriculture in the Chimu Heartland, Peru', Ph.D. dissertation, University of California, Los Angeles.

—— (1984). 'The Chicama–Moche Canal: Failure or Success? An Alternative Explanation for an Incomplete Canal', *American Antiquity*, 49: 408–15.

LAMB, F. BRUCE (1987). 'The Role of Anthropology in Tropical Forest Ecosystem Resource Management and Development', *Journal of Developing Areas*, 21: 429–58.

LARCO HOYLE, RAFAEL (1945). *Los Mochicas*, Buenos Aires: Sociedad Geográfica Americana.

LATCHAM, RICARDO E. (1936). *La agricultura precolombina en Chile y los países vecinos*, Santiago, Chile: Universidad de Chile.

LATHRAP, DONALD W. (1962). 'Yarinachocha: Stratigraphic Excavations in the Peruvian Montaña', Ph.D. dissertation, Harvard University.

—— (1968a). 'Aboriginal Occupation and Changes in River Channel on the Central Ucayali, Peru', *American Antiquity*, 33: 62–79.

—— (1968b). 'The Hunting Economies of the Tropical Forest Zone of South America: An Attempt at Historical Perspective', in Richard B. Lee and Irven DeVore (eds.), *Man the Hunter*, Chicago: Aldine, 23–9.

—— (1970). *The Upper Amazon*, New York: Praeger.

—— (1977). 'Our Father the Cayman, Our Mother the Gourd: Spinden Revisited, or a Unitary Model for the Emergence of Agriculture in the New World', in Charles A. Reed (ed.), *Origins of Agriculture*, The Hague: Mouton, 713–51.

——, ANGELIKA GEBHART-SAYER, and ANN M. MESTER (1985). 'The Roots of the Shipibo Art Style: Three Waves on Imiríacocha or There Were "Incas" Before the Incas', *Journal of Latin American Lore*, 11: 31–119.

LE COINTE, PAUL (1947). *Amazônia Brasileira III: Arvores e Plantas Úteis (Indígenas e Aclimadas)*, São Paulo: Companhia Editora Nacional.

LEE, KENNETH (1977). '7,000 años de historia del hombre de Mojos: Agricultura en pampas esteriles—Informe preliminar', *Panorama Universitario*, Trinidad, Bolivia: 1: 23–6.

LEEDS, ANTHONY (1961). 'Yaruro Incipient Tropical Forest Horticulture: Possibilities and Limits', in Wilbert (1961: 13–46).

LENNON, THOMAS J. (1982). 'Raised Fields of Lake Titicaca, Peru: A Pre-Hispanic Water Management System', Ph.D. dissertation, University of Colorado.

—— (1983). 'Pattern Analysis of Pre-Hispanic Raised Fields of Lake Titicaca, Peru', *British Archaeological Reports, International Series*, 189: 183–99.

——, WARREN B. CHURCH, and MIGUEL CORNEJO G. (1989). 'Investigaciones arqueológicas en el Parque Nacional Río Abiseo, San Martin', *Boletín de Lima*, 62: 43–56.

LENTZ, DAVID L. (ed.) (2000). *Imperfect Balance: Landscape Transformations in the Precolumbian Americas*, New York: Columbia University Press.

LEÓN, JORGE (1994). 'Plant Genetic Resources of the New World', in J. E. Hernández Bermejo and J. León (eds.), *Neglected Crops: 1492 from a Different Perspective*, FAO Plant Production and Protection Series 26, Rome: Jardín Botanico de Córdoba, 3–21.

LÉRY, JEAN DE (1990) [1578]. *History of a Voyage to the Land of Brazil, Otherwise Called America*, Janet Whatley (trans.), Berkeley: University of California Press.

LÉVI-STRAUSS, CLAUDE (1963) [1952]. 'The Concept of Archaism in Anthropology', in C. Lévi-Strauss, *Structural Anthropology*, New York: Anchor Books, 97–115.

LEWENSTEIN, SUZANNE M. (1987). *Stone Tool Use at Cerros: The Ethnoarchaeological and Use-Wear Evidence*, Austin: University of Texas Press.

LIGHTFOOT, DALE (1994). 'Morphology and Ecology of Lithic-Mulch Agriculture', *The Geographical Review*, 84: 172–85.

LIZARAZU, JUAN DE (1906) [1638]. 'Informaciones hechas por Don Juan de Lizarazu sobre el descubrimiento de los Mojos', in Victor M. Maurtua (ed.), *Juicio de límites entre el Perú y Bolivia*, Madrid: Imprenta de los Hijos de M. G. Hernández., 9: 124–216.

LIZÁRRAGA, REGINALDO DE (1968) [c. 1600]. *Descripción breve de toda la tierra del Perú, Tucumán, Río de la Plata y Chile*, Madrid.

LIZOT, JACQUES (1980). 'La agricultura Yanomami', *Antropológica*, Caracas, 53: 3–93.

LORANT, STEFAN (ed.) (1946). *The New World: The First Pictures of America*, New York: Duell, Sloan and Pearce.

LUIS DE CABRERA, GERÓNIMO (1965) [c. 1573]. 'Relación en suma de la tierra y poblazones', in Marcos Jiménez de la Espada (ed.), *Relaciones geográficas de Indias*, Biblioteca de Autores Españoles, Madrid: Ediciones Atlas, 185/1: 388–9.

MALDONADO, ANGEL and LUIS GAMARRA DULANTO (1978) [1945]. 'Significado arqueológico, agrológico y geográfico de los andenes abandonados de Santa Inés y Chosica en el valle del Rimac', in Ravines (1978c: 157–71).

MALPASS, MICHAEL A. (1987). 'Prehistoric Agricultural Terracing at Chijra in the Colca Valley, Peru: Preliminary Report II', *British Archaeological Reports, International Series*, 359: 45–66.

MALPASS, MICHAEL A. (cont.) (1988). 'Irrigated versus Non-Irrigated Terracing in the Andes: Environmental Considerations', unpublished paper presented at the Northeast Conference of Andean Archaeology and Ethnohistory Meeting, Amherst, MA.

——and PABLO DE LA VERA CRUZ CHÁVEZ (1988). 'Ceramic Sequence from Chijra, Coporaque', in Denevan (1986–1988: 2: 204–33).

MAMPEL GONZÁLEZ, ELENA and NEUS ESCANDELL TUR (eds.) (1981). *Lope de Aguirre Crónicas, 1559–1561*, Barcelona: Ediciones Universidad.

MARBÁN, PEDRO (1898) [*c*. 1676]. 'Relación de la provincia de la vírgen del Pilar de Mojos', *Boletín de la Sociedad Geográfica de La Paz*, 1: 120–67.

MARCOS, JORGE G. (1987). 'Los campos elevados de la Cuenca del Guayas, Ecuador: El proyecto Peñón del Río', *British Archaeological Reports, International Series*, 359: 217–24.

MARTIN, DEBORAH (1986). 'Archaeology of Terraces and Settlement at Chilacota, Coporaque', in Denevan (1986–1988:1: 221–34).

MARTIN, M. KAY (1969). 'South American Foragers: A Case Study in Cultural Devolution', *American Anthropologist*, 71: 243–60.

MARTÍNEZ, HÉCTOR (1962). El indígena y el mestizo de Taraco, *Revista del Museo Nacional*, Lima, 30: 172–244.

MARTÍNEZ, VALENTINA L. (1987). 'Campos elevados al norte del sitio arqueológico Peñón del Río, Guayas, Ecuador', *British Archaeological Reports, International Series*, 359: 267–77.

MASSON M., LUIS (1984). *La recuperación de los andenes para la ampliación de la frontera agrícola en la sierra*, Lima: COFIDE.

——(1986). 'Rehabilitación de andenes en la comunidad de San Pedro de Casta, Lima', in Carlos de la Torre and Manuel Burga (eds.), *Andenes y camellones en el Perú andino: Historia presente y futuro*, Lima: Consejo Nacional de Ciencia y Tecnología, 207–16.

——(1987). 'La ocupación de andenes en el Perú', *Pensamiento Iberoamericano*, 12: 179–200.

MASUDA, SHOZO, IZUMI SHIMADA, and CRAIG MORRIS (eds.) (1985). *Andean Ecology and Civilization: An Interdisciplinary Perspective on Andean Ecological Complementarity*, Tokyo: University of Tokyo Press.

MATHENY, RAY T. and DEANNE L. GURR (1983). 'Variation in Prehistoric Agricultural Systems in the New World', *Annual Review of Anthropology*, 12: 79–103.

MATHEWSON, KENT (1986). 'Alexander von Humboldt and the Origins of Landscape Archaeology', *Journal of Geography*, 85: 50–6.

——(1987*a*). 'Landscape Change and Cultural Persistence in the Guayas Wetlands, Ecuador', Ph. D. dissertation, University of Wisconsin–Madison.

——(1987*b*). 'Estimating Labor Inputs in Raised Field Complexes of the Guayas Basin, Ecuador', *British Archaeological Reports, International Series*, 359: 217–24.

MATIENZO, JUAN DE (1910) [1567]. *Gobierno del Perú: Obra escrita en el siglo XVI por el Licenciado Don Juan Matienzo, Oidor de la Real Audiencia de Charcas*, Buenos Aires: Universidad de Buenos Aires.

MAURTUA, VÍCTOR M. (ed.) (1906). *Juicio de límites entre el Perú y Bolivia*, 12 vols, Madrid: Los Hijos de M. G. Hernández.

MAYER, ENRIQUE (1979). *Land Use in the Andes: Ecology and Agriculture in the Mantaro Valley of Peru, with Special Reference to Potatoes*, Lima: Centro Internacional de la Papa.

McCAMANT, KRIS ANN (1986). 'The Organization of Agricultural Production in Coporaque, Peru', Master's thesis, University of California, Berkeley.

McCANN, JOSEPH M., WILLIAM I. WOODS, and DONALD W. MEYER (2001). 'Organic Matter and Anthrosols in Amazonia: Interpreting the Amerindian Legacy', in R. M. Rees *et al.* (eds.), *Sustainable Management of Soil Organic Matter*, New York: CAB International, 180–9.

McDOUGLE, EUGENE J. (1967). 'Water Use and Settlements in the Changing Environments of the Southern Ecuadorian Coast', Master's thesis, Columbia University.

MEGGERS, BETTY J. (1954). 'Environmental Limitation on the Development of Culture', *American Anthropologist*, 56: 801–24.

—— (1957). 'Environment and Culture in the Amazon Basin: An Appraisal of the Theory of Environmental Determinism', *Studies in Human Ecology*, Social Science Monograph, Washington, DC: Pan American Union, 3: 71–89.

—— (1966). *Ecuador*, New York: Praeger.

—— (1971). *Amazonia: Man and Culture in a Counterfeit Paradise*, Chicago: Aldine.

—— (1984). 'The Indigenous Peoples of Amazonia, Their Cultures, Land Use Patterns and Effects on the Landscape and Biota', in Harold Sioli (ed.), *The Amazon: Limnology and Landscape Ecology of a Mighty Tropical River and its Basin*, Dordrecht: W. Junk, 627–48.

—— (1987). 'The Early History of Man in Amazonia', in T. C. Whitmore and G. T. Prance (eds.), *Biogeography and Quaternary History in Tropical America*, Oxford: Oxford Scientific Publications, 151–74.

—— (1992*a*). 'Prehistoric Population Density in the Amazon Basin', in John W. Verano and Douglas H. Ubelaker (eds.), *Disease and Demography in the Americas*, Washington, DC: Smithsonian Institution Press, 197–205.

—— (1992*b*). 'Amazonia: Real or Counterfeit Paradise?', *The Review of Archaeology*, 13/2: 25–40.

—— (1993–5). 'Amazonia on the Eve of European Contact: Ethnohistorical, Ecological, and Anthropological Perspectives', *Revista de Arqueología Americana*, 8: 91–115.

—— (1994). 'Prehistoric Cultural Development in Amazonia: An Archaeological Perspective', *Research and Exploration*, 10: 398–421.

—— (1995). 'Judging the Future by the Past: The Impact of Environmental Instability on Prehistoric Amazonian Populations', in Sponsel (1995: 15–43).

—— (1996). *Amazonia: Man and Culture in a Counterfeit Paradise* (2nd edn), Washington, DC: Smithsonian Institution Press.

—— and CLIFFORD EVANS (1957). *Archeological Investigations at the Mouth of the Amazon*, Bureau of American Ethnology, Bulletin 167, Washington, DC: Smithsonian Institution.

———— (1983). 'Lowland South America and the Antilles', in Jesse D. Jennings (ed.), *Ancient South Americans*, San Francisco: W. H. Freeman, 287–335.

—— ONDEMAR F. DIAS, EURICO T. MILLER, and CELSO PEROTA (1988). 'Implications of Archeological Distributions in Amazonia', in W. R. Heyer and P. E. Vanzolini (eds.), *Proceedings of a Workshop on Neotropical Distribution Patterns*, Rio de Janeiro: Academia Brasileira de Ciências, 275–94.

MÉTRAUX, ALFRED (1942). *The Native Tribes of Eastern Bolivia and Western Matto Grosso*, Bureau of American Ethnology, Bulletin 134, Washington, DC: Smithsonian Institution.

——(1959). 'The Revolution of the Ax', *Diogenes*, 25: 28–40.

—— and PAUL KIRCHHOFF (1948). 'The Northeastern Extension of Andean Culture', in Steward (1946–1959: 4: 349–68).

MEYER, WILLIAM B. and B. L. TURNER II (eds.) (1994). *Changes in Land Use and Land Cover: A Global Perspective*, Cambridge: Cambridge University Press.

MILLER, NAOMI F. and KATHRYN L. GLEASON (eds.) (1994*a*). *The Archaeology of Garden and Field*, Philadelphia: University of Pennsylvania Press.

————(1994*b*). 'Fertilizer in the Identification and Analysis of Cultivated Soil', in Miller and Gleason (1994*a*: 25–43).

MILTON, KAY (1997). 'Ecologies: Anthropology, Culture and the Environment', *International Social Science Journal*, 154: 477–95.

MITCHELL, WILLIAM P. and DAVID GUILLET (eds.) (1994). *Irrigation at High Altitudes: The Social Organization of Water Control Systems in the Andes*, Society for Latin American Anthropology Publication Series 12.

MOLINA, CRISTÓBAL DE (1968) [1552]. 'Relación de muchas cosas acaescidas en el Perú', in *Crónicas peruanos de interés indígena*, Madrid: Biblioteca de Autores Españoles, 209: 56–95.

MOMSEN JR., RICHARD P. (1964). 'The Isconahua Indians: A Case Study of Change and Diversity in the Peruvian Amazon', *Revista Geográfica*, 60: 59–82.

MONHEIM, FELIX (1956). *Beiträge zur Klimatologie und Hydrologie des Titicacabeckens*, Heidelberger Geographische Arbeiten 1.

MOORE, JERRY D. (1988). 'Prehistoric Raised Field Agriculture in the Casma Valley, Peru', *Journal of Field Archaeology*, 15: 265–76.

MORA C., SANTIAGO, LUISA FERNANDA HERRERA, INÉS CAVELIER F., and CAMILIO RODRÍGUEZ (1991). *Cultivars, Anthropic Soils and Stability: A Preliminary Report of Archaeological Research in Araracuara, Colombian Amazonia*, Pittsburgh: University of Pittsburgh Latin American Archaeology Reports 2.

MORAN, EMILIO F. (1991). 'Human Adaptive Strategies in Amazonian Blackwater Ecosystems', *American Anthropologist*, 93: 361–82.

——(1993). *Through Amazonian Eyes: The Human Ecology of Amazonian Populations*, Iowa City: University of Iowa Press.

——(1995). 'Disaggregating Amazonia: A Strategy for Understanding Biological and Cultural Diversity', in Sponsel (1995: 71–95).

MOREY, NANCY C. (1975). 'Ethnohistory of the Colombian and Venezuelan Llanos', Ph.D. dissertation, University of Utah.

MOSELEY, MICHAEL E. (1977). 'Waterways of Ancient Peru', *Field Museum of Natural History Bulletin*, 48/3: 10–15.

——(1983). 'The Good Old Days *Were* Better: Agrarian Collapse and Tectonics', *American Anthropologist*, 85: 773–99.

—— and KENT C. DAY (eds.) (1982). *Chan Chan: Andean Desert City*, Albuquerque: University of New Mexico Press.

—— and ERIC E. DEEDS (1982). 'The Land in Front of Chan Chan: Agrarian Expansion, Reform, and Collapse in the Moche Valley', in M. E. Moseley and K. C. Day (eds.), *Chan Chan: Andean Desert City*, Albuquerque: University of New Mexico Press, 25–53.

MOSELEY, MICHAEL E. and ROBERT A. FELDMAN (1984). 'Hydrological Dynamics and the Evolution of Field Form and Use: Resolving the Knapp-Smith Controversy', *American Antiquity*, 49: 403–8.

——, CHARLES R. ORTLOFF, and ALFREDO NARVAEZ (1983). 'Principles of Agrarian Collapse in the Cordillera Negra, Peru', *Annals of Carnegie Museum*, 52: 299–327.

MOTHES, PATRICIA ANN (1986). 'Pimampiro's Canal: Adaptation and Infrastructure in Northern Ecuador', Master's thesis, University of Texas at Austin.

MULLEN, WILLIAM (1986). 'Secrets of Tiwanaku: How Ingenious Farmers Built an Empire that Inspired the Incas and Rivaled Rome', *Chicago Tribune Magazine*, 23 November, 10–32.

MURRA, JOHN V. (1972). 'El "control vertical" de un máximo de pisos ecológicos en la economía de las sociedades Andinas', in J. V. Murra (ed.), *Visita de la Provincia de León de Huánuco en 1562, Iñigo Ortiz de Zúñiga, visitador*, Huánuco, Perú: Universidad Nacional Hermilio Valdizán, 2: 427–76.

——(1983). 'La capacidad gerencial y macroorganizadora de la sociedad andina antigua', in Ana María Fries (ed.), *Evolución y tecnología de la agricultura andina*, Cuzco: Instituto Indigenísta Interamericano, 5–10.

MUSE, MICHAEL and FAUSTO QUINTERO (1987). 'Experimentos de reactivación de campos elevados, Peñón del Río, Guayas, Ecuador', *British Archaeological Reports, International Series*, 359: 249–66.

MYERS, THOMAS P. (1973). 'Toward the Reconstruction of Prehistoric Community Patterns in the Amazon Basin', in Donald W. Lathrap and Jody Douglas (eds.), *Variation in Anthropology*, Urbana: Illinois Archaeological Survey, 233–52.

——(1974*a*). 'Evidence of Prehistoric Irrigation in Northern Ecuador', *Journal of Field Archaeology*, 1: 309–13.

——(1974*b*). 'Spanish Contacts and Social Change on the Ucayali River, Peru', *Ethnohistory*, 21: 135–58.

——(1976). 'Defended Territories and No-Man's-Lands', *American Anthropologist*, 78: 354–5.

——(1990). *Sarayacu: Ethnohistorical and Archaeological Investigations of a Nineteenth-Century Franciscan Mission in the Peruvian Montaña*, Lincoln: University of Nebraska Studies, New Series 68.

——(1992*a*). 'Agricultural Limitations of the Amazon in Theory and Practice', *World Archaeology*, 24: 82–97.

——(1992*b*). 'The Expansion and Collapse of the Omagua', *Journal of the Steward Anthropological Society*, 20: 129–52.

NEIRA AVENDAÑO, MÁXIMO (1990). 'Arequipa prehispánica', in M. Neira Avendaño *et al.* (eds.), *Historia general de Arequipa*, Arequipa: Fundación M. J. Bustamente de la Fuente, 5–184.

NELSON, BRUCE W. and MARILANE NASCIMENTO IRMÃOL (1998). 'Fire Penetration in Standing Amazon Forests', *Proceedings, Ninth Brazilian Remote Sensing Symposium*, Santos, 13–18.

—— VALERIE KAPOS, JOHN B. ADAMS, WILSON J. OLIVEIRA, OSCAR P. G. BRAUN, and IĔDA L. DO AMARAL (1994). 'Forest Disturbance by Large Blowdowns in the Brazilian Amazon', *Ecology*, 75: 853–8.

NETHERLY, PATRICIA J. (1984). 'The Management of Late Andean Irrigation Systems on the North Coast of Peru', *American Antiquity*, 49: 227–54.

NETHERLY, PATRICIA J. (cont.) (1988). 'From Event to Process: The Recovery of Late Andean Organizational Structure by Means of Spanish Colonial Written Records', in Richard W. Keatinge (ed.), *Peruvian Prehistory: An Overview of Pre-Inca and Inca Society*, Cambridge: Cambridge University Press, 257–75.

NEVES, EDUARDO GÓES (1998). 'Twenty Years of Amazonian Archaeology in Brazil (1977–1997), *Antiquity*, 72: 625–32.

NIALS, FRED L., ERIC E. DEEDS, MICHAEL E. MOSELEY, S. G. POZORSKI, T. G. POZORSKI, and ROBERT FELDMAN (1979). 'El Niño: The Catastrophic Flooding of Coastal Peru', *Bulletin of the Field Museum of Natural History*, 50/7: 4–14, 50/8: 4–10.

NIETSCHMANN, BERNARD Q. (1973). *Between Land and Water: The Subsistence Ecology of the Miskito Indians, Eastern Nicaragua*, New York: Seminar Press.

NILES, SUSAN A. (1987). *Callachaca: Style and Status in an Inca Community*, Iowa City: University of Iowa Press.

——(1999). *The Shape of Inca History: Narrative and Architecture in an Andean Empire*, Iowa City: University of Iowa Press.

NIMUENDAJÚ, CURT (1952) [1939]. 'The Tapajó', John H. Rowe (trans. and ed.), *Kroeber Anthropological Society Papers*, 6: 1–15.

NORDENSKIÖLD, ERLAND (1916). 'Die Anpassung der Indianer an die Verhältnisse in den Uberschwemmungsgebieten in Súdamerika', *Ymer*, Stockholm, 36: 138–55.

——(1919). *An Ethno-Geographical Analysis of the Material Culture of Two Indian Tribes in the Gran Chaco*, Göteborg: Comparative Ethnographical Studies 1.

——(1920). *The Changes in the Material Culture of Two Indian Tribes Under the Influence of New Surroundings*, Göteborg: Comparative Ethnological Studies 2.

——(1924). *The Ethnology of South America Seen from Mojos in Bolivia*, Göteborg: Comparative Ethnological Studies 3.

NRC (National Research Council) (1989). *Lost Crops of the Incas: Little-Known Plants of the Andes with Promise for Worldwide Cultivation*, Washington, DC: National Academy Press.

NÚÑEZ JIMÉNEZ, ANTONIO (1987). *Petróglifos del Perú*, 4 vols, Havana: UNESCO.

NÚÑEZ REGUEIRO, VICTOR A. (1978). 'Considerations on the Periodizations of Northwest Argentina', in David L. Browman (ed.), *Advances in Andean Archaeology*, The Hague: Mouton, 453–84.

NYE, P. H. and D. J. GREENLAND (1960). *The Soil Under Shifting Cultivation*, Commonwealth Agricultural Bureaux, Farnham Royal, UK.

OCHOA, CARLOS M. (1984). '*Solanum hygrothermicum*, New Potato Species Cultivated in the Lowlands of Peru', *Economic Botany*, 38: 128–33.

——(1990). *The Potatoes of South America: Bolivia*, Cambridge: Cambridge University Press.

OHLY, J. J. and WOLFGANG J. JUNK (1999). 'Multiple Use of Central Amazon Floodplains: Combining Ecological Conditions, Requirements for Environmental Protection, and Socioeconomic Needs', in Padoch *et al.* (1999: 283–99).

ONERN (Oficina Nacional de Evaluación de Recursos Naturales) (1965). *Inventario y evaluación de los recursos naturales del Departamento de Puno*, 6 vols, Lima.

——(1976). *Inventario, evaluación, y uso racional de los recursos naturales de la costa: Cuencas de los ríos Chilca, Mala, y Asia*, 2 vols, Lima.

——(1987). 'Inventario Nacional de Andenes', incomplete manuscript, Lima.

ORELLANA, ANTONIO DE (1906). 'Carta del Padre Antonio de Orellana, sobre el origen de las misiones de Mojos, 18 octubre 1687', in *Juicio de límites entre el Perú y Bolivia*, Madrid: Los Hijos de M. G. Hernández, 10: 1–24.

ORLOVE, BENJAMIN S. (1977). 'Integration through Production: The Use of Zonation in Espinar', *American Ethnologist*, 4: 84–101.

——and RICARDO GODOY (1986). 'Sectoral Fallowing Systems in the Central Andes', *Journal of Ethnobiology*, 6: 169–204.

ORTLOFF, CHARLES R. (1988). 'Canal Builders of Pre-Inca Peru', *Scientific American*, 259/6: 100–7.

——(1993). 'Chimu Hydraulic Technology and Statecraft on the North Coast of Peru, AD 1000–1470', in Vernon L. Scarborough and Barry L. Isaac (eds.), *Economic Aspects of Water Management in the Prehispanic New World*, Research in Economic Anthropology, Supplement 7, Greenwich, CT: JAI Press, 327–67.

——(1995). 'Surveying and Hydraulic Engineering of the Pre-Columbian Chimú State, AD 900–1450', *Cambridge Archaeological Journal*, 5: 55–74.

——and ALAN L. KOLATA (1989). 'Hydraulic Analysis of Tiwanaku Aqueduct Structures at Lukurmata and Pajchiri, Bolivia', *Journal of Archaeological Science*, 16: 513–35.

————(1993). 'Climate and Collapse: Agro-Ecological Perspectives on the Decline of the Tiwanaku State', *Journal of Archaeological Science*, 20: 195–221.

——, MICHAEL E. MOSELEY, and ROBERT A. FELDMAN (1982). 'Hydraulic Engineering Aspects of the Chimu Chicama–Moche Intervalley Canal', *American Antiquity*, 47: 572–95.

——————(1983). 'The Chicama–Moche Intervalley Canal: Social Explanations and Physical Paradigms', *American Antiquity*, 48: 375–89.

——, ROBERT A. FELDMAN, and MICHAEL E. MOSELEY (1985). 'Hydraulic Engineering and Historical Aspects of the Pre-Columbian Intravalley Canal Systems of the Moche Valley, Peru', *Journal of Field Archaeology*, 12: 77–98.

OVIEDO y VALDÉS, GONZALO FERNÁNDEZ DE (1934) [1543–1548]. 'Oviedo's Description of Gonzalo Pizarro's Expedition to the Land of Cinnamon . . .', in H. C. Heaton (ed.), *The Discovery of the Amazon*, Special Publication 17, New York: American Geographical Society, 390–404.

——(1959) [1526]. *Natural History of the West Indies*, Sterling A. Stoudemire (trans. and ed.), University of North Carolina Studies in the Romance Languages and Literatures 32, Chapel Hill: University of North Carolina Press.

PABST, EIJE E. (1993). 'Terra Preta: Ein Beitrag zur Genese-Diskussion auf der Basis von Geländearbeiten bei Tupí-Völkern Amazoniens', Doctoral dissertation, Kassel University, Germany.

PADOCH, CHRISTINE and WIL DE JONG (1992). 'Diversity, Variation, and Change in Ribereño Agriculture', in Kent H. Redford and C. Padoch (eds.), *Conservation of Neotropical Forests: Working from Traditional Resource Use*, New York: Columbia University Press, 158–74.

——, JOSÉ MÁRCIO AYRES, MIGUEL PINEDO-VASQUEZ, and ANDREW HENDERSON (eds.) (1999). *Várzea: Diversity, Development, and Conservation of Amazonia's Whitewater Floodplains*, Advances in Economic Botany 13.

PALERM, ANGEL (1973). *Obras hidraúlicas prehispánicas en el sistema lacustre del Valle de México*, México: Instituto Nacional de Antropología e Historia.

PARK, CHRIS C. (1983). 'Water Resources and Irrigation Agriculture in Pre-Hispanic Peru', *Geographical Journal*, 149: 153–66.

PARKER, EUGENE P. (1981). 'Cultural Ecology and Change: A Caboclo Várzea Community in the Brazilian Amazon', Ph.D. dissertation, University of Colorado.

——(1992). 'Forest Islands and Kayapó Resource Management in Amazonia: A Reappraisal of the Apêtê', *American Anthropologist*, 94: 406–28.

——(1993). 'Fact and Fiction in Amazonia: The Case of the Apêtê', *American Anthropologist*, 95: 715–23.

——, DARRELL A. POSEY, JOHN FRECHIONE, and LUIZ FRANCELINO DA SILVA (1983). 'Resource Exploitation in Amazonia: Ethnoecological Examples from Four Populations', *Annals of Carnegie Museum*, 52/8: 163–203.

PARSONS, JAMES J. (1949). *Antioqueño Colonization in Western Colombia*, Ibero-Americana 32, Berkeley: University of California Press.

——(1969). 'Ridged Fields in the Río Guayas Valley, Ecuador', *American Antiquity*, 34: 76–80.

——and WILLIAM A. BOWEN (1966). 'Ancient Ridged Fields of the San Jorge River Floodplain, Colombia', *Geographical Review*, 56: 317–43.

——and WILLIAM M. DENEVAN (1967). 'Pre-Columbian Ridged Fields', *Scientific American*, 217/1: 92–100.

——and ROY SHLEMON (1982). 'Nuevo informe sobre los campos elevados prehistoricos de la Cuenca del Guayas, Ecuador', *Miscelánea Antropológica Ecuatoriana*, 2: 31–7.

————(1987). 'Mapping and Dating the Prehistoric Raised Fields of the Guayas Basin, Ecuador', *British Archaeological Reports, International Series*, 359: 207–16.

PARSONS, JEFFREY R. (1968). 'The Archaeological Significance of Mahamaes Cultivation on the Coast of Peru', *American Antiquity*, 33: 80–9.

——(1991). 'Political Implications of Prehispanic Chinampa Agriculture in the Valley of Mexico', in Herbert R. Harvey (ed.), *Land and Politics in the Valley of Mexico: A Two Thousand Year Perspective*, Albuquerque: University of New Mexico Press, 17–41.

——and NORBERT P. PSUTY (1975). 'Sunken Fields and Prehispanic Subsistence on the Peruvian Coast', *American Antiquity*, 40: 259–82.

PATIÑO, VÍCTOR M. (1963–1969). *Plantas cultivadas y animales domesticos en América equinoccial*, 4 vols, Cali, Colombia: Imprenta Departamental.

PATRICK, LARRY L. (1980). 'Los orígenes de las terrazas de cultivo', *América Indígena*, 40: 757–72.

PAZ SOLDÁN, CARLOS ENRIQUE and MAXIME KUCZYNSKI-GODARD (1939). *La selva peruana: Sus pobladores y su colonización en seguridad sanitaria*, Lima: Ediciones de La Reforma Médica.

PEARSALL, DEBORAH M. (1987). 'Evidence for Prehistoric Maize Cultivation on Raised Fields at Peñón del Río, Guayas, Ecuador', *British Archaeological Reports, International Series*, 359: 279–95.

——(1992). 'The Origins of Plant Cultivation in South America', in C. Wesley Cowan and Patty Jo Watson (eds.), *The Origins of Agriculture: An International Perspective*, Washington, DC: Smithsonian Institution Press, 173–205.

PETERS, CHARLES M. (2000). 'Pre-Columbian Silviculture and Indigenous Management of Neotropical Forests, in Lentz (2000): 203–23.

PETRICK, CARSTEN (1978). 'The Complementary Function of Floodlands for Agricultural Utilization: The Várzea of the Brazilian Amazon Region', *Applied Sciences and Development*, 12: 24–46.

PIERPONT MORGAN LIBRARY (1996) [1590s]. *Histoire Naturelle des Indes: The Drake Manuscript in the Pierpont Morgan Library*, New York: W. W. Norton.

PINTO PARADA, RODOLFO (1987). *Pueblo de leyenda*, Trinidad, Bolivia: Editorial Tiempo del Beni.

PIPERNO, DOLORES R. and DEBORAH M. PEARSALL (1998). *The Origins of Agriculture in the Lowland Neotropics*, San Diego, CA: Academic Press.

PLAFKER, GEORGE (1963). 'Observations on Archaeological Remains in Northeastern Bolivia', *American Antiquity*, 28: 372–8.

PLAZAS, CLEMENCIA and ANA MARÍA FALCHETTI (1981). *Asentamientos prehispánicos en el bajo Río San Jorge*, Publicación de la Fundación de Investigaciones Arqueológicas Nacionales 11, Bogotá: Banco de la República.

——————(1987). 'Poblamiento y adecuación hidráulica en el bajo Río San Jorge, Costa Atlantica, Colombia', *British Archaeological Reports, International Series*, 359: 483–503.

——————(1990). 'Manejo hidráulico Zenú', in Santiago Mora Camargo (ed.), *Ingenierias prehispánicas*, Bogotá: Fondo FEN Colombia, 151–71.

——————, Juanita Saenz Samper, and Sonia Archila (1993). *La sociedad hidráulica Zenú: Estudio arqueológico de 2,000 años de historia en las llanuras del Caribe colombiano*, Bogotá: Banco de la República.

PLOTKIN, MARK J. (1993). *Tales of a Shaman's Apprentice*, New York: Penguin Books.

POHL, MARY D. and PAUL BLOOM (1996). 'Prehistoric Maya Farming in the Wetlands of Northern Belize: More Data from Albion Island and Beyond', in Scott L. Fedick (ed.), *The Managed Mosaic: Ancient Maya Agriculture and Resource Use*, Salt Lake City: University of Utah Press, 145–64.

——————, and KEVIN O. POPE (1990). 'Interpretation of Wetland Farming in Northern Belize: Excavations at San Antonio Río Hondo', in M. D. Pohl (ed.), *Ancient Maya Wetland Agriculture: Excavations on Albion Island, Northern Belize*, Boulder, CO: Westview Press, 187–278.

POLITIS, GUSTAVO G. (1996). 'Moving to Produce: Nukak Mobility and Settlement Patterns in Amazonia', *World Archaeology*, 27: 492–511.

PORRO, ANTONIO (1994). 'Social Organization and Political Power in the Amazon Floodplain: The Ethnohistorical Sources', in Roosevelt (1994: 79–94).

PORTER, PHILIP W. (1965). 'Environmental Potentials and Economic Opportunities: A Background for Cultural Adaptation', *American Anthropologist*, 67: 409–20.

POSEY, DARRELL A. (1982). 'The Keepers of the Forest', *Garden*, 6/1: 18–24.

——(1984a). 'Keepers of the Campo', *Garden*, 8/6: 8–12, 32.

——(1984b). 'A Preliminary Report on Diversified Management of Tropical Forest by the Kayapó Indians of the Brazilian Amazon', *Advances in Economic Botany*, 1: 112–26.

——(1985a). 'Indigenous Management of Tropical Forest Ecosystems: The Case of the Kayapó Indians of the Brazilian Amazon', *Agroforestry Systems*, 3: 139–58.

——(1985b). 'Native and Indigenous Guidelines for New Amazonian Development Strategies: Understanding Biological Diversity Through Ethnoecology', in John Hemming (ed.), *Change in the Amazon Basin*, Manchester: Manchester University Press, 1: 156–81.

354 *References*

Posey, Darrell A. (cont.) (1987). 'Contact before Contact: Typology of Post-Columbian Interaction with Northern Kayapó of the Amazon Basin', *Boletim do Museu Paraense Emílio Goeldi, Série Antropologia*, 3/2: 135–54.

—— (1992). 'Reply to Parker', *American Anthropologist*, 94: 441–3.

—— (1998). 'Diachronic Ecotones and Anthropogenic Landscapes in Amazonia: Contesting the Consciousness of Conservation', in Balée (1998: 104–18).

—— and William Balée (eds.) (1989). *Resource Management in Amazonia: Indigenous and Folk Strategies*, Advances in Economic Botany 7.

Pozorski, Thomas (1987). 'Changing Priorities within the Chimu State: The Role of Irrigation Agriculture', in Jonathan Haas, Shelia Pozorski, and T. Pozorski (eds.), *The Origins and Development of the Andean State*, Cambridge: Cambridge University Press, 111–20.

—— and Shelia Pozorski (1982). 'Reassessing the Chicama–Moche Intervalley Canal: Comments on "Hydraulic Engineering Aspects of the Chimu Chicama–Moche Intervalley Canal"', *American Antiquity*, 47: 851–68.

———, Carol J. Mackey, and Alexandra M. Ulana Klymyshyn (1983). 'Pre-Hispanic Ridged Fields of the Casma Valley, Peru', *The Geographical Review*, 73: 407–16.

Prance, Ghillean T. (1979). 'Notes on the Vegetation of Amazonia III: The Terminology of Amazonian Forest Types Subject to Inundation', *Brittonia*, 31: 26–38.

Preston, David A. (1996). 'People on the Move: Migrations Past and Present', in D. Preston (ed.), *Latin American Development: Geographical Perspectives*, Essex, UK: Longman, 165–87.

Price, Barbara J. (1971). 'Prehispanic Irrigation Agriculture in Nuclear America', *Latin American Research Review*, 6/3: 3–60.

Price, David (1989). *Before the Bulldozer: The Nambiquara Indians and the World Bank*, Cabin John, MD: Seven Locks Press.

Puleston, Dennis E. (1977). 'The Art and Archaeology of Hydraulic Agriculture in the Maya Lowlands', in Norman Hammond (ed.), *Social Process in Maya Prehistory: Studies in Memory of Sir Eric Thompson*, London: Academic Press, 449–67.

Raffles, Hugh (1999). 'Exploring the Anthropogenic Amazon: Estuarine Landscape Transformations in Amapá, Brazil', in Padoch *et al.* (1999: 355–70).

Ralegh, Walter (1997) [1596]. *The Discoverie of the Large, Rich and Bewtiful Empyre of Guiana*, Neil L. Whitehead (ed.), Norman: University of Oklahoma Press.

Räsänen, Matti, Risto Kalliola, and Maarit Puhakka (1993). 'Mapa geoecológico de la selva baja peruana: Explicaciones', in R. Kalliola, M. Puhakka, and Walter Danjoy (eds.), *Amazonia peruana: Vegetación humeda tropical en el llano subandino*, Turku, Finland and Lima: PAUT/ONERN, 207–16.

Ravines, Rogger (1978*a*). 'Recursos naturales de los Andes', in Ravines (1978*c*: 1–90).

—— (1978*b*). 'Agricultura y riego', in Ravines (1978*c*: 91–106).

—— (ed.) (1978*c*). *Tecnología andina*, Instituto de Estudios Peruanos, Lima.

—— and Félix Solar la Cruz (1980). 'Hidráulica agrícola prehispánica', *Allpanchis*, 14: 69–81.

Raymond, J. Scott (1988). 'A View from the Tropical Forest', in William Keatinge (ed.), *Peruvian Prehistory*, Cambridge: Cambridge University Press, 279–300.

—— (1994). 'The Intellectual Legacy of Donald W. Lathrap', in Augusto Oyuela-

Caycedo (ed.), *History of Latin American Archaeology*, Aldershot, UK: Avebury, 173–182.

REDMOND, ELSA M. and CHARLES S. SPENCER (1994). 'Savanna Chiefdoms of Venezuela', *Research and Exploration*, 10: 422–39.

REEVES, R. G. (ed.) (1975). *Manual of Remote Sensing*, 2 vols, Falls Church: American Society of Photogrammetry.

REGAL, ALBERTO (1945). 'Política hidráulica del imperio incaico', *Revista de la Universidad Católica del Perú*, 13: 75–110.

——(1970). *Los trabajos hidráulicos del Inca en el antiguo Perú*, Lima: Imprenta Gráf. Industrial.

REICHEL-DOLMATOFF, GERARDO (1961). 'The Agricultural Basis of the Sub-Andean Chiefdoms of Colombia', in Wilbert (1961: 83–100).

——and ALICIA REICHEL-DOLMATOFF (1974). 'Un sistema de agricultura prehistórica de los Llanos Orientales', *Revista Colombiana de Antropología*, 17: 189–200.

RENÉ-MORENO, GABRIEL (1888). *Biblioteca boliviana: Catálogo del archivo de Mojos y Chiquitos*, Santiago: Imprenta Gutenberg.

RENGIFO VÁSQUEZ, GRIMALDO (1987). *La agricultura tradicional en los Andes*, Lima: Editorial Horizonte.

RICHARDS, PAUL W. (1996) [1952]. *The Tropical Rain Forest: An Ecological Study* (2nd edn), Cambridge: Cambridge University Press.

RINDOS, DAVID (1984). *The Origins of Agriculture: An Evolutionary Perspective*, London: Academic Press.

RISWAN, S. and K. KARTAWINITA (1988). 'A Lowland Dipterocarp Forest 35 Years after Pepper Plantation in East Kalimantan, Indonesia', in S. Soemodihardjo (ed.), *Some Ecological Aspects of Tropical Forests of East Kalimantan*, Jakarta: Indonesian Institute of Sciences, 1–31.

RIVERA DÍAZ, MARIO A. (1987). 'Land Use Patterns in the Azapa Valley, Northern Chile', in Browman (1987*a*: 225–50).

RIVERO L., VÍCTOR (1983). 'Herramientas agrícolas andinas', in Ana María Fries (ed.), *Evolución y tecnología de la agricultura andina*, Cuzco: Instituto Indigenísta Interamericano, 123–42 plus 16 figures.

ROBINSON, DAVID A. (1964). *Peru in Four Dimensions*, Lima: American Studies Press.

ROCHE, MICHEL A., J. BOURGES, J. CORTES, and R. MATTOS (1992). 'Climatology and Hydrology of the Lake Titicaca Basin', in C. Dejoux and A. Iltis (eds.), *Lake Titicaca: A Synthesis of Limnological Knowledge*, Dordrecht: Kluwer, 63–88.

RODRIGUES, ARLINDO JOSÉ (1993). 'Ecology of the Kayabí Indians of Xingú, Brazil: Soil and Agroforestry Management', Ph.D. dissertation, Cambridge University.

ROE, PETER G. (1994). 'Ethnology and Archaeology: Symbolic and Systemic Disjunction or Continuity?', in Augusto Oyuela-Caycedo (ed.), *History of Latin American Archaeology*, Aldershot, UK: Avebury, 183–208.

ROJAS RABIELA, TERESA (ed.) (1983). *La agricultura chinampera: Compilación histórica*, México: Universidad Autónoma Chapingo.

——(1984). 'Agricultural Implements in Mesoamerica', in H. R. Harvey and Hanns J. Prem (eds.), *Explorations in Ethnohistory: Indians of Central Mexico in the Sixteenth Century*, Albuquerque: University of New Mexico Press, 175–204.

——(1987). 'Prospects for Collaboration Between Archaeology and Ethnohistory', *British Archaeological Reports, International Series*, 349: 197–205.

Rojas Rabiela, Teresa (1988). *Las siembras de ayer: La agricultura indígena del siglo XVI*, México: Secretaría de Educación Publica.

Romero, Emilio (1961). *Geografía económica del Perú*, Lima.

Roosevelt, Anna C. (1980). *Parmana: Prehistoric Maize and Manioc Subsistence along the Amazon and Orinoco*, New York: Academic Press.

——(1987). 'Chiefdoms in the Amazon and Orinoco', in Robert D. Drennan and Carlos A. Uribe (eds.), *Chiefdoms in the Americas*, Lanham, MD: University Press of America, 153–85.

——(1989*a*). 'Lost Civilizations of the Lower Amazon', *Natural History*, 66/2: 74–83.

——(1989*b*). 'Resource Management in Amazonia before the Conquest: Beyond Ethnographic Projection', *Advances in Economic Botany*, 7: 30–62.

——(1991*a*). *Moundbuilders of the Amazon: Geophysical Archaeology on Marajó Island, Brazil*, San Diego, CA: Academic Press.

——(1991*b*). 'Determinismo ecológico na interpretação do desenvolvimento social indígena da Amazônia', in Walter A. Neves (ed.), *Origens, adaptações e diversidade biológica do homen nativo de Amazônica*, Belém: Museu Paraense Emilio Goeldi, 103–41.

——(ed.) (1994). *Amazonian Indians from Prehistory to the Present: Anthropological Perspectives*, Tucson: University of Arizona Press.

——(1998). 'Ancient and Modern Hunter-Gatherers of Lowland South America: An Evolutionary Problem', in Balée (1998: 190–212).

——(1999). 'Twelve Thousand Years of Human-Environment Interaction in the Amazon Floodplain', in Padoch *et al.* (1999: 371–92).

——(2000). 'The Lower Amazon: A Dynamic Human Habitat', in Lentz (2000): 455–91.

Rostain, Stéphen (1991). *Les champs surélevés amérindiens de la Guyane*, Paris: Centre ORSTOM de Cayenne.

——(1994). 'L'occupation amérindienne ancienne du littoral de Guyane', Thèse de Doctorat, 2 vols, Université de Paris–Pantheon/Sorbonne.

——(1995). 'La mise en culture des marécages littoraux de Guyane à la période pré-colombienne récente', in Alain Marliac (ed.), *Milieux, Sociétés et Archéologues*, Paris: KARTHALA/ORSTOM, 119–60.

Rostworowski de Diez Canseco, María (1962). 'Nuevos datos sobre tenencia de tierras reales en el incario', *Revista del Museo Nacional*, Lima, 31: 130–64.

——(1972). 'Las etnías del valle del Chillón', *Revista del Museo Nacional*, Lima, 38: 250–314.

——(1987). 'El uso de fertilizanates agrícolas en tiempos prehispánicos', unpublished paper presented at the Primera Reunión-Seminario sobre 'Manejo de Suelos y Aguas en la Sociedad Andina', Lima.

Rouse, Irving (1948). 'The Arawak', in Steward (1946–1959: 4: 507–46).

Rowe, John H. (1946). 'Inca Culture at the Time of the Spanish Conquest', in Steward (1946–1959: 2: 183–330).

——(1969). 'The Sunken Gardens of the Peruvian Coast', *American Antiquity*, 34: 320–5.

Ruddle, Kenneth (1974). *The Yukpa Cultivation System: A Study of Shifting Cultivation in Colombia and Venezuela*, Ibero Americana 52, Berkeley: University of California Press.

Ryder, Roy (1970). 'El valor de la fotografía aerea en los estudios históricos y arqueológicos del Ecuador', *Revista Geográfica del IGM*, Quito, 6: 40–2.

SALAMAN, REDCLIFFE N. (1949). *The History and Social Influence of the Potato*, Cambridge: Cambridge University Press.

SALDARRIAGA, JUAN G. and DARRELL C. WEST (1986). 'Holocene Fires in the Northern Amazon Basin', *Quaternary Research*, 26: 358–66.

SALICK, JAN M. (1989). 'Ecological Basis of Amuesha Agriculture, Peruvian Upper Amazon', *Advances in Economic Botany*, 7: 189–212.

——and MATS LUNDBERG (1990). 'Variation and Change in Amuesha Agriculture, Peruvian Upper Amazon', *Advances in Economic Botany*, 8: 199–223.

SALINAS DE LOYOLA, JUAN DE (1965) [1560s]. 'Descubrimientos, conquistas y poblaciones de Juan de Salinas', in Marcos Jiménez de la Espada (ed.), *Relaciones geográficas de Indias*, Biblioteca de Autores Españoles, Madrid: Ediciones Atlas, 185/3: 197–232.

SALISBURY, RICHARD F. (1962). *From Stone to Steel: Economic Consequences of a Technological Change in New Guinea*, Cambridge: Cambridge University Press.

SALO, JUKKA S., RISTO J. KALLIOLA, ILMARI HÄKKINEN, YRJÖ MÄKINEN, PEKKA NIEMELÄ, MAARIT PUHAKKA, and PHYLLIS D. COLEY (1986). 'River Dynamics and the Diversity of Amazon Lowland Forest', *Nature*, 322: 254–8.

SÁNCHEZ FARFAN, JORGE (1983). 'Pampallaqta, centro productor de semilla de papa', in Ana María Fries (ed.), *Evolución y tecnología de la agricultura andina*, Cuzco: Instituto Indigenísta Interamericano, 163–75.

SANCHO DE LA HOZ, PEDRO (1917) [1535]. *An Account of the Conquest of Peru*, New York: Cortés Society.

SANDERS, WILLIAM T., JEFFREY PARSONS, and ROBERT S. SANTLEY (1979). *The Basin of Mexico: Ecological Processes in the Evolution of a Civilization*, New York: Academic Press.

SANDOR, JONATHAN A. (1987a). 'Initial Investigation of Soils in Agricultural Terraces in the Colca Valley, Peru', *British Archaeological Reports, International Series*, 359: 163–92.

——(1987b). 'Soil Conservation and Redevelopment of Agricultural Terraces in the Colca Valley, Peru', *Journal of the Washington Academy of Sciences*, 77: 149–54.

——and N. S. EASH (1995). 'Ancient Agricultural Soils in the Andes of Southern Peru', *Soil Science Society of America Journal*, 59: 170–9.

SARAYDAR, STEPHEN and IZUMI SHIMADA (1971). 'A Quantitative Comparison of Efficiency Between a Stone Axe and a Steel Axe', *American Antiquity*, 36: 216–17.

————(1973). 'Experimental Archaeology: A New Outlook', *American Antiquity*, 38: 344–50.

SARMIENTO DE GAMBOA, PEDRO (1907) [1572]. *History of the Incas*, Clements R. Markham (ed.), London: Hakluyt Society 22.

SAUER, CARL O. (1950). 'Cultivated Plants of South and Central America', in Steward (1946–1959: 6: 487–543).

——(1952). *Agricultural Origins and Dispersals*, New York: American Geographical Society.

——(1958). 'Man in the Ecology of Tropical America', *Proceedings of the Ninth Pacific Science Congress*, Bangkok, 20: 104–10.

——(1966). *The Early Spanish Main*, Berkeley: University of California Press.

——(1987) [1964]. ' "Now This Matter of Cultural Geography": Notes from Carl Sauer's Last Seminar at Berkeley', in Martin S. Kenzer (ed.), *Carl O. Sauer: A Tribute*, Corvallis: Oregon State University Press, 153–63.

SAUER, JONATHAN D. (1993). *Historical Geography of Crop Plants: A Select Roster*, Boca Raton: CRC Press.

SCHAEDEL, RICHARD P. (1986). 'Paleohidrologías y política agraria en el Perú', *América Indígena*, 46: 319–29.

SCHERY, ROBERT W. (1972). *Plants for Man* (2nd edn), Englewood Cliffs, NJ: Prentice Hall.

SCHMIDT, MAX (1974) [1914]. 'Comments on Cultivated Plants and Agricultural Methods of South American Indians', in Patricia J. Lyon (ed.), *Native South Americans: Ethnology of the Least Known Continent*, Boston: Little, Brown and Co., 60–8.

SCHOMBURGK, ROBERT H. (1841). 'Report of the Third Expedition into the Interior of Guyana . . .', *Journal of the Royal Geographical Society*, 10: 159–267.

SCHREIBER, KATHARINA J. and JOSUÉ LANCHO ROJAS (1995). 'The Puquios of Nasca', *Latin American Antiquity*, 6: 229–54.

SCHULTES, RICHARD E. and ROBERT F. RAFFAUF (1990). *The Healing Forest: Medicinal and Toxic Plants of the Northwest Amazonia*, Portland: Dioscorides Press.

SCHWERIN, KARL H. (1966). *Oil and Steel: Processes of Karinya Culture Change in Response to Industrial Development*, Latin American Studies 4, Los Angeles: University of California.

SELTZER, GEOFFREY O. and CHRISTINE A. HASTORF (1990). 'Climatic Change and its Effect on Prehispanic Agriculture in the Central Peruvian Andes', *Journal of Field Archaeology*, 17: 397–414.

SHARP, LAURISTON (1952). 'Steel Axes for Stone-Age Australians', *Human Organization*, 11: 17–22.

SHEA, DANIEL E. (1987). 'Preliminary Discussion of Prehistoric Settlement and Terracing at Achoma in the Colca Valley, Peru', *British Archaeological Reports, International Series*, 359: 67–88.

——(ed.) (1997). *Achoma Archaeology: A Study of Terrace Irrigation in Peru*, Logan Museum of Anthropology, Museums of Beloit College, Beloit, WI.

——and MARIO A. RIVERA (1995–6). 'Reticulate Irrigation in the Atacama', *Diálogo Andino*, Arica, Chile, 14–15: 281–9.

SHERBONDY, JEANETTE E. (1969). 'El regadío en el área andina central: Ensayo de distribución geográfica', *Revista Española de Antropología Americana*, 4: 113–43.

——(1982). 'The Canal Systems of Hanan Cuzco', Ph.D. dissertation, University of Illinois.

——(1987). 'The Incaic Organization of Terraced Irrigation in Cuzco, Peru', *British Archaeological Reports, International Series*, 359: 365–71.

SHIMADA, IZUMI (1982). 'Horizontal Archipelago and Coast–Highland Interaction in North Peru', in Luis Millones and H. Tomoeda (eds.), *El hombre y su ambiente en los Andes centrales*, Senri Ethnological Studies 10, Osaka: National Museum of Ethnology, 137–210.

SHIPPEE, ROBERT (1932*a*). 'The "Great Wall of Peru" and Other Aerial Photographic Studies by the Shippee–Johnson Peruvian Expedition', *The Geographical Review*, 22: 1–29.

——(1932*b*). 'Lost Valleys of Peru: Results of the Shippee–Johnson Peruvian Expedition', *The Geographical Review*, 22: 562–81.

——(1933). 'Air Adventures in Peru', *National Geographic Magazine*, 63: 80–120.

——(1934). 'A Forgotten Valley of Peru', *National Geographic Magazine*, 65: 110–32.

SIEMENS, ALFRED H. (1998). *A Favored Place: San Juan River Wetlands, Central Veracruz, AD 500 to the Present*, Austin: University of Texas Press.

SILLITOE, PAUL (1996). *A Place Against Time: Land and Environment in the Papua New Guinea Highlands*, Amsterdam: Harwood Academic Publishers.

SIPPEL, SUZANNE J., STEPHEN K. HAMILTON, and JOHN M. MELACK (1992). 'Inundation Area and Morphometry of Lakes on the Amazon River Floodplain, Brazil', *Archiv für Hydrobiologie*, 123: 385–400.

SLUYTER, ANDREW (1994). 'Intensive Wetland Agriculture in Mesoamerica: Space, Time, and Form', *Annals of the Association of American Geographers*, 84: 557–84.

——(1999). 'The Making of the Myth in Postcolonial Development: The Material-Conceptual Landscape Transformation in Sixteenth-Century Veracruz', *Annals of the Association of American Geographers*, 89: 377–401.

SMITH, CLIFFORD T. (1970). 'Depopulation of the Central Andes in the 16th Century', *Current Anthropology*, 11: 453–64.

——, WILLIAM M. DENEVAN, and PATRICK HAMILTON (1968). 'Ancient Ridged Fields in the Region of Lake Titicaca', *The Geographical Journal*, 134: 353–67.

SMITH, HERBERT H. (1879). *Brazil: The Amazons and the Coast*, New York: Scribner's.

SMITH, NIGEL J. H. (1980). 'Anthrosols and Human Carrying Capacity in Amazonia', *Annals of the Association of American Geographers*, 70: 553–66.

——(1999). *The Amazon River Forest: A Natural History of Plants, Animals, and People*, New York: Oxford University Press.

——, J. T. WILLIAMS, DONALD L. PLUCKNETT, and JENNIFER P. TALBOT (1992). *Tropical Forests and Their Crops.*, Ithaca, NY: Cornell University Press.

SMITH, RICHARD T. (1979). 'The Development and Role of Sunken Field Agriculture on the Peruvian Coast', *The Geographical Journal*, 145: 387–400.

——(1983). 'Making a Meal Out of Mahamaes: A Reply to Knapp's "Prehistoric Flood Management on the Peruvian Coast"', *American Antiquity*, 48: 147–9.

——(1987). 'Indigenous Agriculture in the Americas: Origins, Techniques and Contemporary Evidence', in David Preston (ed.), *Latin American Development: Geographical Perspectives*, Harlow, UK: Longman, 34–69.

SMOLE, WILLIAM (1976). *The Yanoama Indians: A Cultural Geography*, Austin: University of Texas Press.

SOLDI, ANA MARÍA (1982). *La agricultura tradicional en hoyas*, Lima: Pontificia Universidad Católica del Perú.

SOMBROEK, W. G. (1966). *Amazon Soils: A Reconnaissance of the Soils of the Brazilian Amazon Region*, Wageningen: Centre for Agricultural Publications and Documentation.

SORIA LENS, LUIS (1954). 'La ciencia agrícola de los antiguos Aymaras', *Boletín de la Sociedad Geográfica de La Paz*, 64: 85–101.

SOTELO NARVÁEZ, PEDRO (1965) [1583]. 'Relación de las provincias de Tucumán', in Marcos Jiménez de la Espada (ed.), *Relaciones geográficas de Indias*, Biblioteca de Autores Españoles, Madrid: Ediciones Atlas, 185/1: 390–6.

SPENCER, CHARLES S., ELSA M. REDMOND, and MILAGRO RINALDI (1994). 'Drained Fields at La Tigra, Venezuelan Llanos: A Regional Perspective', *Latin American Antiquity*, 5: 119–43.

SPENCER, JOSEPH E. and G. A. HALE (1961). 'The Origin, Nature, and Distribution of Agricultural Terracing', *Pacific Viewpoint*, 2: 1–40.

SPONSEL, LESLIE E. (1983). 'Yanomama Warfare, Protein Capture, and Cultural Ecology: A Critical Analysis of the Arguments of the Opponents', *Interciencia*, 8: 204–10.

——(1986). 'Amazon Ecology and Adaptation', *Annual Review of Anthropology*, 15: 67–97.

——(ed.) (1995). *Indigenous Peoples and the Future of Amazonia: An Ecological Anthropology of an Endangered World*, Tucson: University of Arizona Press.

SQUIER, EPHRAIM G. (1877). *Incidents of Travel and Exploration in the Land of the Incas*, New York: Harper.

STADEN, HANS (1928) [1557]. *Hans Staden: The True History of His Captivity*, Malcolm Letts (ed.), London: Routledge and Sons.

STAHL, PETER W. (ed.) (1995). *Archaeology in the Lowland American Tropics: Current Analytical Methods and Applications*, Cambridge: Cambridge University Press.

STANISH, CHARLES (1987). 'Agroengineering Dynamics of Post-Tiwanaku Settlements in the Otora Valley, Peru', *British Archaeological Reports, International Series*, 359: 337–64.

——(1994). 'The Hydraulic Hypothesis Revisited: Lake Titicaca Basin Raised Fields in Theoretical Perspective', *Latin American Antiquity*, 5: 312–32.

——(n.d.). *Prehispanic Sierra Agrarian Economics and Interzonal Exchange in the Moquegua Drainage, Southern Peru: Preliminary Report*, Department of Anthropology, University of Chicago.

Stearman, Allyn M. (1987). *No Longer Nomads: The Sirionó Revisited*, Lanham, MD: Hamilton Press.

——(1989). *Yuquí: Forest Nomads in a Changing World*, Orlando: Holt, Rinehart and Winston.

STEENSBERG, AXEL (1980). *New Guinea Gardens: A Study of Husbandry with Parallels in Prehistoric Europe*, London: Academic Press.

STEMPER, DAVID M. (1993). *The Persistence of Prehistoric Chiefdoms on the Río Daule, Coastal Ecuador*, University of Pittsburgh Memoirs in Latin American Archaeology 7.

STERNBERG, HILGARD O'R. (1960). 'Radiocarbon Dating as Applied to a Problem of Amazonian Morphology', *Comptes Rendus du XVIII Congrès International de Géographie*, Rio de Janeiro: International Geographical Union, 2: 399–424.

——(1964). 'Land and Man in the Tropics', *Proceedings of the Academy of Political Science*, 27: 319–29.

——(1975). *The Amazon River of Brazil*, Erdkundliches Wissen 40, Beihefte zur Geographischen Zeitschrift, Wiesbaden: Franz Steiner.

——(1995). 'Waters and Wetlands of Brazilian Amazonia: An Uncertain Future', in Toshie Nishizawa and Juha I. Uitto (eds.), *The Fragile Tropics of Latin America: Sustainable Management of Changing Environments*, Tokyo: United Nations University Press, 113–79.

STEVENS, WILLIAM K. (1988). 'Scientists Revive a Lost Secret of Farming', *New York Times* (Science Times), 22 November, 19, 22.

STEWARD, JULIAN H. (ed.) (1946–1959). *Handbook of South American Indians*, Bureau of American Ethnology, Bulletin 143, 7 vols, Washington, DC: Smithsonian Institution.

——(1948). 'Culture Areas of the Tropical Forests', in Steward (1946–1959: 3: 883–99).

STEWARD, JULIAN H. (cont.) (1949*a*). 'The Native Population of South America', in Steward (1946–1959: 5: 655–68).

——(1949*b*). 'South American Cultures: An Interpretative Summary', in Steward (1946–1959: 5: 669–772).

——(1955). *Theory of Culture Change: The Methodology of Multilinear Evolution*, Urbana: University of Illinois Press.

——and LOUIS C. FARON (1959). *Native Peoples of South America*, New York: McGraw-Hill.

——and ALFRED MÉTRAUX (1948). 'Tribes of the Peruvian and Ecuadorian Montaña', in Steward (1946–1959: 5: 535–656).

STOCKS, ANTHONY (1983*a*). 'Candoshi and Cocamilla Swiddens in Eastern Peru', *Human Ecology*, 11: 69–84.

——(1983*b*). 'Cocamilla Fishing: Patch Modification and Environmental Buffering in the Amazon', in Hames and Vickers (1983*a*: 239–67).

STODDART, DAVID R. (1987). 'To Claim the High Ground: Geography for the End of the Century', *Transactions, The Institute of British Geographers*, New Series, 12: 327–36.

STOTHERT, KAREN E. (1995). 'Las albarradas tradicionales y el manejo de aguas en la península de Santa Elena', *Miscelánea Antropológica Ecuatoriana*, 8: 131–60.

STURTEVANT, WILLIAM C. (1961). 'Taino Agriculture', in Wilbert (1961: 69–82).

——(general ed.) (1978–). *Handbook of North American Indians*, c. 20 vols, Washington, DC: Smithsonian Institution.

SWANSON, E. (1955). 'Terrace Agriculture in the Central Andes', *Davidson Journal of Anthropology*, 1: 123–32.

TESSMAN, GÜNTER (1930). *Die Indianer Nordost-Perus*, Hamburg: De Gruyter.

THOMPSON, LONNIE G., MARY E. DAVIS, ELLEN MOSLEY-THOMPSON, and K-B. LIU (1988). 'Pre-Incan Agricultural Activity Recorded in Dust Layers in Two Tropical Ice Cores', *Nature*, 336: 763–5.

—————— (1994). 'Glacial Records of Global Climate: A 1500-Year Tropical Ice Core Record of Climate', *Human Ecology*, 22: 83–95.

THURSTON, H. DAVID (1997). *Slash/Mulch Systems: Sustainable Methods for Tropical Agriculture*, Boulder, CO: Westview Press.

——and JOANNE M. PARKER (1995). 'Raised Beds and Plant Disease Management', in D. Michael Warren, L. Jan Slikkerveer, and David Brokensha (eds.), *The Cultural Dimensions of Development: Indigenous Knowledge Systems*, London: Intermediate Technology Publications, 140–6.

TORMO SANZ, LEANDRO and JAVIER TERCERO (1966). 'El sistema comunalista indiano en la región comunera de Mojos–Chiquitos, II: La organización del trabajo', *Communidades*, Spain, 1/2: 89–117.

TORRE, CARLOS DE LA and MANUEL BURGA (eds.) (1986). *Andenes y camellones en el Perú andino: Historia presente y futuro*, Lima: Consejo Nacional de Ciencia y Tecnología.

TOSI, JOSEPH A. (1960). *Zonas de vida natural en el Perú*, Instituto Interamericano de Ciencias Agrícolas de la OEA, Zona Andina, Lima: Boletín Técnico 5.

TOURNON, JACQUES (1988). 'Las inundaciones y los patrones de ocupación de las orillas del Ucayali por los Shipibo-Conibo', *Amazonía Peruana*, 8/16: 43–66.

TOWLE, MARGARET A. (1961). *The Ethnobotany of Pre-Columbian Peru*, Viking Fund Publications in Anthropology 30, Chicago: Aldine.

TOWNSEND, WILLIAM H. (1969). 'Stone and Steel Tool Use in a New Guinea Society', *Ethnology*, 8: 199–205.

TREACY, JOHN M. (1987a). 'An Ecological Model for Estimating Prehistoric Population at Coporaque, Colca Valley, Peru', *British Archaeological Reports, International Series*, 359: 147–62.

—— (1987b). 'Building and Rebuilding Agricultural Terraces in the Colca Valley of Peru', *Yearbook, Conference of Latin Americanist Geographers*, 13: 51–7.

—— (1989a). 'The Fields of Coporaque: Agricultural Terracing and Water Management in the Colca Valley, Arequipa, Peru', Ph.D. dissertation, University of Wisconsin–Madison.

—— (1989b). 'Agricultural Terraces in Peru's Colca Valley: Promises and Problems of an Ancient Technology', in John O. Browder (ed.), *Fragile Lands of Latin America: Strategies for Sustainable Development*, Boulder, CO: Westview Press, 209–29.

—— (1994a). 'Teaching Water: Hydraulic Management and Terracing in Coporaque, the Colca Valley, Peru', in Mitchell and Guillet (1994: 99–114).

—— (1994b). *Las chacras de Coporaque: Andenería y riego en el valle del Colca*, Lima: Instituto de Estudios Peruanos.

—— and WILLIAM M. DENEVAN (1986). 'Survey of Abandoned Terraces, Canals, and Houses at Chijra, Coporaque', in Denevan (1986–1988 :1: 198–220).

—— —— (1994). 'The Creation of Cultivable Land through Terracing', in Miller and Gleason (1994a: 91–110).

TROLL, CARL (1958) [1931]. 'Las culturas superiores andinas y el medio geográfico', *Publicaciones del Instituto de Geografía*, Lima, series 1: 1.

—— (1963). 'Qanat-Bewässerung in der Alten und Neuen Welt', *Mitteilungen der Österreichischen Geographischen Gesellschaft*, 105: 313–30.

—— (1968). 'The Cordilleras of the Tropical Americas: Aspects of Climatic, Phyto-geographical and Agrarian Ecology', in Carl Troll (ed.), *Geo-Ecology of the Mountainous Regions of the Tropical Americas*, Colloquium Geographicum, Bonn, 9: 15–56.

TSCHOPIK JR., HARRY (1946). 'The Aymara', in Steward (1946–1959:2: 501–73).

TSCHUDI, J. J. VON (1852). *Travels in Peru, During the Years 1838–1842*, New York: Putnam.

TURNER II, B. L. (1983). 'The Excavations of Raised and Channelized Fields at Pulltrouser Swamp', in B. L. Turner II and Peter D. Harrison (eds.), *Pulltrouser Swamp: Ancient Maya Habitat, Agriculture, and Settlement in Northern Belize*, Austin: University of Texas Press, 30–51.

—— (1989). 'The Specialist-Synthesis Approach to the Revival of Geography: The Case of Cultural Ecology', *Annals of the Association of American Geographers*, 79: 88–100.

—— (1997). 'Spirals, Bridges and Tunnels: Engaging Human-Environment Perspectives in Geography', *Ecumene*, 4: 196–217.

—— and WILLIAM M. DENEVAN (1985). 'Prehistoric Manipulation of Wetlands in the Americas: A Raised-Field Perspective', *British Archaeological Reports, International Series*, 232: 11–30.

UHL, CHRISTOPHER (1983). 'You Can Keep a Good Forest Down', *Natural History*, 92/4: 69–79.

—— (2000). 'Raised Field Abandonment in the Upper Amazon', *Culture and Agriculture*, 22: 27–31.

UHLE, MAX (1954) [1923]. 'The Aims and Results of Archaeology', in John H. Rowe (ed.), *Max Uhle, 1856–1944: A Memoir of the Father of Peruvian Archaeology*, Berkeley: University of California Publications in American Archaeology and Ethnology, 46/1: 54–100.

UP DE GRAFF, F. W. (1923). *Head Hunters of the Amazon*, Garden City: Garden City Publishing Co.

VALCÁRCEL, LUIS E. (1942). 'La agricultura entre los antiguos peruanos', *Revista del Museo Nacional*, Lima, 12: 1–7.

VASEY D. E., D. R. HARRIS, G. W. OLSON, M. J. T. SPRIGGS, and B. L. TURNER II (1984). 'The Role of Standing Water and Water-Logged Soils in Raised Field, Drained-Field, and Island-Bed Agriculture', *Singapore Journal of Tropical Geography*, 5: 63–72.

VÁSQUEZ, RODOLFO and ALWYN H. GENTRY (1989). 'Use and Misuse of Forest-Harvested Fruits in the Iquitos Area', *Conservation Biology*, 3: 350–61.

VÁZQUEZ DE ESPINOSA, ANTONIO (1942) [1628]. *Compendium and Description of the West Indies*, Charles Upson Clark (ed), Smithsonian Miscellaneous Collections 102, Washington, DC: Smithsonian Institution Press.

——(1948) [1628]. *Compendio y descripción de las Indias Occidentales*, Charles Upson Clark (ed.), Smithsonian Miscellaneous Collections 108, Washington, DC: Smithsonian Institution.

VERA CRUZ CHÁVEZ, PABLO DE LA (1987). 'Cambio en los patrones de asentamiento e el uso y abandono de los andenes en Cabanaconde, Valle del Colca, Perú', *British Archaeological Reports, International Series*, 359: 89–128.

——(1988). 'Estudio Arqueológico en el Valle de Cabanaconde, Arequipa', Tesis de Bachiller en Ciencias Arqueológicas, Universidad Católica Santa María, Arequipa.

——(1989). 'Cronología y corología de la cuenca del Río Camaná–Majes–Colca–Arequipa', Tesis de Licenciado en Arqueología, Universidad Católica Santa María, Arequipa.

VICKERS, WILLIAM T. (1976). 'Cultural Adaptation to Amazonian Habitats: The Siona-Secoya of Eastern Ecuador', Ph.D. dissertation, University of Florida.

—— (1983*a*). 'The Territorial Dimensions of Siona–Secoya and Encabellado Adaptation', in Hames and Vickers (1983*a*: 451–78).

——(1983*b*). 'Tropical Forest Mimicry in Swiddens: A Reassessment of Geertz's Model with Amazonian Data', *Human Ecology*, 11: 35–45.

—— and TIMOTHY PLOWMAN (1984). 'Useful Plants of the Siona and Secoya Indians', *Fieldiana: Botany*, New Series, 15.

VIVEIROS DE CASTRO, EDUARDO (1996). 'Images of Nature and Society in Amazonian Ethnology', *Annual Review of Anthropology*, 25: 179–200.

VREELAND JR., JAMES M. (1986). 'Una perspectiva antropológica de la paleotecnología en el desarrollo agrario del norte de Perú', *América Indígena*, 46: 275–318.

WADDELL, ERIC (1972). *The Mound Builders: Agricultural Practices, Environment, and Society in the Central Highlands of New Guinea*, Seattle: University of Washington Press.

WALKER, JOHN H. (1999). 'Agricultural Change in the Bolivian Amazon,' Ph.D. dissertation, University of Pennsylvania.

——(2000). 'Raised Field Abandonment in the Upper Amazon', unpublished paper presented at the American Anthropological Association Meeting, Philadelphia.

WAUCHOPE, ROBERT (general ed.) (1964–1976). *Handbook of Middle American Indians*, 16 vols, Austin: University of Texas Press.

WAUGH, RICHARD and JOHN TREACY (1986). 'Hydrology of the Coporaque Irrigation System', in Denevan (1986–1988 :1: 116–48).

WEBBER, ELLEN R. (1993). 'Cows in the Colca: Household Cattle Raising in Achoma, Peru', Master's thesis, University of Wisconsin–Madison.

WERNER, DENNIS (1983). 'Why Do the Mekranoti Trek?', in Hames and Vickers (1983a: 225–38).

WEST, MICHAEL (1977). 'Alternatives to Canal Irrigation in a Coastal Peruvian Valley: Past and Present', Los Angeles County Museum of Natural History, unpublished manuscript.

——(1979). 'Early Watertable Farming on the North Coast of Peru', *American Antiquity*, 44: 138–44.

——(1981). 'Agricultural Resource Use in an Andean Coastal Ecosystem', *Human Ecology*, 9: 47–78.

WEST, ROBERT C. (1957). *The Pacific Lowlands of Colombia*, Louisiana State University Studies, Social Science Series 8.

——(1959). 'Ridge or "Era" Agriculture in the Colombian Andes', *Actas del XXXIII Congreso Internacional de Americanistas*, San José, Costa Rica, 1: 279–82.

WHITEHEAD, NEIL L. (1993). 'Ethnic Transformation and Historical Discontinuity in Native Amazonia and Guayana, 1500–1900', *L'Homme*, 126–8: 285–305.

——(1994). 'The Ancient Amerindian Polities of the Amazon, the Orinoco, and the Atlantic Coast: A Preliminary Analysis of Their Passage from Antiquity to Extinction', in Roosevelt (1994: 33–53).

WHITMORE, THOMAS M. and B. L. TURNER II (2001). *Cultivated Landscapes of Middle America on The Eve of Conquest*, Oxford: Oxford University Press, forthcoming.

WHITTEN JR., NORMAN E. (1976). *Sacha Runa: Ethnicity and Adaptation of Ecuadorian Jungle Quichua*, Urbana: University of Illinois Press.

WILBERT, JOHANNES (ed.) (1961). *The Evolution of Horticultural Systems in Native South America, Causes and Consequences: A Symposium*, Caracas: Sociedad de Ciencias Naturales La Salle.

WILK, RICHARD R. (1985). 'Dry Season Agriculture among the Kekchi Maya and its Implications for Prehistory', in Mary Pohl (ed.), *Prehistoric Lowland Maya Environment and Subsistence Economy*, Papers of the Peabody Museum of Archaeology and Ethnology 77, Cambridge: Harvard University, 47–57.

WILKEN, GENE C. (1987). *Good Farmers: Traditional Agricultural Resource Management in Mexico and Central America*, Berkeley: University of California Press.

WILLEY, GORDON R. (1953). *Prehistoric Settlement Patterns in the Virú Valley, Perú*, Bureau of American Ethnology, Bulletin 155, Washington, DC: Smithsonian Institution.

——(1966). *An Introduction to American Archaeology, Volume One: North and Middle America*, Englewood Cliffs, NJ: Prentice Hall.

——(1971). *An Introduction to American Archaeology, Volume Two: South America*, Englewood Cliffs, NJ: Prentice Hall.

WILLIAMS, LYNDEN S. (1986). *Agricultural Terrace Construction*, International Development Program, Clark University, Worcester.

——(1990). 'Agricultural Terrace Evolution in Latin America', *Yearbook, Conference of Latin Americanist Geographers*, 16: 82–93.

——, LESLIE COOPERBAND, and BOB J. WALTER (1986). *Agricultural Terrace Construction: The Valles Altos Project of Venezuela as an Example for AID*, International Development Program, Clark University, Worcester.

WILSON, DAVID J. (1999). *Indigenous South Americans of the Past and Present: An Ecological Perspective*, Boulder, CO: Westview Press.

WILSON, DAVID R. (1982). *Air Photo Interpretation for Archaeologists*, New York: St. Martin's Press.

WINKLERPRINS, ANTOINETTE M. G. A. (1999). 'Between the Floods: Soils and Agriculture on the Lower Amazon Floodplain, Brazil', Ph.D. dissertation, University of Wisconsin–Madison.

WINTERHALDER, BRUCE (1994). 'The Ecological Basis of Water Management in the Central Andes: Rainfall and Temperature in Southern Peru', in Mitchell and Guillet (1994: 21–67).

——, ROBERT LARSEN, and R. BROOKE THOMAS (1974). 'Dung as an Essential Resource in a Highland Peruvian Community', *Human Ecology*, 2: 89–104.

WOLF, ERIC R. (1982). *Europe and the People without History*, Berkeley: University of California Press.

WOODBURY, RICHARD B. and JAMES A. NEELY (1972). 'Water Control Systems of the Tehuacán Valley', in Frederick Johnson (ed.), *The Prehistory of the Tehuacán Valley*, Austin: University of Texas Press, 4: 81–153.

WOODS, WILLIAM I. (1995). 'Comments on the Black Earths of Amazonia', *Papers and Proceedings of Applied Geography Conferences*, Arlington, VA: 18: 159–65.

—— and JOSEPH M. MCCANN (1999). 'The Anthropogenic Origin and Persistence of Amazonian Dark Earths', *Yearbook, Conference of Latin Americanist Geographers*, 25: 7–14.

WORKS, MARTHA A. (1990). 'Continuity and Conversion of House Gardens in Western Amazonia', *Yearbook of the Association of Pacific Coast Geographers*, 52: 31–64.

WRIGHT, A. C. S. (1962). 'Some Terrace Systems of the Western Hemisphere and Pacific Islands', *Pacific Viewpoint*, 3: 97–101.

—— (1963). 'The Soil Process and the Evolution of Agriculture in Northern Chile', *Pacific Viewpoint*, 4: 65–74.

WRIGHT, KENNETH R., R. M. WRIGHT, M. E. JENSEN, and A. VALENCIA Z. (1997). 'Machu Picchu Ancient Agricultural Potential', *Applied Engineering in Agriculture*, 13: 39–47.

WÜST, IRMHILD (1994). 'The Eastern Bororo from an Archaeological Perspective', in Roosevelt (1994: 315–42).

—— and CRISTIANA BARRETO (1999). 'The Ring Villages of Central Brazil: A Challenge for Amazonian Archaeology', *Latin American Antiquity*, 10: 3–23.

YEN, D. E. (1974). *The Sweet Potato and Oceania: An Essay in Ethnobotany*, Honolulu: Bernice P. Bishop Museum Bulletin 236.

ZARIN, DANIEL J. (1999). 'Spatial Heterogeneity and Temporal Variability of Some Amazonian Floodplain Soils', in Padoch *et al.* (1999: 313–21).

ZEGARRA, JORGE M. (1978). 'Irrigación y técnicas de riego en el Perú pre-colombino', in Ravines (1978c: 107–16).

ZENT, STANFORD R. (1992). 'Historical and Ethnographic Ecology of the Upper Cuao River Wõthĩhã: Clues for an Interpretation of Native Guianese Social Organization', Ph.D. dissertation, Columbia University.

——(1998). 'Independent yet Interdependent "Isolde": The Historical Ecology of Traditional Piaroa Settlement Pattern', in Balée (1998: 251–86).

ZEVEN, A. C. and J. M. J. DE WET (1982). *Dictionary of Cultivated Plants and their Regions of Diversity*, Wageningen: Centre for Agricultural Publishing and Documentation.

ZIMMERER, KARL S. (1995). 'The Origins of Andean Irrigation', *Nature*, 378: 481–3.

——(1996a). *Changing Fortunes: Biodiversity and Peasant Livelihood in the Peruvian Andes*, Berkeley: University of California Press.

——(1996b). 'Ecology as Cornerstone and Chimera in Human Geography', in Carville Earle, Kent Mathewson, and Martin S. Kenzer (eds.), *Concepts in Human Geography*, Lanham, MD: Rowman and Littlefield, 161–88.

——(1999). 'Overlapping Patchworks of Mountain Agriculture in Peru and Bolivia: Toward a Regional-Global Landscape Model', *Human Ecology*, 27: 135–65.

ZONNEVELD, J. I. S. (1952). 'Luchtfoto-geografie in Suriname', *De West-Indische Gids*, 33: 35–48.

ZUCCHI, ALBERTA and WILLIAM M. DENEVAN (1979). 'Campos elevados e historia cultural prehispánica en los Llanos Occidentales de Venezuela', *Montalban*, Caracas, 9: 565–736.

INDEX

mounding (non crops):
  habitation and ceremonial 215, 226, 227,
    234, 243
  termite 243, 244
Movima (people) 239
Moxitania, Bolivia 76
Moxos, *see* Mojos
Moyobamba, Peru 71, 302
muck, mucking 37, 38, 49 n. 5
  raised fields 72, 220, 236, 248
mulch 37, 47, 69, 73
  raised fields 248, 285
  see also slash/mulch
Mullen, William 275
Mundurucú (people) 123
muros 18
  see also embankments
Murra, John V. 49 n. 7, 58, 134
Muse, Michael 232, 233, 237, 248
Museo de la Nación, Peru 191
Mutah, Robert 82
Muzo (people) 141
Myers, Thomas P. 31, 55, 74 n. 2, 90, 105, 108,
  110, 113, 129, 130, 142

Nambicuara (people) 29, 67, 72, 125
Nasca Valley, Peru 162
Nascimento Irmãol, Marilane 120
National Geographic 171
native:
  defined 11 n. 1
Native Peoples of South America 11 n. 4
natural environment 55–6
Neely, James A. 22
Negev Desert, Israel 184 n. 2
Neira Avendaño, Máximo 186, 196
Nelson, Bruce W. 120
Nepeña Valley, Peru 161, 162, 169 n. 3
Netherlands 236
Netherly, Patricia J. 25, 138, *145*, 149, 152,
  167
Neves, Eduardo Góes 130, 132 n. 13
New Caledonia 249
New Guinea 132 n. 11
  raised fields 37, 125, 215, 236, 248, 249
New Mexico (state) 21
new synthesis 130–1
New York Times 215, 275
Nials, Fred L. 152, 169 n. 2
Nicaragua 59
Nickerson, Norton H. 16
Nietschmann, Bernard Q. 59
Niles, Susan A. 158, 172, 180
Nimuendajú, Curt 105
nitrates 37, 38

nitrogen fixation 37–8, 40, 41, 220, 248,
  270
nomadic, nomads 67, 68
  agriculture 42
  semi-nomadic 80, 81, 101 n. 6
  see also forest fields; trekking
no-man lands, *see* uninhabited zones
Nordenskiöld, Erland 31, 77, 217, 218,
  253 n. 4, 253 n. 7
North America xv, 26 n. 2, 122, 236
NRC (National Research Council) 320
Nuevo Mamo, Venezuela 282, 285
Nuevo Mundo, Bolivia 247
Nuevo Nazareth, Peru 113
Nukak, *see* Makú
Núñez Jiménez, Antonio 162, *163*, *164*
Núñez Regueiro, Víctor A. 142, 162, 163,
  164
nutrients:
  depletion 85, 86
  hot spots 43
  see also fertilization; soils
Nye, P. H. 39

oca 45, 165, 257, 265
Ocamo, Venezuela 69
Ochoa, Carlos M. 27, 101 n. 3, 317
Oenocarpus spp. 109
Ohly, J. J. 59, 60, 61, 290 n. 1
Oitavo Bec, Brazil 123
Ollantaytambo, Peru 141, 171, 290
Omagua (people) 63, 64, 90
ONERN (Oficina Nacional de Evaluación de
  Recursos Naturales), Peru 135, 173, 262,
  263, 266, 276 n. 2
Orbignya spp. 101 n. 8
orchards, *see* fruit trees
Orellana, Antonio de 77
Orellana, Francisco de 63
Orinoco:
  Llanos 30, 31, 71, 72, 97, 226
  raised fields 217, 219, 223, 287–8
  region 104, 119, 217, 226
  River 41, 99, 105, 217, 281
Orlove, Benjamin S. 37, 45, 47, 49 n. 9, 166,
  281
Ortiz de Zúñiga, Iñigo 171
Ortloff, Charles R. 37, 135, 147, *149*, 152,
  154, 155, 156, 157, 169 n. 2, 255, 269, 270,
  273, 274, 302
Otavalo, Equador 233
Otomac, Otomaco (people) 63, 217
Oviedo y Valdés, Gonzalo Fernández de 30,
  37, 64
oxisols 106, 114 n. 6

Index